Specific Interactions and Biological Recognition Processes

Edited by

Nicolae Voiculetz, Ph.D., D.Sc.

Research Professor of Molecular Biology

Institute of Oncology and University of Bucharest

Bucharest, Romania

Ioan Moţoc, Ph.D.

Assistant Professor of Medicinal Chemistry

University of Georgia

Athens, Georgia

Zeno Simon, Ph.D.

Professor of Biophysics

University of Medicine and Pharmacy

Timişoara, Romania

CRC Press

Boca Raton Ann Arbor London Tokyo

Library of Congress Cataloging-in-Publication Data

Specific interactions and biological recognition processes/editors,
Nicolae Voiculetz, Ioan Moţoc, Zeno Simon.
p.cm.
Includes bibliographical references and index.
ISBN 0-8493-5398-X
1. Molecular recognition. 2. Ligand binding (Biochemistry)
I. Voiculetz, Nicolae. II. Moţoc, Ioan, 1950-1991. III. Simon,
Zeno.
QP517.M67S74 1993
574.8′8—dc20 92-38801
CIP

Direct all inquiries to CRC Press, Inc., 2000 Corporate Blvd., N.W., Boca Raton, Florida 33431.

International Standard Book Number 0-8493-5398-X

Library of Congress Card Number 92-38801

Printed in the United States of America 1 2 3 4 5 6 7 8 9 0

Printed on acid-free paper

PREFACE

In recent years, significant progress has been made in the field of recognition processes of (and binding to) specific DNA sequences, by proteins involved in the regulation of DNA replication, control of gene activity, self-nonself discrimination in immunology, specific interactions in ionic transport through biological membranes, and interaction of chemical messengers with biological receptors. These progresses are due to advances in recombinant DNA-technology and improvements in both X-ray crystallographic techniques and the use of two and three dimensional NMR and potent computers and programs.

In this book, we have tried to present the theoretical basis for specificity of biological recognition and the degree of specificity with data concerning errors in the corresponding recognition processes.

Since this field of molecular biology is so complex, much remains to be discovered and elucidated. But we hope that this book presents a synthesis of data and comprises a good start toward the new discoveries.

THE EDITORS

Nicolae Voiculetz, Ph.D., is Research Professor of Molecular Biology in the Institute of Oncology and the University of Bucharest.

He was IAEA fellow in Belgium (1965) and Eleanor Roosevelt Fellow at Stanford University Medical School, Palo Alto, California (1970 to 1971).

He is the author of five books, nine chapters in monographs, and has published more than 180 scientific papers. For CRC Press, he has contributed a chapter in *Chemistry and Biology of Synthetic Retinoids* (edited by Marcia Dawson and William H. Okamura) and served as co-editor of *Modeling in Cancer Genesis and Prevention* (1991).

He is vice president of the Romanian Oncology Society and Romanian Liga for Cancer, member of the European Association for Cancer Research, and a member of other scientific societies. He has been UNESCO Expert in Carcinogenesis since 1979.

Dr. Voiculetz is interested in DNA lesions and repair, molecular mechanisms of carcinogenesis-anticarcinogenesis, and in oncogenes-antioncogenes.

Ioan Moţoc, Ph.D., was graduated from the Faculty of Industrial Chemistry of the Polytechnical Institute of Timişoara in 1974 and received a Ph.D. in physical organic chemistry with Professor A.T. Balaban in 1978. Since 1982, he worked in the field of QSAR in Germany and the U.S., the last position being as Senior Research Investigator at the Computer-Assisted Drug Design Group of the Bristol-Myers/Squibb Company, and was Assistant Professor of Medicinal Chemistry at the University of Georgia. He made important contributions in 3D-QSAR based upon three-dimensional molecular shape descriptors and various computational chemistry techniques. He was co-founder of the QSAR-group at the University of Timişoara. He published about 140 papers and three books.

Dr. I. Moţoc passed away quite unexpectedly in September 1991.

Zeno Simon, Ph.D., was graduated from the University of Bucharest in 1957 and received a Ph.D. in chemical physics in 1965. He is Professor of Biophysics at the University of Medicine and Pharmacy, Timişoara. He is interested in quantum biochemistry, trigger theory for regulation of gene activity, and QSAR, for which he developed the MTD (minimal steric difference) method. He is author and co-author of more than 200 scientific papers and five books.

CONTRIBUTORS

Adrian Chiriac, Ph.D.
Professor
Department of Chemistry
University of Timişoara
Timişoara
Romania

Maria-Luiza Flonta, Ph.D.
Associate Professor
Department of Biophysics
Faculty of Biology
University of Bucharest
Bucharest
Romania

Ludovic Kurunczi, Ph.D.
Senior Investigator
Chemical Research Center of the Romanian Academy
Timişoara
Romania

Mariana Marin, Ph.D.
Senior Investigator
Department of Chemical Carcinogenesis and Molecular Biology
Institute of Oncology
Bucharest
Romania

Ioan Moţoc, Ph.D.
Senior Investigator
Computer-Assisted Drug Design Group
Bristol Meyers-Squibb Co.
Princeton, New Jersey, and
Assistant Professor of Medicinal Chemistry
University of Georgia
Athens, Georgia

Tudor I. Oprea, M.D., Ph.D.
Assistant Professor of Physiology
Department of Physiology
University of Medicine and Pharmacy
Timişoara
Romania, and
Research Associate
Center for Molecular Design
Washington University in St. Louis
St. Louis, Missouri

Zeno Simon, Ph.D.
Professor
Department of Biophysics
University of Medicine and Pharmacy
Timişoara
Romania

Geza Szegli, Ph.D.
Head
Immunology Department
I. Cantacuzino Institute for Sera and Vaccines
Bucharest
Romania

Simona Vasilescu, Ph.D.
Junior Investigator
Department of Chemical
Carcinogenesis and
Molecular Biology
Institute of Oncology
Bucharest
Romania

Nicolae Voiculetz, Ph.D., D.Sc.
Research Professor of
Molecular Biology
Institute of Oncology, and
University of Bucharest
Bucharest
Romania

TABLE OF CONTENTS

In memory of our colleague and friend
Ioan Moţoc
who was a distinguished scientist

INTRODUCTION

Nicolae Voiculetz and Zeno Simon

Specificity is perhaps the main characteristic of biological processes at the molecular level. Generally, one assumes that this specificity of biological recognition processes is devoid of errors, as in the case of the enzymatic catalysis of the transformation of only a certain substrate — not also of a similar one — the activation of a certain receptor only by its corresponding chemical messenger. However, such errors occur, for example, spontaneous mutations, self auto-aggression in immunology, but relatively little attention is paid to errors in biological recognition processes. The specificity-molecular size relationship was discussed in relation to DNA-RNA hybridization-competition experiments as early as 1966 by Thomas;[1] the probability of false base incorporation in DNA replication was related to binding free enthalpy differences as early as 1961 by Volkenshtein and Eliashevici.[2] Nevertheless, only three books concerned with reliability of biological recognition processes seem to have been published up to now.[3-5] Some review articles on these items[6,7] may also be mentioned.

Important progress has been made recently in the research of various recognition processes and concerning the structure of the implied proteins and nucleic acids, especially in DNA replication, charging of tRNAs translation, control of gene activity, self-nonself recognition in immunology, etc. On the other hand, recent progress in methods for drug design, for quantitative structure-activity relations (QSAR) brings important contributions to the understanding of effector-receptor interaction mechanisms. All these features justify the opportunity of this book.

Two aspects will be separately treated. First, the theoretical basis for specificity in biological recognition, the thermodynamic, chemical equilibrium background, with the relation between size of combining sites and specificity (i.e., capacity to discern, with sufficient precision, the correct partner from a large number of possible partners). The molecular basis for the receptor-effector interactions will be also discussed. Significant advances were made in the last decade in the use of quantum chemistry methods and especially in the use of semiempirical potentials for calculation of interaction energies, in the use of potent computers and programs. The potentialities of quantum chemistry methods in calculating receptor-effector affinities will be assessed. Nevertheless, the possibilities of a more intuitive, semiempiric approach will be also discussed — types of intermolecular forces, various structural parameters of the type used in the, now classical, Hansch-type approach in QSAR.

The second aspect will be an analysis of various recognition processes, such as DNA-replication, transcription, RNA-translation (protein synthesis), enzymatic reactions, transmembrane transport processes, mechanisms of action of hormones and other chemical messengers, and self-nonself recognition in immunology. The topics discussed will be the molecular features of the corresponding receptor-effector interactions, the degree of specificity with data concerning errors in the corresponding recognition processes. Our opinion is that more thorough consideration of the molecular mechanisms of recognition processes, of their intrinsic limited reliability, can contribute to the knowledge of various complex regulation processes, such as self-nonself recognition in immunology.

REFERENCES

1. **Thomas, C. A., Jr.,** Oligonucleotide size and specificity in DNA-RNA Hybridisation, *Progr. Nucleic Acid Res. Mol. Biol.,* 5, 315, 1966.
2. **Volkenshtein, M. V. and Eliashevici, A. M.,** About the theory of mutations, *Dokl. Akad. Nauk SSSR (Moscow),* 136, 1216, 1961.
3. **Simon, Z.,** *Quantum Biochemistry and Specific Interactions,* Abacus Press, Tunbridge Wells, Kent, 1976.
4. **Praga, S., Saxena, K. M. S., and Torres, M.,** *Biomolecular Information Theory,* Elsevier, Amsterdam, 1978.
5. **Kirkwood, T. B. L., Rosenberg, R. F., and Gals, D. J., Eds.,** *Accuracy in Molecular Processing, Its Control and Relevance to Biological Systems,* Chapman and Hall, London, 1986.
6. **Rebeck, Jr., J.,** Model studies in molecular recognition, *Science,* 235, 1478, 1987.
7. **Mueller, K.,** Molecular recognition: its role in chemistry and biochemistry, *J. Mol. Graphics,* 7, 92, 1989.

Chapter 1

RECEPTOR-LIGAND EQUILIBRIA AND BIOLOGICAL RECOGNITION PROCESSES

Zeno Simon and Ioan Moţoc

TABLE OF CONTENTS

0-8493-5398-X/93/$0.00 + $.50

I. INTRODUCTION

Throughout various biological processes, such as elicitation of a hormonal response, a certain hormone as ligand (effector) E_0, recognizes a biological receptor, R_0. Although there are several other ligand species (E_i, i = 1,2, . . . ,N − 1) or receptor species (R_i, i = 1,2, . . . ,N − 1) available, E_0 recognizes only R_0. The ligand (effector) is usually a micromolecule, generally smaller than the biological receptor, which is a macromolecule or complex of macromolecules. The result of the recognition is the formation of a complex, C_{00} with the elicitation of a biological, cellular response (DNA, transcription, enzymatic intracellular synthesis, modification of a transmembranar potential, etc.). The intensity of the biological response increases with the concentration $[C_{00}]$ of the complex or with the fraction $[C_{00}]/([R_0] + [C_{00}])$ of the receptor which has recognized the ligand. The efficiency of the ligand producing the biological response is usually measured as ED_{50}, more exactly as biological activity, $A = -\lg ED_{50}$, in which ED_{50} is the molar concentration producing a half-maximal biologic response.

Recognition of R_0 by a "false" ligand, E_i, thus inducing a biological response not required by the organism or blocking the receptor towards a possible "activation" by E_0, would be harmful to the organism. This is also true for the recognition of an E_0 ligand by a false R_i receptor. Quantitatively, specificity may be defined as follows: consider a certain receptor R_0 and N molecular species (generally similar), E_0, E_1, . . . , whose concentration are or may attain the (maximal) values $[E_0]$, $[E_1]$, . . . ,$[E_{N-1}]$. The receptor should recognize (i.e., form a complex with) only the E_0 ligand under these conditions; alternatively, the probability of complex formation with a false ligand, E_i (i = 1,2, . . . ,N − 1) should, at most, be equal to a certain limit, P_{er} (error probability).

This recognition concept can be applied to all the processes of cell regulation and regulation of the metabolic processes. Activation of a biological response means induction of an active receptor conformation by combining with the ligand, very often the activation of an enzymatic activity of the receptor. This implies that the active receptor conformation is not the normal (lowest energy) conformation of the receptor. Alternatively, the ligand may play the role of a coenzyme. Mechanisms by which receptors are triggered to activity after ligand binding were recently reviewed by Hollenberg.[1] There are rather few such mechanisms: (1) the receptor can be part of a ligand-gated oligomeric ionic channel, as the nicotinic, colinergic receptor; (2) the receptor is an enzyme transmembranarly regulated by the ligand, such as tyrosine kinase receptors for insulin and some oncoproteins; and (3) the receptor interacts in a ligand-regulated manner with a membrane-associated guanine nucleotide binding protein. Mobility of the receptors in the membrane plane is important thereby. Ligand binding, in several cases, produces dimerization or oligomerization of receptor proteins.

Inhibition of a biological response also may be the result of a ligand action (antagonistic drugs). In this case, the ligand fixes an inactive receptor conformation; other ligands and molecules may produce inhibition by fixing other inactive conformations of the receptor.

Enzymatic reactions, and transmembranar transport processes are kinetic processes. Recognition here means the enzymatic reaction or the transport; in this case the "receptor" represents the enzyme (enzymatic transport complex) and the "ligand", the transformed substrate, the transported ion. Several "correct" ligands, substrates which are normally transformed or transported, could exist in this case.

In immunologic processes, antigen recognition has antibody-immunocyte receptor sites as the "receptor", R_0, which must recognize a certain foreign antigenic determinant, a macromolecule, E_0. Such a receptor must not recognize the large number of self-antigenic determinants (false E_i) with which it comes into contact. Surface concentrations of antigenic determinants are to be considered, instead of usual concentrations.

II. EQUILIBRIUM THEORY OF RECEPTOR-LIGAND INTERACTIONS

Already, Ehrlich[2] and Arrhenius[3] (quoted by Albert[4]) have assumed that this recognition implies the formation of a complex, C_{00}, between the correct partners, E_0 and R_0, and not between R_0 and a "false" E_i. This distinction implies that the affinity of R_0 for E_0 is higher than for the false E_i. Generally, since a receptor may combine m ligand molecules, the corresponding equilibrium processes can be described as:

$$R_0 + m\,E_0 \rightleftarrows C_{00} \quad ; \quad \frac{[C_{00}]}{[R_0]\cdot[E_i]^m} = K_{00} = \exp\left(-\frac{\Delta G_{00}}{RT}\right) \qquad (1)$$

$$R_0 + m\,E_i \rightleftarrows C_{0i} \quad ; \quad \frac{[C_{0i}]}{[R_0]\cdot[E_i]^m} = K_{0i} = \exp\left(-\frac{\Delta G_{0i}}{RT}\right) \qquad (2)$$

It is hereby assumed that the m-ligand molecules (in solution) do not associate significantly among themselves.

Such equilibrium interactions adequately describe processes such as antigen recognition and activation of an extracellular membrane receptor by a hormone, but processes such as enzymatic reactions or transmembranar transport are kinetic processes. In terms of Eyring's absolute reaction rate theory[5] for such kinetic processes C_{00} and C_{0i} can be considered as activated complexes, and the free enthalpies, ΔG_{00} and ΔG_{0i}, as activation-free enthalpies ($\Delta G^{\neq}$'s).

A more detailed description of receptor-ligand interactions as stoichiometric and equilibrium equations are given by Dean.[6] In the binding of ligands (drugs) to receptors, some authors distinguish two subsequent steps: one so-called recognition step in which mutual electrostatic fields are involved, followed by a second one, the complex formation per se, in which short-range interactions and conformational changes dominate.[7] From a thermodynamic point of view this does not differ from using a single ΔG for the entire process.

With these equilibrium considerations, the probability, p_{er}, of an erroneous function or blocking of R_0 will be[8]

$$p_{er} = \sum_{i=1}^{N-1} [C_{00}] \Big/ \left([C_{00}] + \sum_{i=1}^{N-1} [C_{0i}] \right) \tag{3}$$

that is, the total concentration of false C_{0i} complexes must be small in comparison to that of the correct C_{00} complex. If we consider a large total (stoichiometric) available amount of ligands in comparison to that of the receptors, and if $[E_0]$, $[E_1]$, . . . , $[E_{N-1}]$, are the concentrations of the correct and false ligands, Equations 1 and 2 yield

$$p_{er} = \sum_{i=1}^{N-1} [E_i]^m \cdot \exp\left(-\frac{\Delta G_{oi}}{RT}\right) \cdot \left[[E_o]^m \cdot \exp\left(-\frac{\Delta G_{oo}}{RT}\right) + \sum_{i=1}^{N-1} [E_i]^m \cdot \exp\left(-\frac{\Delta G_{oi}}{RT}\right) \right]^{-1} \tag{4}$$

Using the notation, $\Delta\Delta G_i = \Delta G_{0i} - \Delta G_{00}$, Equation 4 becomes:

$$p_{er} = \sum_{i=1}^{N-1} [E_i]^m \cdot \exp\left(-\frac{\Delta\Delta Gi}{RT}\right) \cdot \left[[E_o]^m + \sum_{i=1}^{N-1} [E_i]^m \cdot \exp\left(-\frac{\Delta\Delta Gi}{RT}\right) \right]^{-1} \tag{5}$$

where $\Delta\Delta G_i$ is the difference between the affinity of R_0 for the correct E_0 ligand and for the false E_i ligands. The larger the specificity (that is, the larger the number N of ligands among which the correct ligand is to be recognized, the larger the concentrations $[E_i]$ of false ligands; and the smaller the error probability, p_{er}), the larger must be the differences $\Delta\Delta G_i$ between the affinities for the correct and false ligands.

If the receptor-ligand interaction has a highly cooperative character ($m >> 1$), possibly in the case of antigen recognition or an antigen-antibody precipitation reaction, the situation is better described by solubility product relationships than by Equations 1 and 2:[9]

$$[R_0] \cdot [E_0] \geq K_{00}^{-1} = \exp\left(\frac{\Delta G_{00}}{RT}\right) \tag{6}$$

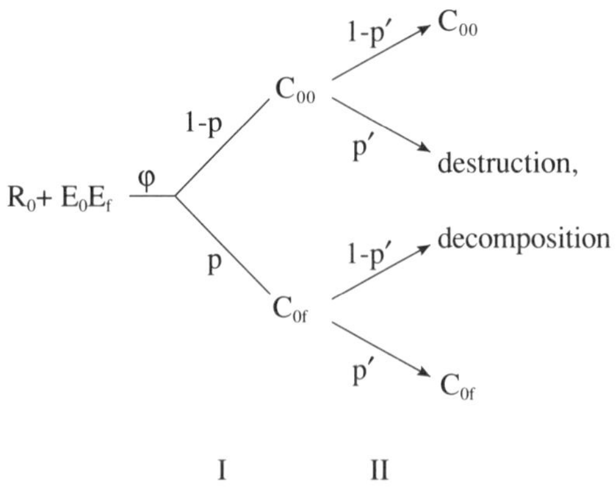

FIGURE 1. Cascade regulation: R_0, receptor; E_0, E_f, correct and false ligands; ϕ, reaction rates; p,p′, probabilities; and C_{00}, C_{0f}, correct and false receptor-ligand complexes.

$$[R_0] \cdot [E_i] \geq K_{0i}^{-1} = \exp\left(\frac{\Delta G_{0i}}{RT}\right) \qquad i = 1,2,\ldots,N - 1 \qquad (7)$$

and the following condition for enthalpy differences results:

$$\Delta\Delta G_i > RT \ln([E_i]/[E_0]) \qquad (8)$$

If phenomena similar to coprecipitation are neglected, no erroneous recognition is possible under such conditions.

A. RECOGNITION WITH SEVERAL SELECTION STEPS

During DNA replication,[10] the precision of a recognition process is verified by a control system which produces the destruction of the false C_{0i} complexes. Such a recognition, with several selection steps, with the goal of decreasing the total error probability, was suggested in 1968 by Scherer[11] under the name of ''cascade regulation'', and by Hopfield[12] as ''kinetic proofreading''. The scheme for such a recognition with two steps is given in Figure 1.

E_f and C_{0f} here represent all false ligands and their complexes. Suppose the rate of the decomposition (step II) to be higher than the total rate, ϕ, of the complex production (step I). With p, the error probability of formation of false complexes, and p′, the error probability for the second step (that is, decomposition of the correct C_{00} or nondecomposition of false C_{0i}), the production rates v_0 and v_f, respectively, of correct and false complexes will be

$$v_0 = \phi \cdot (1 - p) \cdot (1 - p') \quad ; \quad v_f = \phi \cdot p \cdot p' \qquad (9)$$

and the error probability, P, of the two-step recognition process will be

$$P = v_f/(v_o + v_f) \approx P \cdot p' \tag{10}$$

for $\phi << 1$ and $p' << 1$.

Generally, for ν consecutive steps in a recognition process, all with similar error probabilities, p,

$$P \approx p^{\nu} \tag{11}$$

In a recognition process with multiple steps, the first can be exergonic and the following must be endergonic, producing a supplementary energy consumption,[12] but this may not raise special difficulties for the organism.

B. MOLECULAR SIZE AND SPECIFICITY

High specificity of a recognition process implies high molecular size of the R and E partners. Micromolecular ions, S^{2-} for example, precipitates (as sulfides, in this example) tens of cations out of some hundreds of cations in analytical chemistry; ethylenediaminetetraacetic acid (EDTA), with a higher molecular weight, is more selective — it complexes perhaps only 10 or 20 such cations. All molecular species capable of high specific recognition (enzymes, antibodies) are macromolecules.

High specificity implies high $\Delta\Delta G_i$ differences between the affinity of R_0 for the correct E_0, and a high number N of false ligands. One usually supposes that the combining sites of R_0 and E_0 are "complementary", in the sense that they allow the maximal combining affinity. Small structural differences in the combining site of a false E_i will produce only a small $\Delta\Delta G_i$ affinity decrease. Therefore, the combining site of the ligands and the receptor must be large enough to allow a sufficiently large $\Delta\Delta G_i$ for all N, presumably different, combining sites of all false ligands, E_i. The steric shape of such combining sites, their conformational rigidity or flexibility, i.e., ability to accommodate each other in the R–E complexation, will depend also upon steric contacts with other parts of the molecule (of R or E, respectively), which do not come into direct contact upon complexation. The molecular size in the recognition process may also imply the presence of other atoms (from amino acid residues, nucleotides, etc.) than those of the combining site *per se*.

The first quantitative considerations concerning the relation between specificity and molecular size were by Thomas[13] and by Conaughy and McCarthy.[14] They considered the minimum size, n, of RNA oligonucleotides which can be recognized as self or alien in hybridization competition experiments by adsorbed DNA (of viral or bacterial origin). These oligonucleotides are recognized by complementary base pair formation with DNA segments. If the whole DNA has the length of N nucleotides, there are also N overlapping, short oligonucleotidic sequences on this DNA; if their base sequences are more or less randomly distributed, a short RNA oligonucleotide, coded by an "alien" DNA, may have by chance a sequence complementary to sequences contained by the adsorbed (self) DNA. If we want this not to happen, the length n of the "alien" RNA oligonucleotide must be large enough that the total number 4^n of possible "words" in the base-type alphabet with α = four letters should be much higher than the number N of "words" on the adsorbed "self" DNA:

$$\alpha^n >> N \quad ; \quad n > \lg N/\lg\alpha \tag{12}$$

For example, the T4 DNA has $N \doteq 4 \cdot 10^5$ nucleotides; Equation 12 requires $n > 9$; *E. coli* or *Salmonella typhimurium* DNA have $N \doteq 4 \cdot 10^6$ with $n > 11$; and the mammalian genome DNA (haploid complement) has $N = 3 \cdot 10^9$ with $n > 16$. The experimental lengths of oligonucleotides above which species specificity appears, agree with these theoretical length evaluations.[14]

This simple type of quantitative consideration cannot always be applied, because a single difference in a false E_i, compared to the correct E_0 (a single atom or monomer residue substituted in the combining site) may be insufficient to give enough affinity decrease, $\Delta\Delta G_i$ to ascertain the nonrecognition (or with sufficient low p_{er}) of E_i by R_0. In the following we shall present a quantitative evaluation of the size of the site for the general case. Such evaluations have been published by Simon and Ghetie,[9] Simon,[16-18] and by Sadler and Smith.[19] They rely on rather crude, simplifying assumptions, but while Equation 12 of Thomas[13] and Conaughy and McCarthy[14] gives a lower limit for the size n, it is also worthwhile to obtain an upper limit for this size as is required by certain specificity conditions (number N of potential effectors, their concentrations, and the allowed error limit, p_{er}).

1. Notations and Assumptions

We will try to evaluate the size, n, of recognition sites of the R and E partners. The recognition site is to contain, in addition to the combining site (the region upon R and E, respectively, which come into direct contact upon interaction), the part of the molecule (of R and E, respectively) which influences the combining affinity (by interacting with the combining site). We shall assume the same size, n, for the recognition site for both the R and E partners, although this assumption is obvious only for the combining sites.

Receptors are macromolecules; ligands also are often macromolecules. The size usually will be expressed in the number of monomer residues which are of the α type. Alternatively, the number and types of atoms can be used (for micromolecules, small recognition sites).

The "correct" ligand E_0 is considered complementary, as a recognition site, to the receptor R_0. The recognition site of the false ligand E_i differs from that of E_0 by ν_i substitutions (of monomers, atoms).

The assumptions upon which our calculations rely are

1. All ligand recognition sites have the same size, n;

2. The affinity decreases $\Delta\Delta G_i$ are proportional to the number ν_i of substitutions (by which the E_i recognition site differs from that of E_0):

$$\Delta\Delta G_i = \nu_i \cdot \Delta g \tag{13}$$

with Δg being the mean affinity decrease for one substitution; and
3. The ligand recognition sites contain n well-defined and noninterchangeable positions in which monomers are placed, as an n-letter word in an α letter alphabet.

2. Size Evaluation — Precipitation Reaction

Let us consider self-nonself recognition of antigens (antigenic determinants) for elicitation of a primary antibody response of B-lymphocytes.[9] A certain lymphocyte class must recognize a foreign, "strong" antigenic determinant, sometimes at concentrations of $c_0 = 10^{-10}$ *M* or less, but must not recognize antigenic determinants on the surface of blood vessels or of serum globulins (with many of their amino acid sequences identical) at a total concentration of about $c_i = 10^{-3}$ *M*.

The B-lymphocyte antibody receptor is to be considered as R_0, the recognized nonself antigenic determinant as E_0 (with $[E_0] = c_0 = 10^{-10}$ *M*); and the self-antigenic determinants as the false E_i's (maximal concentration, $[E_i] = c_i = 10^{-3}$ *M*) which must be not recognized. According to Equation 8, the maximal affinity difference (at about 300° K) is

$$\Delta\Delta G > 2.3 \cdot RT\ \lg([E_i]/[E_o]) = 1.4 \cdot \lg 10^7 = 10\ \text{kcal/mol} \tag{14}$$

An analysis of affinity decrease, due to amino acid substitution in oligopeptidic antigen-specific antibody interactions, yields a value of about $\Delta g = 2.0$ kcal/mol for the mean affinity decrease per amino acid substitution in the oligopeptide which elicited the antibody.[9,16] The number ν of amino acid substitutions producing the $\Delta\Delta G$ decrease of Equation 14 is thus:

$$\nu = \Delta\Delta G/\Delta g \approx 5 \tag{15}$$

It is more difficult to evaluate the total number, N, of self-antigenic determinants which an antibody must not recognize. If the entire mammalian genome were transcribed, the corresponding amino acid sequence (per haploid genome) would have a length of about $1 \cdot 10^9$ residues. Only a few percent of this genome are transcribed, and several proteins (intracellular ones) do not come into contact with the immune system. A guess of $N = 10^7$ seems reasonable for the number of self-antigenic determinants (short, overlapping, amino acid sequences or oligosaccharide sequences) which the B-lymphocyte must not recognize.

There are $\alpha = 20$ natural amino acid residues, and the lower limit for the size, n, of the recognition site of the antibody receptor, according to

Equation 8, is

$$n > (\lg 10^7)/(\lg 20) \approx 6 \tag{16}$$

We will now obtain an upper limit to the (minimal required) size, n. How large must the recognition site be such that there is only a small probability that neither of the N = 10^7 "words" of length n (written at random in an α = 20-letter alphabet) differs by less than ν = 5 substitutions from the "word" corresponding to the E_0 recognition site?[8] This probability, P, can be evaluated by combinatorics (see Appendix 1) and is

$$P \approx N \cdot C_n^{n-\nu+1} \cdot \alpha^{-(n-\nu+1)} << 1 \tag{17}$$

while for the "word" length, n, the following approximate result is obtained:

$$n > \nu - 1 + \frac{\lg N + (\nu - 1) \cdot \lg\{n/(\nu - 1)\}}{\lg\alpha} \tag{18}$$

Equal frequencies ($1/\alpha$) for each of the α letters (monomer types) within the available pool of all effectors are hereby supposed (Appendix 1). For our special case, with N = 10^7, α = 20, and ν = 5, one obtains n > 12. Our result corresponds to "strong" antigens. X-ray crystallography studies for an antibody indicate a number of n = 17 amino acid residues in the combining site which come into contact with n = 16 residues of the recognized antigenic determinant (of lysozyme).[15]

3. Size Evaluation — General (Noncooperational) Equilibria in Solution

Let us consider the general equilibria as described by Equations 1 and 2 with the error probability, p_{er}, and Equation 5, for the usual noncooperational case, with m = 1.

As an example, we consider the interaction of the lac repressor, considered as R_o, with the lac operator (E_o) and other short, overlapping, nucleotide sequences (the E_i's) on the *E. coli* DNA. The lac repressor-operator interaction is a well-studied process.[20] The *E. coli* DNA has a length of $4.5 \cdot 10^6$ base pairs and this is also the total number, N, of short, overlapping, nucleotide sequences. A 300-fold excess of *E. coli* DNA does not compete favorably with DNA containing the operator gene;[23] an error probability of, at most, p_{er} = 0.1 can be considered for this case. The mean affinity decrease per point mutation is Δg = 2.1 kcal/mol.[23] Here, α = 4, the number of base pair types.

We introduce the notation:

$$r = \exp\left(-\frac{\Delta g}{RT}\right) \quad ; \quad \exp\left(-\frac{\Delta\Delta G_i}{RT}\right) \approx \exp\left(-\frac{\nu_i \cdot \Delta g}{RT}\right) = r^{\nu_i} \tag{19}$$

with which Equation 5 for p_{er} becomes (m = 1):

$$p_{er} = \sum_{i=1}^{N-1} [E_0] \cdot r^{\nu_i}/([E_0] + \sum_{i=1}^{N-1} [E_0] \cdot r^{\nu_i}) \quad (20)$$

For the concentration $[E(\nu)]$ of all ligands, differing by ν substitutions from E_o, we make the natural supposition (similar to Sadler and Smith[15]) that this is proportional to the probability $\pi(n,\nu)$ that a "word" of length n, formed at random in an α-letters alphabet, differs from the word corresponding to (the recognition site of) E_o in ν positions. With $[E_{tot}]$ the total concentration of false effectors:

$$[E(\nu)] = [E_{tot}] \cdot \pi(n,\nu) \quad (21)$$

and Equation 20 becomes, for $p_{er} << 1$:

$$\frac{[E_{tot}]}{[E_0]} \cdot \sum_{\nu=1}^{n} \pi(n,\nu) \cdot r^{\nu} << p_{er} \quad (22)$$

With combinatorics and certain approximations (Appendix 2), one obtains the following result for the site size:

$$n > \frac{\lg\{[E_{tot}]/(p_{er}[E_0])\}}{\lg\alpha - \lg[1 + r \cdot(\alpha - 1)]} \quad (23)$$

With the previously mentioned data, $p_f = 0.1$, $[E_{tot}] = 300 \cdot 4.5 \cdot 10^6 \cdot [E_o]$ (the 300 time excess DNA without lac operator and $4.5 \cdot 10^6$ base pair per *E. coli* DNA) and $r = \exp(-2.1/RT) = 0.027$ (indicated by Smith and Sadler[23]), one obtains $n > 20$.

A theoretical evaluation by Sadler and Smith[15] along similar lines yielded $n > 18$. The lower-limit evaluation of Equation 8 yields $n > 11$. The experimentally determined size by Gilbert and Maxam[24] (length of the *E. coli* DNA fragment protected by the lac repressor from DNase-digestion) is $n = 27$.

APPENDIX 1

The probability P that N "words" of length n in an α-letter alphabet differ with more than ν substitutions from a given "word", corresponding to E_o, is

$$P = 1 - [1 - p(n - \nu + 1)]^N \approx N \cdot p(n - \nu + 1) << 1 \quad (A1)$$

Here $p(n - \nu + 1)$ is the probability that such a word differs by, at most, $\nu - 1$ substitutions from the E_o word, i.e., that an E_i word has the same "letters" in at least $n - \nu + 1$ positions as the E_o word. We consider approximately equal quantities for each monomer type, such that the probability of finding a given "letter" in each position within a "word" is α^{-1}. The probability of finding the same "letters", as in E_o, in $n - \nu + 1$ fixed positions is thus $\alpha^{-(n - \nu + 1)}$; the probability of finding the same letters as in E_o in less than $n - \nu + 1$ positions is $1 - \alpha^{-(n - \nu + 1)}$. The probability of finding the same letters as in E in less than $n - \nu + 1$ arbitrary positions out of a total of n is thus:

$$p(n - \nu + 1) = 1 - [1 - \alpha^{-(n-\nu+1)}]C_n^{n-\nu+1} \approx C_n^{n-\nu+1} \cdot \alpha^{-(n-\nu+1)} \tag{A2}$$

as this $p(n - \nu + 1)$ probability must be small. Thus, the condition is

$$N \cdot C_n^{n-\nu+1} \cdot \alpha^{-(n-\nu+1)} << 1 \tag{A3}$$

With the approximation of Stirling (quoted by Onicescu and Mihoc[25]):

$$\begin{aligned} C_n^{n-\nu+1} &= \frac{n!}{(n - \nu + 1)! \cdot (\nu - 1)!} \\ &\approx \frac{n^n}{(n - \nu + 1)^{(n-\nu+1)} \cdot (\nu - 1)^{(\nu-1)}} \\ &= \left[\frac{1}{1 - \frac{\nu - 1}{n}}\right]^{n-\nu+1} \cdot \left[\frac{n}{\nu - 1}\right]^{\nu-1} \end{aligned} \tag{A4}$$

The condition (Equation 3) now becomes:

$$N \cdot \left[\frac{1}{1 - \frac{\nu - 1}{n}}\right]^{n-\nu+1} \cdot \left[\frac{n}{\nu - 1}\right]^{\nu-1} << \alpha^{(n-\nu+1)} \tag{A5}$$

By logarithmation:

$$\begin{aligned} &\lg N - (n - \nu + 1) \cdot \lg\left(1 - \frac{\nu - 1}{n}\right) \\ &\quad + (\nu - 1) \cdot \lg \frac{n}{\nu - 1} < (n - \nu + 1)\lg\alpha \end{aligned} \tag{A6}$$

and by rearranging, with $\lg(1 - (\nu - 1)/n) << \lg\alpha$, one finally obtains Equation 18 of Section II. B.2:

$$n - \nu + 1 > \frac{\lg N + (\nu - 1) \cdot \lg\{n/(\nu - 1)\}}{\lg\alpha}$$

APPENDIX 2

The probability $\pi(n,\nu)$ that a "word" E_i, in an α-letter alphabet differs from a certain E_0 "word" in ν positions, can be obtained by similar reasonings as in Appendix 1. The probability of having in certain $n - \nu$ positions the same letters as in E_0 and different letters in ν positions is

$$\alpha^{-(n-\gamma)} \cdot (1 - \alpha^{-1})^{\nu}$$

With similar considerations as above:

$$\pi(n,\nu) = 1 - [1 - \alpha^{-(n-\nu)} \cdot (1 - \alpha^{-1})]C_n^{\nu} \approx C_n^{\nu}(1 - 1/\alpha)^{\nu} \cdot (1/\alpha)^{n-\nu} \tag{B1}$$

since the $\pi(n,\nu)$ probabilities are supposed to be very small. Introducing this $\pi(n,\nu)$ into Equation of Section II.B.3 one obtains:

$$\frac{[E_{tot}]}{[E_0]} \sum_{\nu=1}^{n} C_n^{\nu} \cdot \left(1 - \frac{1}{\alpha}\right)^{\nu} \cdot \left(\frac{1}{\alpha}\right)^{n-\nu} \cdot r^{\nu} \leq p_{er} \tag{B2}$$

The sum over ν represents the n-th power of $[1/\alpha + r(1 - 1/\alpha)]$ with the first term, corresponding to $\nu = 0$, missing. One can write:

$$\frac{[E_{tot}]}{[E_0]} \left[\frac{1}{\alpha} + r \cdot \left(1 - \frac{1}{\alpha}\right)\right]^{n} << p_{er} \tag{B3}$$

By logarithmation and some rearrangements, Equation 23 of Section II.B.3. resolves to:

$$n > \frac{\lg([E_{tot}]/p_{er}[E_0])}{\lg\alpha - \lg[1 + r \cdot (\alpha - 1)]} \tag{B4}$$

REFERENCES

1. **Hollenberg, M. D.,** Structure-activity relationship from the receptors point of view, *J. Med. Chem.*, 33, 1275, 1990.
2. **Ehrlich, P.,** *Studies in Immunity,* John Wiley & Sons, New York, 1906.
3. **Arrhenius, S.,** *Immunochemistry,* Macmillan, New York, 1907.
4. **Albert, A.,** *Selective Toxicity,* John Wiley & Sons, New York, 1951.
5. **Lin, S. H., Li, K. P., and Eyring, H.,** Theory of reaction rates in condensed phase, in *Physical Chemistry. An Advanced Treatise,* Vol. VIII, Eyring, H., Henderson, D., and Jost, W., Eds., Academic Press, New York, 1970, chap. 1.
6. **Dean, D. M.,** *Molecular Foundations of Drug-Receptor Interactions,* Cambridge University Press, London, 1987.
7. **Wilting, J., Kremer, J. M. H., Ijzerman, A. P., and Schulman, S.,** The kinetics of the binding of warfarin to human serum albumine as studied by stopped flow spectrophotometry, *Biochim. Biophys. Acta,* 706, 96, 1982.
8. **Simon, Z.,** Recognition between biological macromolecules, *J. Theoret. Biol.*, 9, 414, 1965.
9. **Simon, Z. and Ghetie, V.,** Molecular size and specificity in immunology, *Rev. Roum. Biochim.*, 8, 261, 1971.
10. **Radman, M. and Wagner, R.,** The high fidelity of DNA replication, *Sci. Am.*, 259(2), 24, 1988.
11. **Scherer, K.,** Messenger RNA pattern in animal cells and the concept of cascade regulation, *Abh. Dtsch. Akad. Wiss. Berlin, Kl. Med.*, 259(1), 1968.
12. **Hopfield, J. J.,** Kinetic proofreading: A new mechanism for reducing errors in biosynthetic processes requiring high specificity, *Proc. Natl. Acad. Sci. U.S.A.*, 71, 4135, 1974.
13. **Thomas, C. A., Jr.,** Oligonucleotide size and specificity in DNA-RNA hybridization, *Prog. Nucl. Acid. Res. Mol. Biol.*, 5, 315, 1966.
14. **Conaughy, B. L. and McCarthy, B. J.,** Interactions of oligonucleotides with denaturated DNA, *Biochim. Biophys. Acta,* 149, 180, 1967.
15. **Sadler, J. R. and Smith, T. F.,** Mapping of the lactose operon, *J. Mol. Biol.*, 62, 139, 1971.
16. **Simon, Z.,** Molecular size and specificity of interactions, *Rev. Roum. Biochim.*, 9, 337, 1972.
17. **Simon, Z.,** Theoretic length of the lac operator, *Stud. Biophis. (Berlin),* 26, 179, 1971.
18. **Simon, Z.,** Specific interaction, intermolecular forces, steric requirements and molecular size, *Angew. Chem. (Int. Ed.),* 13, 719, 1974.
19. **Amit, A. G., Mariuzza, R. A., Phillips, S. E., and Pollack, R. J.,** Three dimensional structure of an antigen-antibody complex at 2.8 Å resolution, *Science,* 233, 747, 1986.
20. **Riggs, A. D., Suzuki, H., and Bourgeois, S.,** Lac repressor-operator interactions. I. Equilibrium studies, *J. Mol. Biol.*, 48, 67, 1970.
21. **Riggs, A. D., Newby, R. F., and Bourgeois, S.,** Lac repressor operator interactions. II. Effect of galactosides and other ligands, *J. Mol. Biol.*, 51, 303, 1970.
22. **Riggs, A. D., Bourgeois, S., and Cohn, M.,** Lac repressor operator interactions. III. Kinetic studies, *J. Mol. Biol.*, 53, 401, 1970.
23. **Smith, T. F. and Sadler, J. R.,** The nature of lac operator constitutive mutations, *J. Mol. Biol.*, 59, 273, 1971.
24. **Gilbert, W. and Maxam, A.,** Size of the lac operator, *Proc. Natl. Acad. Sci. U.S.A.*, 70, 3581, 1973.
25. **Onicescu, O. and Mihoc, G.,** *Lectii de Statistica Matematica,* Ed. Tineretului, Bucaresti, 1958, 185.

Chapter 2

MOLECULAR BASIS OF RECEPTOR-LIGAND INTERACTIONS

Zeno Simon, Ioan Moţoc, and Adrian Chiriac

TABLE OF CONTENTS

0-8493-5398-X/93/$0.00 + $.50

I. THE PROBLEM

The aim of this chapter is to discuss the molecular basis of receptor-ligand (effector) interactions, the structural factors of which determine the receptor-ligand affinity. Prediction of biological activities, of structural characteristics pertinent for drug molecule affinities towards a certain receptor also is the key problem for quantitative structure-activity relationships. Elements of quantum chemistry will be discussed in connection with these topics, as well as a more empirical approach to types of intermolecular forces, steric fit, and hydration effects.

Receptor-ligand interactions take place in aqueous solutions. If we want to separate the interactions per se from various hydration-dehydration effects, we obtain the following picture (Figure 1).

The asterics in R* marks the biologically active receptor conformation induced by the ligand. The ΔG free enthalpy change for the combination of the hydrated R and E species can be written (if the PV terms are neglected) as:

$$\Delta G = -\Delta G_1 + \Delta G_2 + \Delta G_3 \quad (1)$$

$$\Delta G_2 = U_{R*E} + \Delta U_{conf} - T\Delta S \quad (2)$$

ΔG_1 and ΔG_3 refer to the hydration of the separated R and E species and of the R*-E complex, respectively; ΔG_2 refers to the R-E interaction per se. ΔU_{R*E} is the potential energy variation for the interaction of the receptor in its biologically active conformation with the ligand; ΔU_{conf} represents the potential energy increase corresponding to the transition of the receptor from

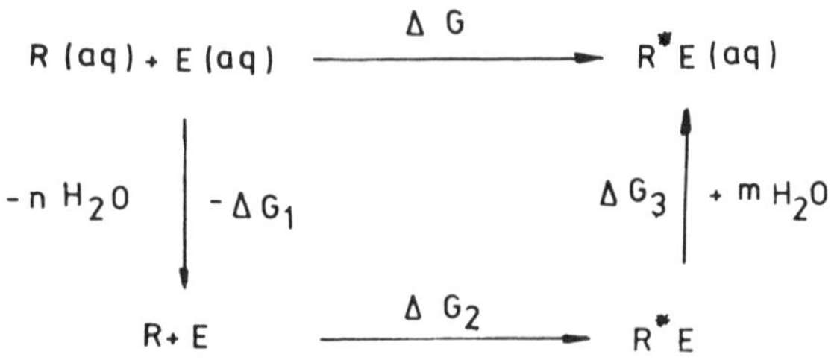

FIGURE 1. Interactions of the receptor, R, and the ligand (effector), E, in aqueous surroundings. The free enthalpies ΔG_1 and DG_3 correspond to hydration effects, ΔG_2 to the interaction of the isolated R and E species.

its "normal", lowest energy, conformation to the active one (found in the R*E complex); ΔS represents the entropy effects for the R-E interaction.

This picture is valid for most biological receptor-chemical messenger interactions, especially for interactions with agonistic ligands. It implies only changes in the intermolecular forces, and does not include rearrangements of chemical bonds. In principle, one complication could appear for antagonistic ligands: binding of E molecules to induce a mixture of several R conformations, all inactive.

In enzymatic reactions, R is the enzyme, E is the substrate, and R*E is the activated complex in the sense of Eyring's absolute reaction rate theory;[1] the ΔG etc. terms of Equations 1 and 2 also refer to the activated complex. Rearrangements of chemical bonds are also to be considered in this case. Transmembrane transport phenomena[2] by ionic channels and carrier enzymes (both in active and passive transport) should be considered in a manner similar to enzymatic processes. Transport by transmembrane diffusion is rather similar to a water:organic solvent partition.

These molecular mechanism problems also are common to the quantitative structure-activity relationship (QSAR) domain. Recent advances in computational chemistry, quantum chemistry, molecular graphics, and problem formulation appear to promise a general solution to QSAR, in what was termed by Marshal and Cramer[2a] as a three-dimensional structure-activity relationship approach. We shall use the term "computational chemistry approach" for it in contrast to the more empirical Hansch-type approach practiced in QSAR since the 1960s.

II. METHODS FOR COMPUTATION OF RECEPTOR-LIGAND AFFINITIES

A. SOME PRINCIPLES OF QUANTUM MECHANICS AND STATISTICAL PHYSICS

Affinities of receptor-ligand binding can be calculated as ΔG differences of free enthalpy, by differences between the G functions of state of the complex and the reactants.[1]

If the difference between G and F is neglected (the PV product is usually negligible for compounds in a condensed phase at the usual pressure, P = 1 atm), we have, for a pure substance, M_i:

$$G_i \cong F_i = -RT \ln Z_i \tag{3}$$

The partition function Z_i, can be calculated from the energy levels ϵ_{ji} of the molecule M_i and the degeneracy degree, g_{ij}, of these levels (the ground state is considered to be of zero energy, $\epsilon_{oi} = 0$):

$$Z_i = e^{-\frac{U_i(o)}{kT}} \cdot \sum_j g_{ji} \cdot e^{-\frac{\epsilon_{ji}}{kT}} \tag{4}$$

$U_i(o)$ is the energy of M_i at 0°K, as related to a certain standard (usually the chemical elements in their natural state at 25°C and 1 atm). The energy of a molecule, U_i (in a certain energetic state), can be written in a fairly good approximation as the sum of energies of various degree of freedom:

$$U_i = \sum_1 (U_{li} + U^i_{el}) \quad ; \quad Z_i = Z_i^{el} \cdot \prod_1 Z_{li} \tag{5}$$

The partition function Z_i results as a product over partition functions for these degrees of freedom, 1i. Another common approach is the Born-Oppenheimer approximation, the separate consideration of the electronic and the nuclear degrees of freedom.

In principle, the ϵ_{ji} energy levels and their degeneracy degrees, g_{ij}, can be calculated by solving the Schrödinger equation for the molecule:

$$\hat{H}_i \cdot \psi_{ij} = (\epsilon_{ij} + U_i(o)) \cdot \psi_{ij} \tag{6}$$

with $\epsilon_{ij} + U_i(o)$ being the energy relative to the component electrons and nuclei in the free state.

For a solution formed of N_1 molecules of type 1, N_2 molecules of type 2, . . . , the partition function is given by:

$$Z = \frac{(N_1 + N_2 + \ldots)!}{N_1!N_2!\ldots} Z_{Z_1}^{N_1} \cdot Z_2^{N_2} \ldots . e^{-\frac{U_{int}}{kT}} \tag{7}$$

In the first approximation of the perturbation theory, the interaction energy between two molecules (in a given reciprocal spatial configuration), U_{12}, will be

$$U_{12} = \int \psi_o \hat{W} \psi_o \, dv \quad ; \quad \psi_o = \psi_{10} \cdot \psi_{20} \tag{8}$$

with $\hat{W}$ describing the interactions (of electrons and nuclei) between molecule M_1 and molecule M_2. The interaction energy of the solution, U_{int}, will be obtained by a mediation over different reciprocal spatial configurations of different types of pairs of molecules (1-2, 2-3, . . .) and also over different energy states (not only the ground state, $j = 0$), taking into account their population at the temperature T.

For the receptor-ligand interaction in aqueous solution, the ΔU terms of the corresponding ΔG (see Equations 1 and 2) can be calculated from quantum mechanical principles, at least in the first approximation, when perturbation of molecular wave functions, ψ_{ij}, by the intermolecular interaction are neglected. This is because in Figure 1 only intermolecular interactions take place, without chemical bond rearrangements (formation of new molecules). Eventually, in the mediation over U_{12}-type terms, one must account for the population changes — an increase of occupation degrees of molecular conformations, which allow a favorable interaction energy U_{int}.

The whole interaction affinity, ΔG of Figure 1, thus can be calculated from quantum mechanical principles. Other thermodynamic functions can be also calculated from the partition function, Z:

$$U = RT^2 \frac{d(\ln Z)}{dT} \quad ; \quad S = R \cdot \ln Z + RT \frac{d(\ln Z)}{dT} \qquad (9)$$

B. QUANTUM MECHANICAL COMPUTATIONS OF INTERMOLECULAR INTERACTION ENERGIES

As discussed in the previous section, the intermolecular interaction energies implied in receptor-ligand complex formation can, at least in principle, be computed from quantum mechanics. The difficulties appear due to the complexity of the problem, the required computing time and capacity, etc.

Recent progress in the use of supercomputers in chemical research was reviewed in two excellent articles by Clementi.[3] The accent is upon the *ab initio* methods. The problem of intermolecular interaction energy calculation implies three steps in this case: computation of molecular wave functions, ψ_{io}; computation of interaction energies of the U_{12} type for fixed reciprocal molecular orientations (Equation 8); and computation of U_{int} by mediation of U_{12} type energy terms.

For calculation of the molecular ψ_{ij} by *ab initio* methods, Clementi[3] recommends the use of Hartree-Fock (HF) functions and the calculation of correlation energies out of HF densities for the eigenvalues. The HF technique is an algorithm of N^4 to N^3 complexity, i.e., the time required for computations increases as the fourth to third power of the number N of basis functions (roughly, of atoms). Density functionals are of N^2 complexity, but any *ab initio* algorithm which tries to improve the HF technique increases the N^4 to N^3 complexity and thus is not feasible for large molecules. Clementi quotes *ab initio* computations for molecules as large as the C–G_{12}-stacked base pairs

(348 atoms, 1632 electrons, ca. 16 h computing time). It is not easy to attest to the precision of the obtained molecular ψ. The calculated atomization energies for molecules of the size of MgO, C_2H_6, and C_6H_6 agree within about 1 eV with experimental values. More directly related to ψ are dipole moments calculated by HF functions for molecules up to the size of SO_2, CH_3NH_2, and pyridine, which agree within about 10% ($0.5 \cdot 10^{-30}$ Cm) with the experimental values (quoted by Naray-Szabo et al.[4]).

Calculation of U_{12}-type terms, starting from calculated molecular ψ by HF functions corrected for electron correlation, are stated by Clementi[3] for H_2O molecules and for $(H_2O)_2$ to $(H_2O)_8$ clusters with various (stable) geometries. In order to obtain a potential energy hypersurface (a set of generalized coordinates to describe the geometry of the interacting H_2O molecules), the calculated interaction energy points are fitted to an analytical expression within the guidelines of some simple physical model. Such interaction energy points also were calculated for interactions between H_2O and amino acid molecules with DNA fragments (up to 30,000 points) or for interaction of H_2O molecules and Li^+ or K^+ counterions with DNA fragments.

The mediation of such point-to-point interaction energies to determine the U_{int} interaction energy within the solution is performed either by Monte Carlo techniques for equilibrium statistical mechanics or by molecular dynamics. Clementi[3] quotes such computations for almost macroscopic batches of H_2O molecules or for thousands of H_2O molecules with a protein (bovine pancreatic trypsin inhibitor, BPTI). The intramolecular potentials used are of a more empirical type, as will be discussed below. The calculated X-ray or neutron scattering intensities of liquid water, as well as the enthalpy, are in good agreement with the experiment. For BPTI in water solution, an average root-mean-square displacement of 1.25 Å is obtained, as compared to the protein structure based upon X-ray data.

The general feeling is that for modeling of interactions of interest to the receptor-ligand problem, the more empirical molecular mechanics methods are the only feasible ones at this time.[5] Significant progress in computing techniques and methods seems to be required before *ab initio* results at the molecular and molecular-pair level can be used directly in processes in water solution of the type depicted in Figure 1.

It is of interest to see in some detail how the U_{12} interaction energy terms are calculated.[6] In terms of the first- and second-order approximation of perturbation theory:

$$U_{12} = \Delta E' + \Delta E'' \ ; \ \Delta E' = \langle \psi_o | \hat{W} | \psi_o \rangle \ ; \ \Delta E'' = \sum_e \frac{[\langle \psi_o \hat{W} \psi_e \rangle]^2}{E_o - E_e} \qquad (10)$$

The $\Delta E'$ term corresponds to the electrostatic interaction and $\Delta E''$ to the polarization or dispersion term, since the excited states in $\bar{\psi}_e$ belong to one molecule or to both molecules, respectively. $\hat{W}$ is the operator for electron-

$$E_{el} = \sum_{\alpha_1} \sum_{\beta_2} \frac{\rho_\alpha \rho_\beta e^2}{R_{\alpha,\beta}} \qquad (12b)$$

$$E_{pol} = 2\sum_{s_2=1}^{n} \sum_{l_1=n'+1}^{\infty} \left[\frac{\left(\sum_{\beta_2} \sum_{\mu_1} c_{s\mu} c_{l\mu} \frac{\rho_\beta e^2}{R_{\beta\mu}} \right)^2}{E_o^{(2)} - E_{s \to l}^{(1)}} \right] +$$

$$+ 2\sum_{k_1=1}^{n'} \sum_{p_2=n+1}^{\infty} \left[\frac{\left(\sum_{\alpha_1} \sum_{\mu_2} c_{k\mu'} c_{p\mu'} \frac{\rho_\alpha e^2}{R_{\alpha\mu}} \right)^2}{E_o^{(1)} - E_{k \to p}^{(2)}} \right] \qquad (12c)$$

$$E_{dis} = 4\sum_{s_2=1}^{n} \sum_{l_2=n+1}^{\infty} \sum_{k_1=1}^{n'} \sum_{l_1=n'+1}^{\infty} \left[\frac{\left(\sum_{\mu} \sum_{\mu'} \frac{c_{s\mu} c_{l\mu} c_{k\mu'} c_{p\mu'} e^2}{R_{\mu\mu'}} \right)^2}{E_o^{(1)} - E_{s \to 2}^{(1)} + E_o^{(2)} - E_{k \to p}^{(2)}} \right] \qquad (12d)$$

FIGURE 2. Expressions of the electrostatic interaction terms, E_{el}, the polarization term, E_{pol}, and the dispersion term, E_{disp} in the intermolecular interaction energy, U_{12} of molecules M_1 and M_2. Here, o_α is the total electric charge of atom α on molecule M_1, o_β of an atom β on molecule M_2. E_o and $E_{s \to l}$ are the energies of the ground state and the excited state by transition from the occupied (in the ground state). ϕ_s molecular orbital upon the unoccupied (in the ground state); ϕ_l molecular orbital; the number of occupied molecular orbitals in the M_1 and M_2 ground states is n and n′, respectively. $R_{\alpha\beta}$, $R_{\alpha\mu}'$, and $R_{\mu\mu}'$ are distances between atomic centers on molecules M_1 and M_2, respectively.

electron, electron-nuclei, and nuclei-nuclei interactions in which one partner is on molecule M_1 and the other on M_2. The wave functions $\overline{\psi}_o, \overline{\psi}_e$ are products, $\psi_1 \cdot \psi_2$, of wave functions for molecule M_1 and M_2, respectively. In turn, the molecular wave functions, ψ, are antisymmetrical products of molecular orbitals, ϕ_s linear combinations of atomic orbitals, ρ_μ:

$$\phi_s = \sum_{\mu} c_{s\mu} \cdot \rho_\mu \qquad (11)$$

The interaction energy, U_{12}, thus becomes

$$U_{12} = E_{el} + E_{pol} + E_{dis} \qquad (12)$$

These terms are detailed in Figure 2.

C. INTERMOLECULAR FORCE POTENTIALS — THE MOLECULAR MECHANICS APPROACH

The molecular mechanics method is a more pragmatic but very successful approach to molecular structure.[7,8] The interacting molecules are considered as a set of balls and springs with a series of potential energy functions expressing the molecular force field as a sum of these forces. The nuclear motions are considered as classical within this approach. The potential energy function is of the type:

$$U = E_{stretching} + E_{bending} + E_{dihedral} + E_{van\ der\ Waals} + E_{electrostatic} \quad (13)$$

The forms of these terms are[15]

$$E_{stretching} = \sum k_b \cdot (b - b_o)^2 \quad (14a)$$

$$E_{bending} = \sum k_o \cdot (\theta - \theta_o)^2 \quad (14b)$$

$$E_{dihedral} = {}^1\!/_2 \sum V_\phi \cdot [1 - \cos(n\phi + \sigma)] \quad (14c)$$

$$E_{van\ der\ Waals} = \sum (-A_{ij} \cdot R_{ij}^{-6} + B_{ij} \cdot R_{ij}^{-12}) \quad (14d)$$

$$E_{electrostatic\ and\ H\text{-}bonds} = \sum \frac{1}{4\pi\epsilon_o D} \cdot \frac{\rho_i \cdot \rho_j}{R_{ij}} \quad (14e)$$

Harmonic potentials are considered for the stretching and bending terms. In Equation 14a the sum is performed over the normal vibrational coordinates, b; in Equation 14b it is over the bond angles, θ. For the dihedral terms (Equation 14c), θ are torsional angles and V_ϕ the corresponding rotational barriers. In the van der Waals terms (Equation 14d) ij are all pairs of non-bonded atoms on the same or on different molecules and R_{ij} represents the distances between them. The electrostatic terms (Equation 14e) have ρ_i and ρ_j as electric charges of interacting atoms, D, as effective dielectric constants. Finally, for hydrogen bonds, special potentials are sometimes considered for the pairs of atoms between which such bonds are formed. This potential energy function contains a great number of parameters characterizing various vibrational potentials (k_b, k_θ, V_ϕ) and interatomic interactions (A_{ij}, B_{ij}, ρ_i, ρ_j, D) which are taken from molecular physics — vibrational spectra, experimental data for interactions in very small-molecule systems. Each of these terms has preferential equilibrium positions (b_o, θ_o, etc.) and deviations from these equilibrium positions increase the energy, U. The total energy of a molecule or of a group of interacting molecules can be a sum of several thousands of such individual contributions. This set of potential energy functions and their associated sets of numerical parameters constitutes the ''force field''.

Despite all the successes obtained, force field development remains a major problem because of the large variety of complex functional groups encountered in biochemistry and medicinal chemistry and because not all force fields are readily transferable from one package to another. Of the most extensively used force fields, MM2 is the current standard for small molecules, but is a poor choice for macromolecules. AMBER and CHARM are used for peptides and nucleic acids, and ECEPP for peptides, but these are rather poor choices for other macromolecules. For calculations involving small-molecule/macromolecule interactions, hybrid force fields such as AMBER all-atom force fields are used.[5]

Energy minimization in a molecular mechanism involves successive iterative computations, where an initial conformation is submitted to full geometric optimization. The geometric parameters are modified in small steps until U reaches a local minimum, which is not necessarily the absolute nor the global minimum. Several methods were developed in the search for this global minimum, also based on molecular dynamics, distance geometric restraints for some, and interatomic distances; near-global optimization is stated to have been obtained in some cases.

The molecular mechanics approach is successful in predicting the stereochemistry of small molecules and oligopeptides — even of proteins — when an initial, sufficiently correct conformation is submitted to optimization.[5,9] To date, no successful predictions of tridimensional protein structure from primary structure have been possible, neither by molecular mechanics, nor by pattern-recognition-based programs.[10,11] The great difficulty is given by the astronomic number of the possible interactions between amino acid residues which are at great distances in the primary polypeptidic structures. In principle, for a peptide formed of N amino acid residues there are 9^N possible conformations. A great number of them can be eliminated because of steric superposition, but the number of conformations to be searched is astronomical for a common protein. As an example, the enzyme papain with $N = 212$ amino acid residues is calculated to have $7 \cdot 10^{48}$ possible backbone conformations,[12] according to a formula by Dill[13] (number of conformations $(Z/a)^N = 1.7^N$).

Many more protein sequences than crystal structures are available, and this gap is expected to grow as DNA sequencing methods become faster. The number of structures of proteins and nucleic acids solved by X-ray crystallography was recently evaluated at over 300. The so-called 2D-NMR techniques, where availability of high quality crystals is not a limiting factor, can now provide three-dimensional structure information on small proteins (up to 100 to 150 residues) and DNA in solution.[14] Protein sequences with high sequence homology to proteins with solved tridimensional structure can be predicted rather successfully. Nevertheless, for the majority of protein sequences with little homology to known structures, the problem of predicting secondary and tertiary structure still seems insurmountable (error rates for secondary structure prediction are usually greater than 40%).[5]

To conclude, the molecular mechanics method can be considered suitable for calculating receptor-ligand ΔU_{R*E} interaction energies of the ΔG receptor-ligand affinities of Equations 1 and 2. This method will probably not be suitable for calculating the ΔU_{conf} energy increases induced by conformational changes within receptor sites in the receptor-ligand complex formation.

D. MOLECULAR DYNAMICS — FREE ENERGY PERTURBATION METHOD — HYDRATION AND ENTROPY EFFECTS

The discussion, in only a single section, of both the effects and computational methods is justified by the fact that theoretical computations of hydration effects imply interactions of the hydrated molecules with large numbers of configurations of water molecules. Water has a flexible molecular structure. Hydration effects rely strongly upon entropy effects. Molecular dynamics is the method used to consider intermolecular interactions and the motions they induce in the interacting molecules. The free energy perturbation method allows one to compute differences in interaction ΔG corresponding to a substitution in one molecule.[9]

Molecular dynamics has considerably enlarged our static view of molecules by treating each atom as a particle responding to Newton's equation of motion. The actual, instantaneous motion of the system is simulated. The acceleration for each atom is calculated[5,15] as follows:

$$\vec{a} = \vec{i}\,\frac{d^2x}{dt^2} + \ldots = \frac{1}{m}\left[\vec{i}\left(\sum_b \frac{\delta u}{\delta b}\cdot\frac{\delta b}{\delta x} + \sum_\theta \frac{\delta u}{\delta \theta}\cdot\frac{\delta \theta}{\delta x} + \ldots\right) + \ldots\right] \tag{15}$$

starting from a given configuration of the N interacting molecules. The displacements of atoms during a time interval t (typically $5 \cdot 10^{-16}$ s) are calculated; then the new positions at step (n + 1) are

$$x_{n+1} = 2x_n - x_{n-1} + {}^1\!/_2\, a_{xn}(\Delta t)^2 \quad ; \quad V_{xn} = (x_{n+1} - x_{n-1})/2\Delta t \tag{16}$$

and new values result for the force components ($\delta U/\delta b \cdot \delta b/\delta x$, . . . , etc.) at the new molecular and atomic configurations. Before starting the computations, the molecules are "heated", thus attributing random speeds to the particles. This "heating" is usually achieved gradually, each step adding speeds corresponding to ΔT = 1°K.

Successive integrations of the Newtonian motion equations lead to trajectories of the atoms in time in the form of a list of positions and velocities, typically through periods of 1 to 100 ps. The motions of atoms, chemical groups, and molecules obtained make it possible to understand phenomena that cannot be explained by the static view. During short time segments, little coherence appears in the displacements, but over longer periods collective motions start to develop, with some groups fluctuating more than others.

Such calculations imply the availability of good computational power and appropriate graphic facilities, without which the huge amounts of data produced by the simulations are difficult to interpret.

Molecular dynamics is useful in searching for energy minima and for preferred conformations for small molecules and macromolecules corresponding to a given force field. It can identify preferred motions of molecules or proteins and provides a high information content picture of the behavior of a molecule interacting with a certain partner.

Restrained molecular dynamics and distance geometry, restraining specific distances and dihedral angles, have been used to generate three-dimensional structures of small molecules, proteins, and nucleic acid consistent with NMR data. As an example, the bovine pancreatic trypsin inhibitor, as studied by molecular dynamics, implies 527 atoms (1581 dimensions), 66,000 successive Δt time intervals, and 111 ps total time.[5,15]

Atom mobility on the surface of 22 different proteins was studied by Tainer et al.[16] Reduced mobility regions are frequently found at the interfaces of protein subunits of oligomeric proteins and at the interface of molecules adjacent in crystal nets. High mobility regions are related to lipophilic regions.

Docking of drug molecules into a receptor site, nevertheless, seems rather complex for pure molecular dynamics simulations and requires interactive procedures with molecular surface displays (''extra radius'' surfaces, hydrophobic or electrostatic potential, color coding) used to guide the fit.[5]

Hydration processes are also simulated by Monte Carlo techniques.[3] Here, N molecules are placed into a volume V at a temperature T and millions of geometric configurations are generated randomly. Interaction energies are calculated by a given force field, Boltzman statistic mediations are performed, and limiting conditions are imposed. In both types of simulations for hydration processes, either the Monte Carlo or the molecular dynamics process, for simulations of hydration of a protein molecule (the bovine pancreatic chymotrypsin molecule with about 3000 H_2O molecules, as an example), the X-ray crystallography conformation of the macromolecules is introduced in the ''equilibrated'' water, and cutoff radii for various intermolecular forces are used. Several such simulations for hydrations of ions, DNA molecules, etc., are reported, but less information is available about computed equilibrium properties.[3]

A major problem is the even qualitative estimate of the binding ΔG between two molecules in aqueous solution.[5] An important advance is the application of the free energy perturbation method which uses the properties of a thermodynamic cycle[9] (Figure 3). The transfer of a drug molecule from solution to the receptor binding site, compared with the transfer of its analogue, is simulated by an equivalent nonphysical process, the ''mutation'' of the drug molecule to its analogue, performed both in solution and in the binding site. The ''mutation'' consists of a gradual change in the parameters

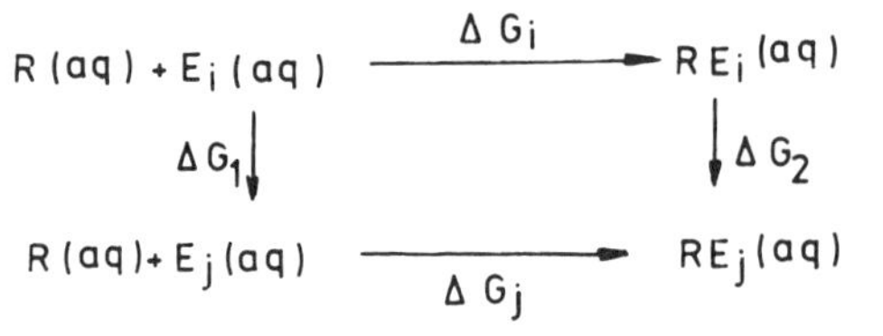

$$\Delta(\Delta G) \equiv \Delta G_i - \Delta G_j = \Delta G_1 - \Delta G_2$$

FIGURE 3. Calculation of the $\Delta(\Delta G)$ difference in the free energy perturbation method; ΔG_1 and ΔG_2 are the free enthalpy changes calculated for the gradual modification of the ligand molecules E_i to the molecule E_j, in aqueous solution and in the complex with the receptor, R, respectively.

of the initial drug molecule to those of the analogue. The change is performed during a molecular dynamic simulation, once in a solution (a box of several hundred water molecules) and, again, in the receptor site. Such simulations require large amounts of supercomputer time, but they offer the possibility of calculating accurate differences in the binding ΔG between related ligands.

Calculations of ΔG differences by the free energy perturbation method were performed for the binding to trypsin of benzamidine vs. *p*fluorobenzamide, the binding of several thermolysin inhibitors to thermolysin, or of a single thermolysin inhibitor to different mutant thermolysins, etc.[5] Although it is difficult to verify when such a simulation has converged, current results suggest that $\Delta\Delta G$ for ligand-macromolecule binding can be calculated to a precision of ± 1.5 to 2.0 kcal/mol.[17]

As a conclusive comparison to our first problem section, the correct calculation of the hydration effects ΔG_1 and ΔG_3 is becoming possible by molecular dynamics techniques with the molecular mechanics force-field approach. At the moment, large amounts of supercomputer time are required for such calculations, and the precision is difficult to estimate. As to the calculation of receptor-ligand affinities in aqueous surroundings, the free energy perturbation method is the only one able to do this. More exactly, it allows one to calculate $\Delta\Delta G$ and to compare a drug molecule's affinity for the receptor with that of similar molecules at a reasonable precision. Here, also, the condition is that various analogues do not induce different local conformational changes (different ΔU_{conf}) within the receptor.

III. INTERMOLECULAR FORCE TYPES — AN INTUITIVE APPROACH

A. INTERMOLECULAR FORCE TYPES — THE COMPLEMENTARITY CONCEPT

The complementarity concept in biomolecular interactions was introduced by Watson and Crick to explain the DNA double helix structure and, later,

TABLE 1
Types of Intermolecular Forces — Potential Energy Function

Interaction	r	Potential function	Typical value (kcal/mol)
Ion-ion	0.92	$305 \frac{Z_1Z_2}{RD}$	87.0
Ion-dipole	0.83	$57.4 \frac{Z \cdot \mu}{R^2D}$	8.5
Dipole-dipole	0.75	$-10.8 \frac{\mu_1\mu_2}{R^3D}$	0.85
Ion-induced dipole	0.67	$-111 \frac{\alpha \cdot Z^2}{R^4D^2}$	1.48
Dipole-induced dipole	0.50	$-7.2 \frac{\alpha \cdot \mu^2}{R^6D^2}$	0.03
Dispersion forces	0.50	$-0.75 \frac{\alpha_1\alpha_2}{R^6} \cdot \frac{I_1I_2}{I_1+I_2}$	0.18

Note: r — correction factor for repulsive forces; Z — charges (e units); μ — dipole moments (D units); α — polarizabilities (Å^3); R — interdistances (Å); I — ionization potentials (eV); and D — dielectric constants.

the code for transcription of mRNA from DNA. One often assumes that the molecule of the natural ligand (effector) for a certain receptor in the molecule has maximal affinity for this receptor (steric fit and matching of groups allowing attractive interactions); i.e., the two partners are complementary.

Complementarity conditions are rarely as clear-cut as in the above-cited two examples. Interaction ΔG can rarely be calculated by additive schemes from simple intermolecular force considerations. Nevertheless, a classification of intermolecular forces in different types is very helpful in understanding the mechanism of specific interactions with biomolecules.

Some types of intermolecular forces and the formulas for interaction energies between pairs of atoms (groups) are listed in Table 1 (data from the treatise of Webb[18]). The types of intermolecular forces to be considered in biomolecular interactions are

1. Ionic groups (electric charges), positive or negative;
2. Hydrogen bonds, proton donating, accepting, or both (the OH group);
3. Hydrophobic interactions, which can be augmented by
4. Stacking interactions (aromatic character); and
5. Charge transfer complexes, electron donating or accepting character.

One possibility for using these force types is the following. In the formation of the receptor-ligand complex, the liberty of movement of the groups (the two partners coming into contact) is reduced, with or without an attractive interaction. The contribution to the ΔG of interaction by substitution of one group by another, upon one of the molecules, can be evaluated by the typical binding energies for the implied intermolecular forces.

The numerical factors which multiply potential functions in Table 1 are for conversion into kcal/mol when Z is the valence of ions, μ the dipole moments in Debye (D); R the distances in Å, α the polarizabilities in $Å^3$, D the effective dielectric constants, and I the ionization potentials in kcal/mol. The correction factor r for the effect of repulsive forces at equilibrium distance is included in the numeric factor, which is given for the most favorable spatial orientations. The potential functions are usually obtained by assuming large R distances compared to the diameter of, supposed, spherical particles. The typical values are for R = 3.5 Å (the usual van der Waals contact distances); D = 1; μ = 1.84 D (H_2O molecule); α = 2.0 $Å^3$ (–CH_2 group); and I = 220 kcal/mol (C_6H_6; 9.20 eV). Absolute values, in kcal/mol, for the corresponding interaction energies are listed in Table 1.

B. POLAR INTERACTIONS

Polar interactions are a principal component of molecular mechanics force fields. Because water is a strong polar medium, their contribution to receptor-effector interactions in an aqueous environment is small, but matching or mismatching of polar groups on the interacting partners is very important for specificity. A glance at the typical interaction energy values of Table 1 shows that an interaction of two ± 0.15 e charges at the van der Waals contact distance (3.5 Å) contributes about 1.5 kcal/mol to interaction energy, i.e, a $10\times$ variation of biological activity (ED_{50}).

The importance of charge distributions within the molecules is obvious. There are no direct experimental methods to determine the total electric charge of an atom in a molecule. As discussed in the previous section (see also References 4,5) advanced quantum chemistry methods have not provided, to date, sufficient reliable results for such charge distributions. Molecular mechanics uses rather semiempirical methods based upon partial equalization of valence orbitals electronegativities.[19,20] Semiempirical methods, calibrated for dipole moments in small molecules, such as the Del Re method for phosphine derivatives (with σ bonds, only)[21] or a combination of partial equalization of electronegativities (σ bonds) and the simple ω-HMO method (π electronic system)[22] for various organic molecules seems to give rather satisfactory results (about 10%, ± 0.20 D for aliphatic phosphines[21] and for various other organic molecules[22]). More elaborate quantum chemistry methods, such as MNDO or HM1 are often used, with special procedures to determine point charges which reproduce the molecular electrostatic potential as calculated from the wave function.[23] The principle of equalization of orbital electronegativities was also used recently; the resulting atomic charges are said to compare satisfactorily with NMR and X-ray crystallography results, as well as with the results of more elaborate quantum chemistry methods.[24]

In the following, we shall consider various types of polar interactions (ion-ion, ion-dipole, hydrogen bonds), especially experimental values for their strengths (ΔG,ΔU,ΔS) in interactions in aqueous solutions.

1. Electrostatic Interactions Between Ionized Groups

Most biological processes take place in neutral solutions ($pH = 7$); groups with $pK > pH \cong 7$ are predominantly protonated and groups with $pK < pH \cong 7$ are predominantly deprotonated.

As regards the ΔG for cation-anion bonds in aqueous solution, they are much smaller than in crystals. Experimental ΔG's for such bonds were obtained from competitive equilibria for a given site (e.g., an antibody) between two molecules of similar shape and size, of which only one bears an ionized group (for example, antigenic determinants, one carrying a CMe_3 group, the other a NMe_3^+ group; one with a NO_2, the other with a COO^- group). For the anion-cation bond with NMe_3^+, a $\Delta G = -1.2$ kcal/mol was calculated and for the bond with COO^-, $\Delta G = -4.8$ kcal/mol was calculated. From binding of various substrates to acetylcholinesterase the $-COO^-$ to $-NMe_3^+$ bond $\Delta G = -1.2$ to -2.1 kcal/mol is inferred. The relative strengths of the ionic bond appear to decrease in the order: guanidinium (arginine) > $-NH_2$(lysine) > imidazolic N(histidine).[25]

The specific bonding of some alkali or alkaline earth ions in kryptates is probably due to the radius of the cation (or hydrated cation) fitting that of the relatively rigid ionic cavity. Metallic ions, especially those of the transitional series, can become specifically bonded to certain groups by means of complexing effects. Cation-ligand interactions in such cases are governed by the principle that cations of the "hard" Lewis acid type prefer ligands of the "hard" Lewis base type, while "mild" cations prefer "mild" ligands. According to increasing "hardness" (in H_2O) the usual cations can be classified from "mild" to "hard": Hg^{2+}, Ag^+, Cu^+, Hg^+, Cu^{2+}, Fe^{2+}, Co^{2+}, Ni^{2+}, Mg^{2+}, and Al^{3+}. The usual ligands, classified from "mild" to "hard", are I^-, HS^-, CN^-, Br^-, Cl^-, pyridine, aliphatic amines, H_2O, HO^-, COO^-, PO_4^{3-}, and F^-. Of course, such interactions are not purely electrostatic.[25]

Bonding of cations to native DNA presents some base specificity: dGC-rich DNA binds Ag^+ stronger than AT-rich DNA. Preference for dAT-rich DNA present the Hg^{2+}, Me_4N^+, and $EtMe_3N^+$ cations, while Na^+, Li, Cs^+, Cu^{2+}, arginine, lysine, and spermine bind equally to DNA, irrespective of base composition. Such preferences are weak, corresponding to a ΔΔG of about 0.2 kcal/mol.[26] Some special problems for polyelectrolytes are reviewed by Brenner and McQuarrie[27] and Katschalski.[28]

2. Hydrogen Bonds

Hydrogen bonds are formed between a sufficiently strongly polarized X-H bond and a pair of free electrons (on a sp^2 or sp^3 orbital) of a sufficiently electronegative Y atom. The enthalpy effect, ΔH, for hydrogen bonds is

TABLE 2
Thermodynamic Data To Hydrogen Bonds (Data Per Bond)

X–H . . . Y	System	−ΔH	−ΔS	−ΔG
O–H . . . O	CH_3COOH	7.0	18	−1.6
	H_2O	5.0	35.8	+2.7
	CH_3OH	4.5	20	+1.5
	C_6H_5OH	4.3	25	+0.2
N–H . . . O	$CH_3CONHCH_3$	3.9	13	0
	Diphenylamine-dioxane	2.3	16	+2.5
	Diphenylamine-$HCONMe_2$	3.6	16	+1.2
	δValerolactame	5.15	11	−1.8
N–H . . . N	NH_3	4.4	26.8	+3.6
	CH_3NH_2	3.4	22.0	+3.2
F–H . . . F	HF	6.8	3	−5.9
C–H . . . N	HCN	4.0	16.5	+1
S–H . . . O	$C_6H_5SH + (CH_3)_2CO$	3.2	13	+0.9
S–H . . . N	C_6H_5SH + pyridine	3.4	11	+0.7
O–H . . . S	$C_6H_5OH + (C_2H_5)_2S$	3.4	8	−1.0

Note: ΔH and ΔG in kcal/mol, ΔS in cal/mol K, T = 300°K.

Data are from References 29 and 30.

negative, but the entropy effect, ΔS, is also negative, so that ΔG is close to zero.

Table 2 lists such thermodynamic values for hydrogen bond formation in nonpolar solvents at 300°K. Hydrogen bonding is favored by conjugation effects which entail formation of partially positive charges on atom X and partially negative charges on Y.

In aqueous media, the hydrogen bonding also implies a rearrangement of hydrogen bonds in water itself:

$$X - H\ldots OH_2 + Y\ldots H_2O \qquad X - H\ldots Y + H_2O\ldots H_2O$$

As examples, the dimerization of $CH_3CONHCH_3$ in aqueous solution (formation of one hydrogen bond) has, at 25°C, ΔH ≅ 0 kcal/mol, ΔS = −10 cal/mol°K, and ΔG = +3.1 kcal/mol. From data for melting curves of polyalanine in water, for the interaction of H_2O with the –CONH– group in polypeptides, ΔH = −2.74 kcal/mol, ΔS = −5.8 cal/mol°K, and ΔG = −1.6 kcal/mol. For the OH (tyrosine)-carboxylate bond ΔH = −7.7 kcal/mol, ΔS = −24 cal/mol°K, and ΔG = −0.55 kcal/mol; and for hydrogen bonding between urea molecules in water ΔH = −1.4 kcal/mol and ΔG = −0.4 kcal/mol.[25]

The hydrogen bond produces a decrease in ΔS, owing, first of all, to the loss of entropy of mixing. The forming of a second and third hydrogen bond between a pair of associated molecules or macromolecules entails a slowdown in the decrease of ΔS. A more negative contribution to ΔG will be given by the second and third hydrogen bonds, rather than by the first. For base pair formation between pyrimidine and purine bases in aqueous solution, the ΔG values range between -0.7 and -2.6 kcal/mol.[34]

Hydrogen bonds can strengthen sensibly the interaction between ionized groups (saline bond). For carboxylate-ammonium group interaction upon copolymers in aqueous solution, $\Delta G \cong -6$ kcal/mol.[25]

In the case of hydrogen bonding between the two strands of a nucleic acid double helix, hydrogen bonding groups are directed towards the interior of the already dehydrated helix. The "crystalline" structure has already reduced the freedom of motion so that the hydrogen bond formation contributes with $\Delta S \cong 0$; $\Delta G \cong \Delta H$ to the ΔG of the double helix formation. For false base pairing in these conditions, if two X-H . . . H-X or Y . . . Y groups are side by side (repulsive interaction between two dipole moments) the contribution to the ΔG of double helix formation will be of about $-2\Delta H$, that is, in the order of $+10$ to $+15$ kcal/mol. This explains the role of hydrogen bonds in observing complementarity.

The data presented in Table 2 refer to interactions in nonpolar solvents; they correspond to the difference between the X-H to Y interaction in the actual hydrogen bond and the interactions of these groups separately with the nonpolar medium. Therefore, the difference between the ΔG for the X-H . . . Y interaction in a cavity and ΔG for interaction of X-H or Y with a nonpolar cavity will be of the order of ΔH for the X-H . . . Y interaction in a hydrophobic medium. The freedom of motion is reduced to approximately the same extent in both cavities (polar and nonpolar), so that the difference between the entropy effects for the two cases will be approximately zero.

A complete description of hydrogen bonding in the framework of a multiplicative approach was given by Raevsky et al.[25a]. The ΔG and ΔH for various hydrogen bonds can be calculated using factors characterizing the proton donating and accepting abilities of various groups.

3. Ion-Dipole and Other Interactions in Water Solution

Ion-dipole interactions appear in hydration of ions. The ΔG for hydration of univalent cations and anions is in the range of -70 to -100 kcal/mol; larger for di- or trivalent ions, and generally large and negative for polar, hydrogen-bonding groups. In receptor-ligand interactions, these ΔG intervene as relatively small differences between large quantities.

If interaction between two atoms or groups takes place through one or more layers of water molecules, all electric interactions are modified by the dielectric constant, D. The short-range dispersion and repulsive forces, negligible at distances larger than direct contact, are not modified. The difficulty is that the macroscopic value, $D = 80$, is only utilizable for distances larger than 20 Å. For distances between 3 and 7 Å one makes use of equations of the type: $D = 6R - 7$.[25] Another complication comes from the ionic atmosphere — the influence of the solution's ionic strengths. All these facts make direct calculations of interaction energies in water solution rather difficult.

C. APOLAR INTERACTIONS

The driving force of receptor-ligand interactions are the so-called hydrophobic interactions or hydrophobic forces. These forces represent a complex effect whose main mechanism is the reduction of the number of water molecules in direct contact with apolar molecules or groups:

$$A \cdot nH_2O + B \cdot mH_2O \rightarrow AB + \frac{n + m}{2}(H_2O \ldots H_2O) \qquad (17)$$

H_2O molecules liberated in this reaction can form an increased number of hydrogen bonds. Between the A and B molecules, or between A and H_2O, or B and H_2O only weak dispersion forces are exerted. Such interactions are sometimes reinforced by stronger dispersion forces between planar surface of superposed aromatic molecules (increased molecular refractions, α) — the so-called stacking interactions. If one of the two partners has an electron-donating character (high polarizability), and the other an electron-accepting character, charge transfer complexes (CT complexes) are formed.

1. Hydrophobic Forces

If an apolar molecule or group, A, comes into contact with the H_2O molecules of water, the H_2O molecules near the apolar, hydrophobic surface become structurally more rigid, more similar to H_2O molecules in ice crystals. The entropy of formation of a hydrophobic bond between two apolar groups (like $-CH_2-$; Equation 17) is positive, $\Delta S = +1.7$ to $+11$ e.u., $\Delta H = +0.3$ to $+1.8$ kcal/mol, and $\Delta G = -0.2$ to -1.5 kcal/mol.[32] Interactions of the same type, solvophobic interactions, also occur in other solvents such as glycerin, formamide, ethylenediamine, propan-1,3-diol, etc., whose molecules have two or more hydrogen-bonding centers (both proton donating and accepting), and so are able to form three-dimensional networks through hydrogen bonds.[33]

The gain in ΔG in hydrophobic interactions comes from the reduction of apolar area in contact with water. This effect was evaluated to be 24 cal/Å^2 for amino acid side chains[34] and 20 to 25 cal/Å^2 for the transfer of various hydrocarbon molecules from water into a hydrocarbon as solvent.[35]

The ΔG for hydrophobic interactions of a molecule, A, is obtained from the study of the equilibrium of partition between water and a hydrophobic solvent, usually octanol (saturated with water):[36]

$$A(aq) \rightleftarrows A(oct) \qquad (18)$$

TABLE 3
Hydrophobic Group Increments[36]

Structural element	$\pi_X + \Delta\pi_X$	$\pi_X^{ar} - \pi_X^{al}$	Amino acid[42] (side chain)	π
–H	0.75		Gly(H)	0
$-CH_2-$	0.50 ± 0.02		Ala(CH_2)	0.52
Ramification	−0.20 ± 0.05		Val(iC_3H_7)	1.21
Ring closure	−0.09 ± 0.02		Leu(Me_2 $CHCH_2$)	1.78
Double bond	−0.30 ± 0.05		Ile($CH_3CH_2CH(Me)-$	2.12
Folding of $-(CH_2)_n$	−0.60 ± 0.05		He($C_6H_5CH_2$)	1.89
Chain (n ≥ 4)			Pro(C_4H_8N)	1.86
Intermolec. H–bond	−0.65 ± 0.10		Met($CH_3SCH_2CH_2$)	0.93
–COOH	−0.65 ± 0.03	0.39	Cys($HSCH_2$)	0.71
–OH	−1.16 ± 0.03	0.49	Tyr($CH_2C_6H_4OH$)	2.12
$-NH_2$	−1.16 ± 0.03	−0.04	His($C_4H_5N_2$)	0.78
C=O	−1.21 ± 0.03	0.16	Ser($HOCH_2$)	0.03
–CN	−0.84 ± 0.04	0.26	AsN(CH_2CONH_2)	−0.01
–O–	−0.98 ± 0.05	0.45	GlN(CH_2 CH_2CONH_2)	−0.07
$-CONH_2$	−1.71 ± 0.05	0.22	Asp(CH_2COOH)	0.38
–F	−0.17 ± 0.03	0.31	Glu(CH_2CH_2COOH)	0.39
–Cl	0.39 ± 0.04	0.32	Lys($C_4H_8NH_2$)	1.07
–Br	0.60 ± 0.04	0.26	Arg($C_3H_6CH_4N_3$)	0.52
–I	1.00 ± 0.05	0.12		
$-NO_2$	−0.85 ± 0.05	0.57		

Note: Group increments refer to groups bonded to aliphatic residue. Uncharged side chains are listed; for a charged chain, an increment of about −4 is added.

The partition coefficient P_{oct} for this equilibrium, defined as the quotient of equilibrium concentrations in octanol and water, respectively, is related to the ΔG for process (Equation 18) at about 300°K by:

$$\Delta G_A(\text{kcal/mol}) = -RT \ln P^A_{oct} \cong -1.4 \lg P^A_{oct} \cong -1.4\pi_A \qquad (19)$$

This hydrophobicity, π_A, of a substance can be calculated with good results as the sum of group contributions (fragmental constants), or at least the hydrophobicity of substitution derivatives can be calculated from the hydrophobicity of the "unsubstituted" derivative and the group contributions of the substituents. Such group contributions and increments for various structural effects were determined by the school of Hansch and Leo[37] and by that of Recker.[38] Some of the most usual increments are listed in Table 3. The partition coefficient for the passage of a molecule from water to a hydrophobic solvent (various alcohols, benzene, CCl_4, diethylether, etc.), does not depend strongly upon the latter. For the passage of CH_4 from C_6H_6, Et_2O, or CCl_4 to H_2O, ΔG is +2.6, +3.3, and +2.9, respectively; for C_2H_6 from C_6H_6 or CCl_4 to H_2O, ΔG is +3.8 and +3.5 kcal/mol, respectively. The partition coefficient of a substance, A, (practically any substance) between water and a hydrophobic solvent "i", can be related to P^A_{oct} by a linear relation of the type:

$$\lg P^A_i = a_i \cdot \lg P^A_{oct} + b_i \tag{20}$$

Such solvent regression equations are usually rather precise, but for molecule A with a hydrogen donor or hydrogen acceptor character, two different relations must be used for most solvents. For a number of 21 different solvents, correlational equations of the Equation 20 type can be used; for numbers between 9 and 65 different molecules, the correlational coefficients are generally between r = 0.95 and 0.99 (with only two exceptions.)[36]

The success of correlational Equations (e.g., Equation 20) is rather surprising if one considers the great difficulties in computing ΔG for interaction of molecules with solvents (see Section II.C and II.D). Also, the process of Equation 18 can be considered similar to the receptor-ligand interactions, the hydrophobic solvent playing the role of the receptor. Solvents have a high conformational mobility and, in interacting with a molecule, A, the solvent molecules will rearrange themselves so as to allow a maximum of attractive interactions. The ΔG for interaction will depend only on the general apolar-polar (hydrogen bond forming) character of molecule A. Actually, an attempt to correlate $\lg P^A_i$ with the steric dissimilarity (MSD, see Section IV.A.1) between the solvent "i" molecule and the A molecule failed completely.[39]

If a hydrophobic group is dissociated from a rigid hydrophobic cavity, the ΔG value may be twice that of the hydrophobic group:

$$\Delta G^{dis}_{cav} \cong 2\Delta G^{HP}_x \tag{21}$$

since, unlike the passing of a hydrophobic molecule from a nonpolar solvent to water, both the group and the cavity come into contact with water.

Ionic strengths of aqueous solutions also influence the hydrophobic interactions, generally reducing the solubilities of hydrophobic groups; in a 2 *M* uni-univalent saline solution ($\mu = 2\ M$), ΔG increases with about 0.1 kcal/mol.[25]

The necessity to reevaluate equilibrium binding parameters for lipophilic drugs in their interactions with biological membranes as compared to their octanol-water partition coefficients was pointed out by Mason et al.[40] For a dozen 1,4-dihydropyridine derivatives and other drugs, the correlation between sarcoplasmatic reticulum:water and octanol:water (buffer) partition coefficient is rather scarce; the first are 100 to 1000 times larger than the second partition coefficients.

The role of ion and water release as a crucial factor in the formation of DNA-protein complexes was discussed by Record and Spolar.[41]

FIGURE 4. Base pairs in stacking interaction. The arrows indicate the 3′ → 5′ directions of the nucleotidic chains.

2. Stacking Interaction

Dispersion forces (see Table 1) between superposed and parallel aromatic cycles should be stronger than between aliphatic groups or between an aliphatic and an aromatic group of similar size: the polarizability perpendicular on aromatic cycles is larger and coplanarity allows atoms to come closer.

Actually, for the passing of a C_6H_6 molecule from water to aliphatic solvents $\Delta G \cong -0.3$ kcal/mol, and to aromatic solvents $\Delta G \cong -1.8$ kcal/mol (25°C); for the Phe-Leu amino acid pair interacting $\Delta G = -0.5$ kcal/mol, while for the Phe-Phe pair $\Delta G = -1.4$ kcal/mol. For various other experimental determinations, the stacking enthalpy for an adenine pair, AA, in a simple chain, $\Delta H = -6.5$ kcal/mol. Stacking energies between different base pairs (Figure 4) as obtained from a semiempirical analysis of phase diagrams are all within -5.5 ± 0.2 kcal/mol. For dimerization of nucleosides in water solution by stacking, $\Delta G = -1.5$ kcal/mol was obtained for deoxyadenosine and $\Delta G = +0.2$ kcal/mol for uridine. The DNA double helix easily binds planar aromatic cycles by intercalation; for example, dyes like acridine and ethidium bromide bind with a rather small overwinding of the DNA double helix (about 13°)[43] and $\Delta H = -6$ to -10 kcal/mol.[25] Association constants of free amino acids in water solution are listed in Table 4.[44]

The energy of stacking interactions should be proportional to the superposition area of the planar cycles. Considering the hexagonal and pentagonal surface areas and $\Delta U_{AR} = -1.5$ kcal/mol for superposed hexacycles (benzene, pyridine, and pyrimidine), the value for superposed pentacycles (pyrrole and imidazole) should be $\Delta U_{AR} = -1.0$ kcal/mol.

For condensed cycles the ΔU_{AR} values should be less than the sum of the component cycles; the -5.5 kcal/mol values for superposed base pairs, quoted before, is somewhat larger than this expectation. Studies on intercalation of aromatic polycyclic hydrocarbons in the polyadenine double helix have shown that the intercalation energy is maximal when the shape and surface area of the hydrocarbon allow a maximal overlap with the diadenine pairs.[45] In the amino acid-nitrogen base interaction from Table 4, Trp presents the strongest interactions, with a certain preference for G over A, and sensibly weaker interactions with the smaller pyrimidine bases (U,C); the preference of Phe for G over A is less pronounced.

An analysis of aromatic-aromatic interactions in some dozens of proteins indicates that 60% of the aromatic side chains are implied in aromatic-aromatic

TABLE 4
Association Constants (K, M^{-1}, 25°C) between Amino Acids, Polynucleotides, Nucleotides, and Nitrogen Bases in Water Solution

Nucleotide / Amino acid	Poly A	GMP	G	I	A	U	C
Trp	7	17.5	8.6	7.5	6.6	3.2	0.4
Tyr		7.1			—		
Phe	2	3.7			2.9		
His	1				1.4		
Arg					1.5		
Asp,Glu,Met,Pro					0.5		
Thr,Ser,Gly,Ala,Val,Leu					0.15		

interactions. Their contributions to the ΔG of protein structures is −0.6 to −1.3 kcal/mol, but for most pairs the geometry has little resemblance to base stacking in the DNA double helix.[46a]

3. Charge Transfer Complexes

Charge transfer complexes are formed between molecules largely differing in electronegativities and usually with planar side surfaces which allow a large contact area. An extra stabilization for such complexes comes from a partial transfer of an electron from the donor, D, molecule to the acceptor, A, molecule.[46]

Some thermodynamic data describing formation of charge transfer complexes in CCl_4 solution are listed in Table 5. The formation enthalpies, H, correspond approximately to the extra stabilization due to charge transfer; the two partners will interact by the usual dispersion forces and also with the solvent molecules. The total interaction energy U for such complexes will be given by half the formation enthalpy for crystalline complexes as compared to gas phase components; for complexes of 1,3,5-trinitrobenzene (as acceptor) with benzene, naphthalene, and anthracene the ΔH/2 values are −17.5, −20.4, and −23.5 kcal/mol, respectively.

For charge transfer complex formation, the acceptor molecule, A, must have a sufficiently low lowest-unoccupied orbital and the donor molecule, D, a sufficiently high highest-occupied orbital. In the framework of simple Hückle molecular orbitals (HMO) this means an acceptor character for $\epsilon_{LE} \lesssim \alpha - 0.5\beta$, while for $\epsilon_{HO} > \alpha + 0.5\beta$ the molecule presents a donor character. Of the natural amino acids, only tryptophane, with $\epsilon_{HO} = \alpha + 0.534\beta$ (according to Reference 47), should present donor character as well as the nucleic acid bases guanine ($\epsilon_{HO} = \alpha + 0.307\beta$), adenine ($\epsilon_{HO} = \alpha + 0.486\beta$), and thymine ($\epsilon_{HO} = \alpha + 0.510\beta$). According to this criterion, neither natural amino acid nor nucleic acid base should present an electron acceptor character. Acceptor character could appear for steroids containing a

TABLE 5
Equilibrium Parameters For Charge Transfer Complexes (20°C)[46]

Complex	Solvent	ΔG	ΔH	ΔS
Hexamethylbenzene-chloranil	CCl_4	−1.30	−5.35	−13.8
	Cyclohexane	−1.97	—	—
Dimethylaniline-chloranil	CCl_4	−0.72	−5.05	−14.8
Hexamethylbenzene-trinitrobenzene	CCl_4	−1.15	−4.71	−12.2
Naphthaline-trinitrobenzene	CCl_4	−0.81	−4.31	−12.0
	$C_2H_5OC_2H_5$	—	−0.99	—
	C_6H_6	—	−1.90	—
Benzene-trinitrobenzene	CCl_4	+0.84	−1.71	−8.7
Benzene-anthracene	CCl_4	—	−4.5	—
	$ClCH_2CH_2Cl$	—	−3.6	—
	$C_2H_5OC_2H_5$	—	−0.98	—
	C_6H_6	—	−0.98	—
Benzene-I_2	CCl_4	+1.11	−1.3	−8.2
$(CH_3)_2C{=}C(CH_3)_2$trinitrobenzene	CCl_4	+1.97	−1.46	−11.7

Note: ΔG, ΔH — kcal/mol; ΔS — cal/mol°K

carbonyl group conjugated with double bonds (for the unconjugated carbonyl group, $\epsilon_{LE} = \alpha - 0.62\beta$). Various coenzyme groups also can have an electron donor or acceptor character for charge transfer.

D. STERIC FIT — STERIC MISFIT

The importance of molecular stereochemistry for biological activity, the fit of the ligand into the receptor like a "key into a lock", was recognized 100 years ago by Emil Fischer.[48] The "lock and key" theory fits perfectly in interactions in crystalline lattices, while in receptor-ligand interactions the "lock", the receptor protein, presents variable degrees of flexibility;[49] the "lock" may be deformed by the "key" and vice versa. At the opposite extreme to crystalline lattices, organic solvents are rather indifferent to the molecular stereochemistry of the solute (see Section III.C.1 and Reference 39).

In structure-activity studies for drug design, series of molecules, sometimes with marked differences in molecular stereochemistry, are allowed to interact with the same receptor. Steric misfit is to be expected in such cases because of partial superpositions of van der Waals volumes of ligand molecules with those of the walls of the receptor binding site. Nevertheless, if the molecule binds, how has this steric misfit been circumvented and what are its energetic (ΔG) costs?

A selection of covalent and van der Waals radii for some usual atoms and groups is listed in Table 6. Distances between nonbonded atoms in direct contact are equal, at a first approximation, to the sum of their van der Waals radii. The short-range repulsion forces (B/r^{12} from the van der Waals potential)

TABLE 6
Covalent and van der Waals Radii[18]

Atom (group)	Covalent single bond	Radius (Å) double bond	van der Waals radius (Å)
H	0.30	—	1.2
C	0.77	0.67	1.57
N	0.70	0.61	1.5
O	0.65	0.57	1.4
S	1.04	0.95	1.85
Cl	0.99	0.90	1.80
CH_3	—	—	2.0
Aromatic ring (half-thickness)			1.7–1.85
–OH			1.60
$-NH_2$			1.85
$-NO_2$			2.02
$-COO^-$			2.10
$-N^+(CH_3)_3$			3.17

are very stiff with respect to compression of contact distances. The same is true for covalent bonds. Bond angles are more easily deformable and dihedral angles even more so. The largest spatial deformations, with a relatively small energy increase, can be obtained by conformational changes within the ligand molecule or within the receptor site. Circumvention of steric misfit will be produced in an inverse order. Figure 5 illustrates a possible case when a rigid ligand molecule induces a local conformational change in a relatively flexible receptor site.

The results of the MTD and SIBIS methods (see Section IV.A.1) suggest that in several cases the circumvention of spatial superposition of effector (or potential drug) molecules with the walls of the receptor combining site has an energy cost roughly proportional to the volume of the superposed region. This energy increase may correspond to conformational changes or dihedral angle changes in both the drug molecule and in the region of the receptor site. If the molecule is ''hard'' (low conformational flexibility, for example, steroids) and the receptor relatively ''soft'', this energy increase corresponds to the ΔU_{conf} of Equation 1. The MTD and SIBIS results suggest that for a spatial superposition of a volume of the size of a small substituent ($-CH_3$, –OH, –Cl, etc.) the corresponding energy increase is about 1.0 kcal/mol.[50]

Another problem is the tolerance of atomic positions in steric fit, i.e., how far can an atom (of a molecule within a receptor) be displaced from its equilibrium position without a sensible energy increase.[51] Let us consider the 6 to 12 van der Waals potential:

$$U(r) = -A/r^6 + B/r^{12} \tag{22}$$

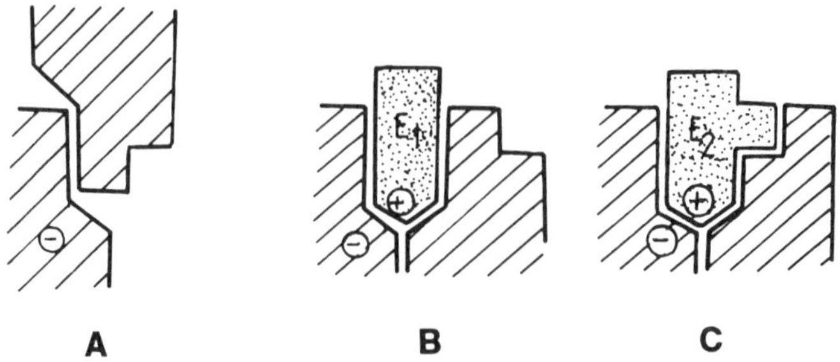

FIGURE 5. Steric misfit in receptor-ligand interactions; (a) receptor in the inactive ground state; (b) perfect fit of the E_1 ligand molecule in the receptor site of the receptor in the biologically active conformation; and (c) steric misfit. The ligand molecule E_2 presents an overlapping volume with the walls of the receptor site of the receptor in active conformation (overlapping of the van der Waals envelopes).

The energy minimum $U_{min} = -B^2/4A$ is obtained for the distance $r_o = (2B/A)^{1/6}$. This energy minimum is rather small — less than ca. 1 kcal/mol for an atom-atom interaction (see Table 1 and Section III.A). We calculate thè shift, Δr, from the equilibrium distance, r_o, which reduces by half the attractive energy, i.e.:

$$U(r_o + \Delta r) = {}^1/_2\, U_{min} = -\frac{B^2}{8A} \tag{22a}$$

The equation can be brought to a quadratic form (Appendix 1) with the solutions:

$$r_o + \Delta r = \begin{cases} r_o \cdot 3.41^{1/6} = 1.227\, r_o \\ r_o \cdot 0.59^{1/6} = 0.916\, r_o \end{cases} \tag{23}$$

With the usual van der Waals contact distance, $r_o = 3.5$ Å, one obtains $\Delta r = +0.79$ Å and -0.33 Å.

Another way is to consider the torsional potential

$$U(\phi) = {}^1/_2\, U_o(1 - \cos n\phi) \tag{24}$$

The shift of atom X (Figure 6) produced by a dihedral angle torsion of ϕ radians is

$$\Delta r \cong l_{XY} \cdot \sin \widehat{XYZ} \cdot \phi \tag{25}$$

We shall calculate the shift, Δr, corresponding to an ϕ radian torsion which produces an energy increase, $U(\phi) - U(o)$, equal to the mean thermal

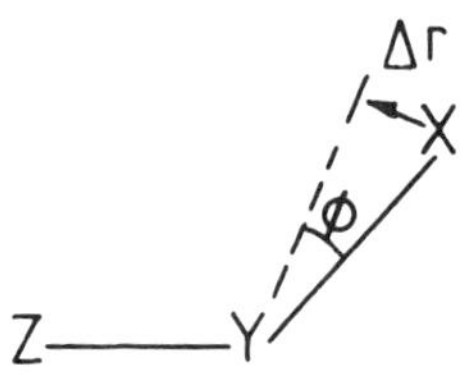

FIGURE 6. Shift Δr of atom X upon a ϕ rad torsion around the ZY bond; $\phi = 0$ corresponds to the energy minimum.

energy (for one freedom degree), $kT = 4 \cdot 10^{-21}$ j at 300°K. If we consider $l_{XY} = 1.5$ Å, $\widehat{XYZ} = 109°30'$ (CC bond, sp^3 hybridization for Y), a potential barrier $U_o = 3$ kcal/mol, $n = 3$ (for rotation around a simple C–C bond), and $\Delta r = 0.22$ Å.

Although the B/r^{12}-A/r^6 van der Waals potential is very resistant toward compression, for small shifts, $\Delta r \lesssim -0.5$ Å, the energy increase is also small. The tolerance of atomic positions for the receptor-ligand steric fit can thus be evaluated to about 0.5 Å, centered around the equilibrium position.

E. SPECIAL EFFECTS

Creation of a small hole — Let us consider the interaction of the enzyme alanyl-AMP-synthetase, implied in tRNA aminoacylation with alanine and glycine, respectively.

$$\text{alanyl-AMP-synthetase} + \text{Ala} \rightarrow \text{Complex} \tag{26a}$$

$$\text{alanyl-AMP-synthetase} + \text{Gly} \rightarrow \text{Complex} \tag{26b}$$

If alanine is accommodated perfectly in the complex with the enzyme in the glycine-enzyme complex, the hole in the enzyme which accommodates the CH_3 group of Ala will be empty or eventually filled with a H_2O molecule. The difference $\Delta G_{Ala} - \Delta G_{Gly}$, for the formation of the two complexes will correspond to the interaction of the CH_3 group with the hole. If the hole is empty, this difference can be calculated using the surface tension, r, and the surface, S, of the hole:

$$\Delta G_{Ala} - \Delta G_{Gly} = r \cdot S \tag{27}$$

For hydrophobic liquids $r \cong 25 \cdot 10^{-3}$ j/m (cyclohexane 20.5 · , benzene 28.9 · and methanol $23 \cdot 10^{-3}$ j/m at 20°C); for the CH_3 group, the hole-surface, a hemisphere with r = 2 Å corresponds to $S \cong 25$ Å^2. Thus, $rS = 6 \cdot 10^{-21}$ j = 0.9 kcal/mol. The inclusion of the H_2O molecule in this hole means its extraction from the hydrogen-bonding environment will be energetically unfavorable.[25]

Entropy effects — For the generation of a new particle in a water solution there is a positive entropy of mixing, $\Delta S_{mix} \cong 10$ cal/mol°K (Appendix 2) corresponding to a contribution of about $-T\Delta S \cong -3.0$ kcal/mol at 300°K. The same amount of entropy will be lost with the formation of a complex of two particles. The entropy variation for a dissociation process will depend upon various other effects, such as modifications of the degree of solvation, etc. For example, for the dissociation of water:

$$2H_2O \rightleftarrows H_3O^+ + OH^- \qquad (28)$$

the generation of two new particles in solution gives a $\Delta S_{mix} \cong +20$ e.u., the hydration entropies are -10.6 e.u. for H_3O^+ and -32.6 e.u. for OH^-. The $\Delta S = -23.2$ e.u. sum is close enough to the experimental value, -18.8 e.u. (Table 29 from Webb[18]). Conformational entropies, due to rotations along simple bonds when a number of g conformations are approximately equi-energetic (as compared to the mean thermal energy, kT), can be calculated by the equation (on a per molecule basis):

$$S_{conf} = k \cdot \ln g \qquad (29)$$

Computation of free-space intramolecular entropies was reported for diphenyl-sulfones.[52]

Let us also mention an empirically derived averaged binding energy of organic molecules to biologic receptors[53] (kcal/mol):

$$-\Delta G_{av} = -14 - 0.7n_{DOF} + 0.7n_{C_{sp2}} + 0.8n_{C_{sp3}} + 2.5n_{OH} + 1.3n_{Hal} \qquad (30)$$

The value -14 stands for the standard loss of translational and rotational entropy in binding; n_{DOF} is the number of freedom degrees of the molecule; and $n_{C_{sp2}}$, $n_{C_{sp3}}$, n_{OH}, and n_{Hal} are the numbers of the corresponding functional groups within the molecule.

Cationic groups seem to interact rather strongly with aromatic environments — the so-called aromatic cage — for example, the trimethylammonium group of acetylcholine in the binding site of acetylcholinesterase.[53a] This can be considered as an interaction of a charge with the highly polarizable π-electronic system (charge-induced dipole interaction).

IV. THE QSAR APPROACH

Quantitative structure activity relations in several cases are approximative relations between receptor-drug molecule affinities and structural features characterizing the drug molecules. Thus, together with the series of experimental data concerning the receptor affinities of rather large numbers of molecules, QSARs are of interest for a characterization of the specificity of interaction with respect to the receptors studied.

It is not our intention to review QSAR methodology, which can be found in several textbooks (for example, those of Martin[54] and Franke[55]), and especially in a recent treatise on medicinal chemistry.[56] We shall only discuss selected topics, especially those related to the computability and predictability of drug-receptor affinity.

We may discern three main types of methods in QSAR. The first is the three-dimensional structure-activity relationship approach,[2a] or what we shall call the computational chemistry approach. These methods rely heavily upon molecular modeling methods and upon calculation of interaction energies of the drug molecule with a receptor of known structure. If the receptor structure is unknown, one uses a model of atom-points with certain interaction characteristics which are optimized by a certain QSAR strategy. This approach requires solid computational technique and programs.

The second method is what we call the classical Hansch approach. The molecules, or various substituents on a molecular core, are described by structural parameters and one picks out a set of such parameters which permits a successful QSAR equation. The demands on computational technique and programs are modest as compared to the first method.

The third method is based upon pattern recognition. In this approach, one tries to find structural features which characterize the biologically active molecules. The corresponding search programs are often very intricate. Methodologies used in the study of artificial neural networks were also introduced into this type of approach.[57] Topologic indices are often used as molecular characteristics.[58] The so-called electron-topologic approach,[59] using atomic charges, bond orders, and atomic interdistances as molecular descriptors, seems to be particularly successful.

A pitfall of the QSAR methodology, in general, is the chance correlation, in the sense of Topliss and Edwards.[60] This can be eliminated by studying the stability of correlational equations (QSARs) toward extensions or reductions in the studied series of molecules, use of a series of test compounds, or cross validation,[61] to ascertain the reliability of predictive results. A review of QSAR publications shows that such procedures are used rather rarely.

In what follows, we shall point out some characteristics of the classical Hansch-type approach and the computational chemistry approach.

An attempt to assess the real predictive power of QSAR methods is not easy. What one can do is to scan QSAR publications. A look at the literature of the last few years attests that the classical Hansch-type approach, by far, is the most popular. The abstracts of publications related to QSAR by Darvas and Darvas[62] provide an idea of the popularity of different procedures. For a 1-year period (mid-1989 to mid-1990), about 460 QSAR titles were recorded, i.e., papers which give a quantitative structure-activity relation, lists calculated vs. experimental affinities, or performance activity predictions which can be compared with experimental activities. From these, about 25 papers are based upon logical, pattern-recognition procedures and 25 on computational chemistry procedures (actually about half of them use parameters such as atomic net charges or molecular electrostatic potentials — calculated by quantum chemistry — in classical QSAR procedures), and about 110 use classical, Hansch-type QSAR. Of the Hansch-type QSAR papers, 14 compare their results with qualitative computational chemistry results (X-ray structures, molecular graphics, and conformational energy calculations). There are also about 100 titles in which qualitative considerations about structure-activity relations are performed by computational chemistry approaches. The use of cross validation or a test series to ascertain reliability of predictive results is rather rare.

In conclusion, while classical Hansch-type approaches, for the moment, are more popular for quantitative prediction (QSAR), there is an increasing use of computational chemistry approaches, especially in qualitative discussions.

A. THE CLASSICAL HANSCH-TYPE APPROACH

The general idea is that biological activity, A_i, which most often reflects the affinity towards a receptor, depends upon several factors which can be classified as hydrophobicity (the receptor, usually, is much less polar than water), electronic effects, and steric effects. The molecules considered for the QSAR series are usually substitution derivatives of a molecular core which also contains the "pharmacophoric constellation", some common group of atoms binding strongly to the receptor. The correlational equation is of the type:

$$\hat{A}_i = a_o + a_1 \cdot \pi_i + a_2 \cdot \sigma_i + a_3 \cdot ES_i + \ldots \tag{31}$$

with π_i the hydrophobic increment of a substituent, σ_i the Hammett constant of the same or another substituent, and ES_i the steric Taft parameter. Hydrophobic increments have been discussed in Section III.C.1. Hammett substituent constants and Taft steric substituent constants are taken from the domain of organic chemistry kinetics (linear free energy relations).[63] The first type of constants refer to electron withdrawing or donating effects as manifested in modulation of the acidity of substituted benzoic or acetic acid. The second type of constants refer to steric screening of some reacting groups, also in rather simple organic molecules. (For a tabulation of such constants see, for example, Reference 37.) More elaborate steric parameters are the L, B_1, B_2, B_3, and B_4 parameters describing five characteristic linear dimensions of a substituent.[64] Molecular refraction increments for substituents are also used as a measure for group volumes (in repulsive or steric misfit interactions) or for group polarizabilities (attractive-dispersive interactions). (A tabulation of the last type is contained in Reference 37.) The molecular geometry is often

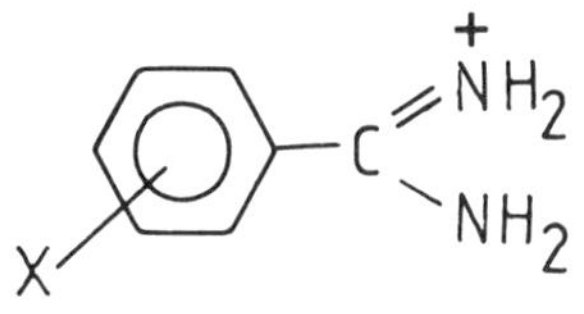

FIGURE 7. Benzamidine derivatives, trypsine inhibitors. General structure.

only implicitly taken into consideration by testing the effect upon the correlation coefficient, r, of the introduction in the correlational equation of constants (of the above-mentioned type) for groups in various positions of the molecular core. Indicator variables reflecting various structural features also are often introduced in such correlational equations.

The possibility of describing the intricate receptor-ligand interactions, at least sometimes, with a small number of physicochemical or structural parameters in a successful and reliable way[63] is rather surprising if one considers the description of these interactions in Sections I and II. This success is to be explained in the same manner as the success of the linear free energy relations (Hammett equation, etc.) in organic chemistry,[65] as compared to the intricate description of organic reactions in quantum chemistry.[10]

This strategy of testing large numbers of various types of group constants for different substituents in order to improve correlational results can certainly produce chance correlations. Special procedures are used in order to ascertain that the correlational improvement, obtained by introduction of a new parameter, reflects some real interaction and not simply chance correlation. We shall illustrate such a procedure on the basis of a relatively recent QSAR and molecular graphics study of the Hansch group upon inhibition of trypsin by amidine derivatives.[66,67] This study considers a number of N = 104 benzamidine derivatives of the general type shown in Figure 7; with substituents, X, in positions 3 and 4 of a large variety of structures such as H, CH_3, C_4H_9, Cl, NO_2, OH, $COOCH_2$, $NHCOC_6H_5$, COC_6H_4-4-Cl, etc. The biological activities, $A = lgK_{inh}$ are related by the inhibition constant, K_{inh}, to the affinity toward the trypsin receptor. Experimental K_{inh} values from three different authors were used in this study.

In obtaining the correlational equation, first a set of ten data from one experimental study was used with the substituents in position 4. The σ Hammett constant alone gives the best uniparametric correlation. Various types of other substituent parameters were added in order to improve the correlation — lgK_{inh} data for new compounds are added. Finally, here is the correlational equation which comprises N = 104 derivatives, both meta- and para-substituted:

$$\hat{A}_4 = -0.59(\pm 0.49)MR_4 + 0.88(\pm 0.52)\lg(\beta.10^{MR_4} + 1)$$
$$+ 0.23(\pm 0.07)\pi'_3 - 0.74(\pm 0.20)\sigma + 0.20(\pm 0.30)IM$$
$$+ 0.65(\pm 0.22)I_1 + 0.43(\pm 0.19)I_2 + 0.51(\pm 0.15)I_3$$
$$+ 1.38(\pm 0.28) \qquad (32)$$
$$r^2 = 0.854 \ ; \ N = 104 \ ; \ \text{optimal } MR_4 = 1.03$$

The IM indicator variable is 1 for meta-substituted benzamidines and 0 for para-substituted ones: I_1, I_2, I_3 is 1 for different types of carbonyl-containing substituents in the meta and para position, and zero otherwise. MR_4 is the molecular refractivity of para substituents, π'_3 the hydrophobic increment for substituents in the meta position (calibrated with respect to substituted nitrobenzenes), and σ Hammett is the substituent constant, irrespective of the substituent position.

What argues against pure chance correlation is the approximative constancy, at least as to sign and within the mean standard error, of the regressional coefficients for π'_3, σ, and MR_4 in the different equations, and of the optimal MR_4 value. More than that, since coordinates for benzamidine and inhibitor-bovine trypsin complex have been determined by X-ray crystallography, the correlational results allow a credible interpretation of the enzyme-inhibitor interactions.

Through the use of molecular graphics, the inhibitor molecule within the enzymatic receptor is visualized by coloring the receptor pocket surfaces in red for hydrophobic zones, blue for polar zones, and yellow for the semi-hydrophilic regions of divalent sulfur atoms. The benzamidine inhibitor is bound into a hydrophobic pocket in the active site. An electrostatic bond is formed between the positively charged amidine group and the negative carboxylate of Asp-189 at the back of the hydrophilic pocket. The OH group of the catalytic Ser-195 is in close contact with substituents in position 4. The general effect of the σ Hammett substituent constant could consist in the improvement of the hydrophobic character of the phenyl group, which is decreased due to the strong electron withdrawing effect of the positively charged amidine groups. The MR_4 character of substituents in para and the π' character of these groups are highly colinear for substituents larger than ethyl; very small groups interact with the Ser-195 OH group, while groups larger than ethyl penetrate into a hydrophobic region. This explains the rather complicated dependence of inhibitory activity upon MR_4. Characteristics for interactions with larger and more complicated substituents, as those characterized by the indicator variables I_1, I_2, and I_3 are also explained by the molecular graphics visualization of the enzyme-inhibitor interaction.

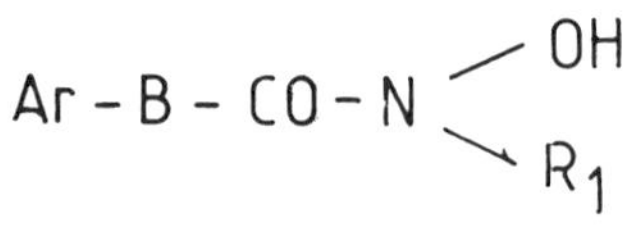

FIGURE 8. General structure of hydroxamic acid derivative, 5-lypoxygenase inhibitors.

An interesting and remarkable success of the Hansch-type approach is a recent study upon hydroxamic acid inhibitors of 5-lipoxygenase.[68] For a number of N = 111 molecules with large structural variations, a successful QSAR with only four structural parameters was obtained. The regressional coefficients for four different correlational equations for subgroups of these molecules are rather similar between these and the final equation, speaking in favor of the robustness of the results. The general structure for these molecules is shown in Figure 8, with Ar a substituted aromatic ring (benzenic, naphthalenic, anthracenic, acenaphthenic), B signifying a direct bond of the hydroxamic group to the aryl ring or a small alkanic, alkenic, etheric, or thioetheric spacer, and R_1 as H, CH_3, or a larger alkylic or arylic substituent. The correlational equation is

$$\log(1/IC_{50}) = 4.30 + 0.57\pi' - 1.16I_{NH} - 0.69I_2 - 0.64I_1$$

$$r^2 = 0.884 \quad ; \quad N = 111 \tag{33}$$

IC_{50} is the 50% inhibitory concentration, π' the general hydrophobicity, corrected for neglecting some larger substituents, and I_{NH}, I_2, and I_1 are indicator variables for the presence of certain substituents in some special positions.

The success of this QSAR, practically only hydrophobicity with few conditions of steric restraints, may be explained by the ability of the drug molecules attaching to a very "soft" (conformationally flexible) and hydrophobic region of the enzyme. Only the hydroxamic acid moiety is essential for the inhibition. The enzyme, 5-lipoxygenase, should have an extremely "soft" binding site for inhibitors.

Successful QSARs based upon hydrophobicity and a few other parameters are relatively frequent for drugs acting on the central nervous system, especially for general, nonspecific central nervous system depressants.[69] This may be explained by a disorganization of nervous structures due to dissolution of the drug in nervous cell membranes.

1. The MTD and SIBIS Methods

This type of method, as a computational requirement, is similar to the classical Hansch-type approach, with which it can also be included by its general strategy and the use of structural parameters as hydrophobicity, etc. The advantage of the MTD and SIBIS methods is that they allow a per

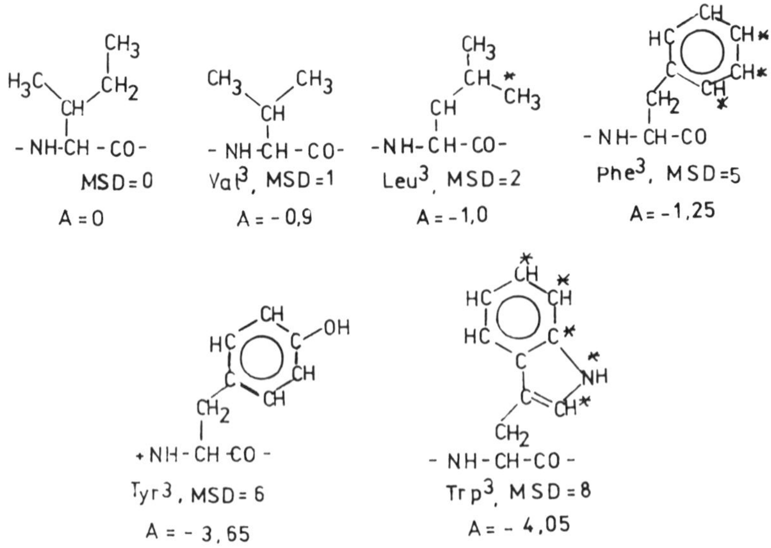

FIGURE 9. Atom per atom superposition and MSD calculations for the oxytocin derivatives with amino acid substitutions in position 3. Biologic activities, A, refer to reduction of oxytocitic activity with respect to oxytocin (logarithmic scale). MSD minimal steric difference of the side chain with respect to that of isoleucine.[71] Atoms marked by asterisk (*) are not superposable upon the Ile[3] side chain. One CH_3 group of Ile[3] is superposable only upon the Val[3] side chain.

molecule account of spatial structure, instead of a per substituent account, which is used by the groups of Hansch[67] or Verloop.[64]

The initial version of the MTD method, in fact, is the MSD (minimal steric difference) method.[70] The minimal steric difference, MSD_i of a molecule, M_i, with respect to a standard molecule, M_o, is equal to the number of unsuperposable atoms, when M_i and M_o are superimposed, atom per atom (with neglect of the H atoms). The relative orientation of M_i with respect M_o, and the conformation chosen for M_i (if M_i has several low-energy conformations) is such as to allow a maximal superposition of the two molecules, i.e., to obtain a minimum for the steric difference value. Figure 9 gives an example of computations of MSD values, for substitution of position 3-Ile by other amino acids in oxytocin, correlated with biologic activity decrease in logarithmic units.[71]

In the MTD[72] and SIBIS[73] methods, the N molecules of the series for QSAR are first superposed, atom by atom (neglecting H atoms and small differences in bond lengths and bond angles). The relative orientation is given by a pharmacophoric constellation or a common molecular core. The result of the superposition is a network, the hypermolecule (investigated receptor space [IRS] for SIBIS) with M vertices j describing the approximative atomic positions for the molecules in the complex with the receptor. This network also has a topological character, not simply a stereochemical one. The vertices j are ascribed to the receptor cavity (beneficial, $\epsilon_j = -1$), receptor walls (detrimental, $\epsilon_j = +1$), or to the exterior space (irrelevant, $\epsilon_j = 0$), according to an optimization procedure (which starts from an initial receptor map, S_o). These attributions are the so-called optimized receptor map (S*) which provides an idea of the stereochemistry of the receptor site. Optimization is performed on the basis of a correlational equation:

$$A_i = a_o - \beta \cdot MTD_i + a_i \cdot \sigma_{li} + \ldots \tag{34}$$

where $\sigma_{li}, \ldots$ are various Hansch-type parameters (hydrophobicity, etc.), and MTD_i, the minimal topological (steric) difference for molecule M_i, is a measure for the steric misfit of the molecule with respect to the receptor site:

$$MTD_i = s + \sum_{j=1}^{M} \epsilon_j \cdot x_{ij} \tag{35}$$

Here, s is the total number of cavity vertices; $x_{ij} = 1$ or 0 if vertex j is occupied or not by molecule M_i.

In the SIBIS procedure, instead of $-\beta.MTD_i$, there are two terms, b.OV + c.NOV, describing OV, the cavity volume and, respectively, NOV, the wall volume occupied by M_i; OV and NOV are the sums of atomic volume increments of M_i, situated within the two regions.

There are certain specific difficulties with these methods. First, the construction of the hypermolecule — a reliable atom per atom superposition, using only what one may call "chemical intuition", is not possible if there are molecules with different condensed rings or with large variations in bond lengths and angles. In such cases, atomic coordinates (X-ray crystallography results) and some docking techniques are to be used to minimize deviations between equivalent atoms in superposed pairs of molecules. A pair of equivalent atoms is considered to be superposable upon the same vertex j, if their interdistance upon superposition is less than 0.5 Å.[51] Also, distances between nonbonded atoms of a molecule, M_i, in the hypermolecule, should be not less than the van der Waals contact distance. If a molecule M_i has several low energy conformations (i.e., with energies differing by not more than kT — the mean thermal energy for one freedom degree), the conformation which best fits the receptor (corresponding to the lowest MTD value) should be considered:

$$MTD_i = \underset{k}{\text{Minimum}} \left(s + \sum_{j=1}^{M} \epsilon_j \cdot x_{ijk} \right) \tag{36}$$

Here k represents conformations.

The optimization of the ϵ_j attribution procedure, as in any optimization procedure, may yield a local minimum (i.e., maximum in the correlation coefficient, r, between experimental and calculated activities). The SIBIS method recommends that all cavity vertices and, separately, all exterior vertices of the IRS should form in each group a single connected network, in the topological sense. Thus, one avoids an optimized receptor map with interspersed cavity, exterior, and wall vertices; it is quite unlikely, for example, that the receptor site consists of several cavities.[71]

More exact computations of MSD, OV, and NOV as volumes of intersecting (or nonintersecting) molecular van der Waals envelopes are possible by the use of Monte Carlo techniques:[74,75] the two molecules, M_o and M_i, for which such volumes are to be calculated in the proper superposition, are included in a parallelepiped. A sufficient number of points is dispersed at random within the parallelepiped and one counts the number of points within the intersecting molecular envelopes. (For the use of such parameters for quantitative shape-comparison of molecules see References 76 and 77.) This Monte Carlo variant of MSD (MCD) allows a relatively simple calculation of overlapping molecular volumes which avoids the problem of multiple intersecting atomic van der Waals volumes required in geometric calculations.

2. Hard Molecules and Soft Receptors — QSAR by MTD for Steroids

Steroids are a domain where successful QSAR applications are rather rare.[78,79] A possible reason is the importance of steric factors. Steroids are rather "hard" molecules, with low conformational flexibility; in "soft" receptors, with sufficient conformational flexibility, different steroids of a series for QSAR may induce different local conformations. Returning to Equations 1 and 2 at the beginning of this chapter, the ΔU_{conf} term will differ within the same series for QSAR while the U_{R*E} term will also refer to interactions with different local receptor conformations. There are some successful QSARs with the MTD method for steroids.[51,80-83] In the MTD method, the receptor-ligand affinity is supposed to decrease linearly with steric misfit, which is approximately equal to the sum of the unoccupied receptor cavity volume and of the overlapping volume of the effector molecule and the receptor walls. This overlap is avoided by local conformational changes and distortions of dihedral angles. It is therefore of interest to discuss at some length such a QSAR example — here, for gestagenic steroids, to assess the robustness of results (i.e., to ascertain that they are not the result of chance correlation), and to discuss the intervening types of intermolecular forces.

Relative binding affinities for the progesterone receptor of the rabbit uterus, as resulted from competition experiments with progesterone, were reported for a large number of steroid derivatives including various degrees of unsaturation of the steroid skeleton. A molecular superposition procedure was used, based upon X-ray crystallography atomic coordinates, with minimization of the sum of squares of interdistances between equivalent atoms

TABLE 7
Distances between Equivalent Atoms for Superposition of Steroid Molecules (Å Units)[51]

Atom no.	Distance (Å) between equivalent atoms in superposition			
	NE/T5	P/T5	P/TD	P/T
C_1	0.112	0.122	—	0.419
C_2	0.097	0.251	—	0.414
C_3	0.117	0.193	0.282	0.333
C_4	0.584	0.551	0.140	0.297
C_5	0.241	0.323	0.041	0.311
C_6	0.081	0.204	0.081	0.269
C_7	0.244	0.059	0.096	0.335
C_8	0.036	0.074	0.143	0.383
C_9	0.056	0.127	0.402	0.417
C_{10}	0.037	0.038	0.364	0.376
C_{11}	0.062	0.185	0.237	0.475
C_{12}	0.059	0.146	0.120	0.587
C_{13}	0.040	0.041	0.105	0.559
C_{14}	0.015	0.028	0.060	0.495
C_{15}	0.086	0.128	0.055	0.514
C_{16}	0.033	0.325	0.048	0.658
C_{17}	0.039	0.052	0.226	0.651
C_{18}	0.092	0.111	0.151	0.500
C_{19}	—	0.231	—	0.328
O_3	0.256	0.279	0.159	0.320
O_{20}	—	—	—	0.728

Note: Distances larger than 0.5 Å are underlined. Abbreviations refer to P: progesterone; T: testosterone; NE: 19-norethisterone; T5: 17-β-hydroxy-5αH-androstane-3-one; TD: 17-β-hydroxy-4,9-estradione-3-one. Superpositions for the NE/T5, P/T5 and P/TD pairs, according to the MOLFIT procedure; for the P/T-pair, according to the SUPRAMOL procedure.

of the steroid rings.[51,82,83] A maximal interdistance of 0.5 Å was allowed between equivalent atoms from different rings, if they are to be superposed upon the same vertex, j, of the hypermolecule. Interdistances upon superposition are listed in Table 7.

On this basis, if testosterone and progesterone derivatives are to be treated within the same series for QSAR, the strong hydrogen-binding 0–17β atom of testosterone is to be superposed upon the 0–20 atom of the 17β acetyl side chain of progesterone. In a testosterone-progesterone superposition on such principles, the atoms of the A and B rings are superposable, but most of the

C- and D-ring atoms fall apart by more than 0.5 Å. Therefore, progesterone and testosterone derivatives were treated separately, in two different series for QSAR (Table 8 and Table 9). At most, only three atoms (of the A ring) fall apart by more than 0.5 Å when the steroid skeleton of progesterone is superposed over that of 5-αH-androstane-3-one and upon that of 4,9-estradien-3-one. The two QSAR series, therefore, also contain derivatives without the 4-5 double bond in the A ring and those with a supplementary 9-10 double bond. The correspondence between substituent position and vertex numerotation is listed in Table 10. The hypermolecules for the two series are depicted in Figure 10.

Figure 10 illustrates the hypermolecules for the testosterone and progesterone series, respectively.[51,84] For a series of N = 41 testosterone derivatives, the receptor site and correlational equation are

$$S_T^* \begin{cases} j(\epsilon = -1)\text{: } 1,13,19 \\ j(\epsilon = 0)\text{: } 4,6,7,9,14,16,18,23\text{–}27,41 \\ j(\epsilon = +1)\text{: } 2,5,8,10,15,17,28 \end{cases}$$

$$\hat{A}_1 = 3.000 - 0.942\text{MTD} + 0.555f \; ; \; r^2 = 0.848 \qquad (37)$$

while for a reduced series of N = 25 testosterone series (only those with one 4-Δ double bond within the steroid cycle)[82,84] the ϵ_j vertex attributions for all common vertices are identical and the correlational equation is similar:

$$\hat{A}_1' = 3.213 - 1.092\text{MTD} + 0.639f \; ; \; r^2 = 0.919 \qquad (38)$$

For a series of N = 55 progesterone derivatives, the receptor site and correlational equation are

$$S_P^* \begin{cases} j(\epsilon = -1)\text{: } 1,14,20,39 \\ j(\epsilon = 0)\text{: } 3,6,8,13,15,16,22\text{–}27,30,33\text{–}36,40 \\ j(\epsilon = +1)\text{: } 2,4,5,9,11,28,29,31,32,37,38 \end{cases}$$

$$\hat{A}_2 = 3.917 - 0.744\text{MTD} + 0.696f \; ; \; r^2 = 0.874 \qquad (39)$$

In a cross-validation-like procedure $r^2 = 0.755$ was obtained for the progesterone derivatives.[84]

For a reduced series, N = 34 progesterone derivatives (only those with one 4-Δ double bond within the steroid cycle) the ϵ_j vertex attributions are also identical for the common vertices; the corresponding correlational equation is[82,84]

$$\hat{A}_2' = 3.924 - 0.725\text{MTD} + 0.854f \; ; \; r^2 = 0.925 \qquad (40)$$

TABLE 8
Data For Testosterone Series[51,84]

No.	Steroid	A_{exp}	A_1	MTD	f	Ref.
1	18–Me–NE	2.550	2.231	1	0.31	13
2	11=CH$_2$,18–Me–NE	2.540	2.320	1	0.47	14
3	7α,17α-diMe-NT	2.330	2.348	1	0.52	15
4	11β-Cl-NE	2.332	1.876	1	−0.33	16
5	NE	2.190	1.942	1	−0.21	15
6	17α-Me-NT	2.000	2.059	1	0.00	17
7	17α-CH$_2$Cl-NT	1.930	2.148	1	0.16	18
8	17α-Et-NT	1.890	2.348	1	1.52	19
9	18-Et-NE	1.860	1.578	2	0.83	15
10	17α-Pr-18norT	1.830	1.695	2	1.04	19
11	17α-CHCH$_2$-NT	1.810	2.148	1	0.16	20
12	17α-CH$_2$-C(Me)=C$_2$-NT	1.680	1.778	2	1.19	20
13	18-Me-NT	1.530	1.147	2	0.00	15
14	7α-Me-NT	1.340	1.147	2	0.00	15
15	3-deoxy-17α-Me-NT	1.300	1.147	2	0.00	16
16	E	1.280	1.290	2	0.31	20
17	3-deoxy-11=CH$_2$-18-Me-NE	1.200	1.378	2	0.47	14
18	3-deoxy-7α-Me-NE	0.950	1.290	2	0.31	16
19	3-deoxy-NE	0.850	1.001	2	−0.21	21
20	NT	0.780	0.827	2	−0.52	20
21	18-Et-NT	0.650	0.464	3	0.52	15
22	4-Cl-17α-Me-NT	0.110	0.553	3	0.68	18
23	T	−0.220	0.176	3	0.00	16
24	17α-CH$_2$CN-T	−0.520	−0.057	3	0.42	18
25	4Cl-11β-OH-17-Me-T	−1.000	−0.363	3	−0.97	18
26	11=CH$_2$-NE	2.480	2.031	1	−0.05	14
27	11β,18-diMe-N	2.180	2.520	1	0.85	22
28	11-CH$_2$-18-Me-N	1.990	1.578	2	0.83	14
29	NE5	1.230	1.201	2	0.15	16
30	E5	0.700	0.548	3	0.67	20
31	T5	−0.300	−0.566	4	0.36	18
32	7α,17α-diMe-TD	2.300	2.148	1	0.16	15
33	17α-Me-TD	1.850	1.859	1	−0.36	15
34	18-Me-ED	1.830	2.031	1	−0.05	15
35	ED	1.600	1.742	1	−0.57	15
36	18-Me-TD	1.420	0.917	2	−0.36	15
37	TD	1.230	0.623	2	−0.88	15
38	18–Et–ED	1.040	1.378	2	0.47	15
39	18-Et-TD	0.660	0.265	3	0.16	15
40	17α-C≡CMe-TD	1.176	2.031	1	−0.05	23
41	11β-Me-17α-C≡CMe-TD	2.371	2.320	1	0.47	23

Note: T: testosterone; E: ethisterone; NE: 19-norethisterone; NE5,E5 and T5 have the 5αH-androstane-3-one skeleton; TD and ED have the 4,9-estradione-3-one skeleton; Me: methyl; Et: ethyl; Pr: propyl; OH: hydroxy; Cl: chloro; CN: cyan; CH$_2$: methylene.

TABLE 9
Data for Progesterone Series[84]

No.	Steroid	A_{exp}	A_2	MTD	f	Ref.
1	17α–Me–P	2.670	2.057	3	0.52	24
2	17α–Me–19–nor–P	2.500	2.454	2	0.00	15
3	6α-Me-17α-OAc-P	2.490	2.510	2	0.08	15
4	16α-Et-21-OH-19-nor-P	2.380	2.057	3	0.52	15
5	19-nor-P	2.360	2.091	2	−0.52	17
6	6α-Me-P	2.060	2.057	3	0.52	19
7	P	2.000	1.694	3	0.00	24
8	16α-Et-P	1.980	1.661	4	1.04	16
9	18-Me-19-nor-P	1.960	1.694	3	0.00	24
10	6α-Cl-P	1.880	1.604	3	−0.13	24
11	17α-F-P	1.860	1.436	3	−0.37	24
12	17α-OMe-P	1.800	1.903	2	−0.79	24
13	17α-Br-P	1.780	1.946	3	0.36	24
14	17α-OAc-P	1.780	2.147	2	−0.44	17
15	6α-F-17α-OAc-P	1.760	1.686	2	−1.10	19
16	7α-Me-P	1.610	1.298	4	0.52	16
17	21-F-P	1.670	1.694	3	0.00	24
18	21-I-P	1.520	1.694	3	0.00	24
19	18,19-dinor-P	1.520	1.728	2	−1.04	24
20	18-Me-P	1.510	1.298	4	0.52	16
21	16α-Me-P	1.420	1.298	4	0.52	24
22	21-OH-P	1.400	1.694	3	0.00	16
23	4-Me-P	1.360	1.298	4	0.52	24
24	3-deoxy-P	1.040	0.935	4	0.00	24
25	6α-Me-17α-OH-P	0.850	1.108	3	−0.84	19
26	11β,21-di-OH-P	0.600	0.543	3	−1.65	16
27	20-deoxy-P	0.600	0.935	4	0.00	24
28	6α-OH-P	0.480	0.543	3	−1.65	24
29	17α-OH-P	−0.050	0.745	3	−1.36	26
30	18-OH-P	−0.160	−0.217	4	−1.65	24
31	7α-OH-P	−0.160	−0.217	4	−1.65	24
32	16α-OH-P	−0.160	−0.217	4	−1.65	24
33	4-OH-P	−0.300	−0.015	3	−1.36	24
34	11β,17α,21-tri-OH-P	−1.000	−0.414	3	−3.02	16
35	16α,17α-CMe_2-P	2.380	2.525	3	1.19	16
36	16α-Et-21-F-19-nor-P	2.360	2.057	3	0.52	16
37	6β-Me-P	1.900	2.057	3	0.52	24
38	15β-Me-P	1.850	2.057	3	0.52	24
39	15α-Me-P	1.720	2.057	3	0.52	24
40	7β-Me-P	1.000	1.298	4	0.52	24
41	15β-OH-P	0.900	0.543	3	−1.65	24

Note: P: progesterone; P5: 17β-acetyl-5αH-androstane-3-one; PD: 17β-acetyl-4,9-estradien-3-one; Me: methyl; Et: ethyl; OH: hydroxy; OMe: methoxy; OAc: acetoxy; F: fluoro; Cl: chloro, Br: bromo; I: iodo.

TABLE 9 (continued)
Data for Progesterone Series[84]

No.	Steroid	A_{exp}	A_2	MTD	f	Ref.
42	2α-Me-P	0.700	1.298	4	0.52	24
43	14α-OH-P	0.520	0.543	3	−1.65	24
44	16β-Me-P	0.340	1.298	4	0.52	24
45	11α-OH-P	0.340	0.543	3	−1.65	24
46	16β-OH-P	0.280	−0.217	4	−1.65	24
47	2α-OH-P	0.150	−0.015	4	−1.36	24
48	16α,17α-epoxy-P	−0.100	−0.231	4	−1.76	16
49	19-OH-P	−0.300	−0.217	4	−1.65	16
50	17α-Et-P5	1.920	2.672	3	1.40	27
51	P5	1.740	1.186	4	0.36	16
52	17α,21-di-Me-PD	2.410	2.203	2	−0.36	24
53	17α-Me-PD	2.360	2.203	2	−0.36	24
54	PD	2.260	1.840	2	−0.88	15
55	16α-Me-PD	1.570	1.443	3	−0.36	24

TABLE 10
Correspondence between Substituent Position and Numerotation of Vertices

Substituent position	Vertex number, j	Substituent position	Vertex number, j
2–α(=CH_3)	29	17-α(-C≡CH,-CN)	13,18
3–β(O_3=)	1	17-α(-C≡C-CH_3)	13,18,41
4-β and 4-(CH_3)	2	17-α(-$OOCCH_3$) –$CH_2C(Me)$=CH_2	13,14,15,16
6-α(-CH_3,-OH)	30	17-α(-CH_2CN)	13,14,17
7-α(-OH,-CH_3)	4	16-α,17-α(epoxy)	38
7-β(-CH_3)	31	16-α,17-α(CMe_2)	38,39,40
11-β(-Cl,CH_3)	6	17-β(-OH)	19
11-β(11=CH_2)	7	17-β(CH_3CO-)	19,20,21
14-α(-OH)	34	17-β(CH_3CH_2CO–FCH_2CO-)	
15-α-(-CH_3)	35		19,20,21,22
15-β(-OH)	36	A-ring(4-en-3-one,T)	23,24,25
16-α(Me,Et)	11,12	A-ring(5Hα-3-one,T5)	23,24,28
16-β(-CH_3,-OH)	37	A-ring(4,9-dien,-3-one,TD)	25,26,27
17-α(-CH_3,-OMe,-CH_2Cl)	13		

Note: Common vertices (occupied by all testosterone derivatives or all progesterone derivatives) are not numerotated. The 4-β-substituent in the 5Hα-3-one(T5)-skeleton, occupies approximately the same position as the 4 substituent in the 4-en-3-one(T) and 4,9-dien-3-one(TD)-skeleton.

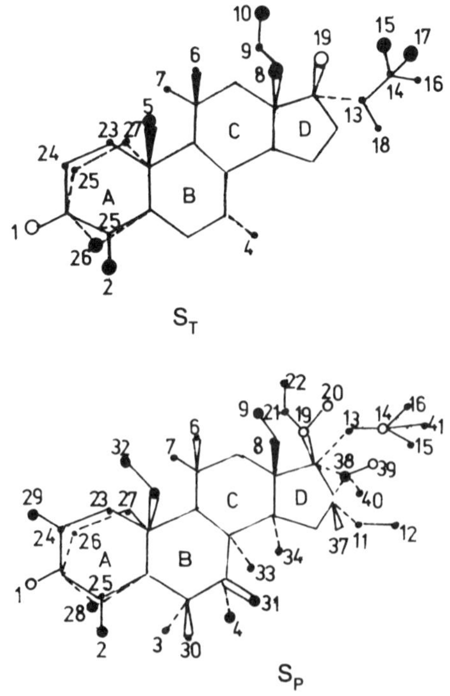

FIGURE 10. Hypermolecule and receptor maps, S_T, for testosterone derivatives, and S_P for progesterone derivatives. Numerotation, j of vertices, o, are wall vertices ($\epsilon_j = +1$); dots, · , represent exterior vertices ($\epsilon_j = 0$).

Here f is a corrected hydrophobicity (relative to progesterone, for which f = 0): the hydrogen bonding 0–3 atom and 17-β-OH, respectively, and 17β-acetyl group (and its substitution derivatives) are excluded from the calculation of f.

The regressional coefficients for MTD and f are rather similar for the four correlational equations. The receptor maps, S_T^* and S_P^* also have identical attributions (except for j:4) for the vertices corresponding to atoms of the superposable A and B cycles (in progesterone/testosterone superpositions) and for substituents at these atoms. These peculiarities of the correlational results are good arguments for their real (not chance) correlational character.

The results can be interpreted as interactions with a hydrophobic receptor site, except hydrogen bonding groups fixing the O-3- and O-17β- or O-20-oxygen atoms. The attractive interactions are given mainly by the hydrophobic, f, parameter; the irrelevant, $\epsilon_j = 0$ vertices are not to be considered as exterior. There are few $\epsilon_j = -1$ (cavity) vertices, most of them (for j:1, 19, and 20) expressing the presence of hydrogen bonding atoms. The MTD parameter expresses mainly the number of steroid-molecule atoms ''penetrating'' into the receptor-wall regions. As steroid molecules are rather rigid these ''penetrations'' are circumvented, probably by local conformational changes within the receptor of a rather ''soft'' character.

The same approach — a hydrophobic receptor with two hydrogen bonding groups also works for aromatase inhibitors.[83] For a series of N = 46 different steroids, the following correlational equation was obtained:

$$\hat{A} = 1.621 - 0.566\text{MTD} + 0.535f + 3.120R \; ; \; r^2 = 0.835 \quad (41)$$

Here R is the resonance energy, in β units, for the conjugated π-electronic systems in the region of the A and B rings.

The receptor for steroidal aglycones (of the type in cardiotonic drugs) on the Na/K-ATPase (sodium-potassium pump) seems much less hydrophobic. For a series of N = 20 substituted steroids (of the type digitoxygenin, digoxygenin, gitoxygenin) the correlational equation is

$$\hat{A} = 7.383 - 0.753\text{MTD} - 0.923\,\delta_7 \; ; \; r^2 = 0.839 \quad (42)$$

Here, $\delta_7 = 1$ for the presence of an OH substituent at C-15. Similarly, in a previous note concerning cardiotoxic steroids (the cardiotoxic action is probably due to the inhibition of the Na/K-pump), one obtains, for a series of N = 30 molecules[80]

$$\hat{A} = 0.70 - 1.15\text{MTD} \; ; \; r^2 = 0.792 \quad (43)$$

For a test series, $r^2 = 0.627$ was obtained in the quoted study.[80] Introduction of the hydrophobicity, π, increases the correlation coefficient only to $r^2 = 0.839$.

The receptor for the Na/K pump inhibiting cardiotoxic steroids should be less hydrophobic than the gestagenic receptor, but also relatively soft, thus allowing different orientations of polar and apolar groups with the result of an optimal accommodation of steroid molecules with different polar and apolar substituents.

3. MTD Vs. Computational Chemistry — Limits of the Simple Approach

We shall consider the limitation of the classical, simple approach in QSAR based upon an example — a class of HMG-CoA-reductase inhibitors.[85] The general structure of the 9,9-bis-(4-fluorophenyl)3,5-dihydroxy-8(substituted)-5,8-nonadienoic acid analogues is shown in Figure 11, which also illustrates the hypermolecule for the MTD approach for 8-substituted derivatives. Biological activity and structure data for the N = 13 derivatives are listed in Table 11.

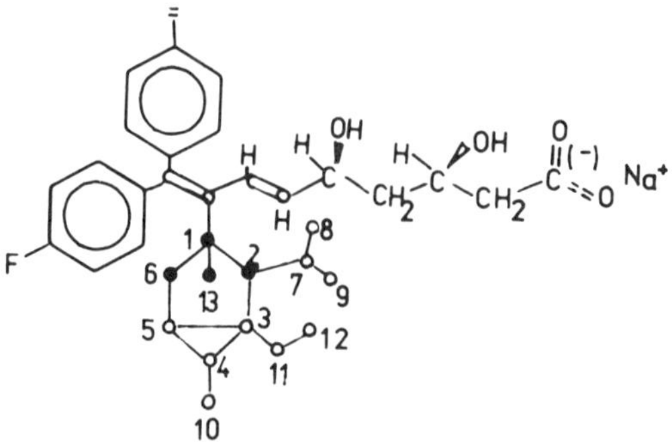

FIGURE 11. General structure of HMG-CoA reductase inhibitors, substituted in position 8. Hypermolecule and vertex rotation for the substituent in position 8. Also see legend of Figure 5.

For the MTD method, the optimized receptor map and the corresponding correlation coefficient are

$$S^* \begin{cases} j(\epsilon = -1)\text{: } 1,2,6 \\ j(\epsilon = 0)\text{: } 3,5,7,13 \\ j(\epsilon = +1)\text{: } 4,8,12 \end{cases}$$

$$\hat{A} = 6.91 - 1.320\text{MTD} \ ; \ r^2 = 0.709 \ ; \ N = 13 \qquad (44)$$

This series also was considered within the so-called 3D-QSAR methodology.[85] In order to determine a logical candidate for the bioactive conformation and the reference space of the C_8-substituent (a receptor cavity), a computational procedure was used,[87] based upon the structure of inhibitors numbered 1 to 3 in Table 11, and of four other different strong inhibitors with a more rigid structure in position 8. The correlational result is

$$\hat{A} = 0.0608 \cdot \text{OV} - 0.114 \cdot \text{NOV} + 0.1667$$

$$r^2 = 0.937 \ ; \ N = 13 \qquad (45)$$

Here, OV represents the overlapping van der Waals volume (Å^3) of the reference space and the C_8 substituent, NOV is the nonoverlapping volume of the C_8 substituent (volume overlapping with the receptor cavity walls). This equation allows us to calculate with excellent results the pIC_{50} values for five 8-substituted derivatives (H, Et, t-Bu, CHF_2, and CF_3). Also, for a series of $N = 9$ derivatives (see Table 11 and this new series) with NOV = 0, i.e., without steric repulsion with the receptor cavity, the pIC_{50} values correlate well with the van der Waals dispersion interaction energy (E_{pol}) calculated with a methane cavity in the shape and size of the receptor cavity ($r^2 = 0.974$).

TABLE 11
Biological Activity and Structural Data for HMG-CoA-Reductase Inhibition

No.	8-(substituent)	pIC_{50}	$j(x_{ij} = 1)$	MTD	OV	NOV
1	1–Me–Tet	7.43	1–3,5–7	0	102.5	0.0
2	tetrazol-3-yl	6.48	1–3,5–6	0	88.6	0.0
3	iso-Pr	6.40	1,2,6	0	90.7	0.0
4	Ph	5.48	1–6	1	96.9	12.0
5	2,4-di-Cl-Ph	3.82	1–7,10	2	103.5	34.5
6	2-Me-Ph	5.02	1–7	1	104.3	18.6
7	3-Me-Ph	4.60	1–6,11	2	99.9	22.8
8	CN	5.10	1,13	2	65.4	0.0
9	CH_3	5.15	1	2	63.4	0.0
10	1-El-Tet	5.95	1–3,5–8	1	98.5	16.9
11	1-iso-Pr-Tet	4.85	1–3,5–9	2	105.8	25.3
12	2-Me-Tet	4.52	1–3,5,6,11	1	87.7	14.6
13	2-Et-Tet	3.70	1–3,5,6,11,12	2	90.3	25.8

Note: $j(X_{ij} = 1)$, vertices j, occupied by molecule "i" in the hypermolecule of Figure 6; MTD minimal steric difference; OV and NOV: molecule-receptor cavity and molecule-receptor wall overlapping volumes ($Å^3$).

These results show clearly the relative character of the MTD method. In this example, the 1-substituent of the 5-tetrazolyl ring is superposed within the hypermolecule construction upon the 2-substituent of the phenyl ring, the vertex j:7. In the first case, the Me-1-substituent increases pIC_{50}, in the second, the Me-2-substituent produces a strong decrease. The more exact calculation of the OV volume, overlapping with the receptor cavity and NOV overlapping with the receptor walls, allows a much better correlational result in this case. Even more, if there is no steric repulsion the calculated attractive dispersion interaction correlates exceptionally well with the inhibitory power of the derivative.

B. THE COMPUTATIONAL CHEMISTRY APPROACH

Recent advances in computational chemistry, molecular graphics, and also in problem formulation appears to promise, according to Marshall and Cramer,[2a] a general solution to the problem of QSAR. These authors termed this type of method, which implies efficient computer techniques, the three-dimensional structure-activity relationship. We prefer here the term "computational chemistry approach" in order to stress the relatedness to methods used in computational chemistry and also because the three-dimensional character of receptor-effector interactions is accounted for in methods which belong rather to the classical approach. Two major problem classes may be distinguished in the computational chemistry approach: those in which the three-dimensional structure of the receptor site has been determined by X-ray crystallography, or inferred from known similar structures, and those in which receptor structure has not been determined.

The computational chemistry approach methods require a knowledge of molecular structures and conformation energies: access to three-dimensional structure, databanks such as the Cambridge Crystallograph Database,[68] programs for molecular mechanics calculations, and programs for real-time interactive stereo computer graphics.[89] In some methods, such as the relatively popular molecular shape analysis of Hopfinger,[90] a molecular parameter — the overlapping volume of drug molecules (in lowest energy conformation) with a standard, very active, molecule — is calculated and used in a Hansch-type QSAR. Very often, these techniques are used in qualitative discussions for the design of hypothetical receptor models.[62]

In order to obtain structure-activity relationships for receptors of unknown structure, an operational model of the receptor must be deduced, that is able to offer a consistent explanation for the known data and, ideally, to predict activity values for new compounds.[3] The features of such receptor models are equivalent to a multiparametric description and require statistical techniques which are capable of dealing with large sets of variables and limited sets of molecules, such as the partial-least-squares (PLS) method.[91]

1. The Three-Dimensional QSAR

If there is no specific information about the three-dimensional structure of a receptor, one must deduce an operational model for the active receptor site. A simplifying hypothesis is the assumption that the geometry of the site is constant and the different effector molecules bind in conformations which present similar steric and electronic patterns, the pharmacophore. Due to conformational freedom of most molecules, each is capable of presenting a multitude of three-dimensional patterns to the site. The first step is to determine the relative importance for recognition of different functional groups of the molecules of interest by chemical modifications, which can give some information about the nature of the functional groups in the active receptor site responsible for binding the set of molecules. The underlying concept is complementarity between the site and the molecules which are bound. Second, a hypothesis concerning the correspondence between functional groups in the series of studied molecules or of the recognition groups within the site must be checked in order to see if one or more geometrical arrangements of the postulated groups is common for the set of active molecules.[92,93]

This active analogue approach thus offers a geometrical arrangement, a candidate pharmacophore or receptor model, which can be evaluated for predictive merit; but a conceptually and computationally integrated approach for validation of such a model is necessary. The pharmacophore model, once validated, can be regarded as a topographical and chemicophysical specification of the receptor. Such a specification allows meaningful comparison of structurally diverse molecules exhibiting the same biological action. The approach is based on a method (the so-called Search method) for systematic analysis of the conformational hyperspace available to the studied molecule. Stable conformations of large molecules are thereby identified as well as is the probable binding mode of the molecule in the active receptor site.[87]

The systematic analysis of the conformational hyperspace[87] starts with the description of the topology of each interesting molecule by a graph $G = (V,E)$, in which the non-empty set V of p vertices describes the atoms and the set E of g edges collects the bonds constituting the molecule.[95] To those vertices, atomic coordinates are added such that to the topological specification $(\{G_l\}, E'')$ of a molecule, for each conformation one adds a set A of Cartesian coordinates. This vector formulation of atomic positions allows various rotational transformations which transform one molecular conformation into another. A differential distance function:

$$\delta_{ij}(\omega) = d_{ij}^2(\omega) - c_{ij}^2 \tag{46}$$

is also introduced, with d_{ij} the Euclidian distance between the nonbonded atoms i and j, c_{ij} the sum of their van der Waals radii, and ω the rotational displacement with respect to some standard bond-containing planes.

The computational search procedure[96] generates and records all points in discrete hyperspace which correspond to sterically allowed conformations and yields their complete geometric description. Search can be used with additional constraints on the allowed distance between pairs of atoms belonging to different aggregates; a constraint is specified by the minimum and maximum allowed distances for a pair of atom.

The pharmacophore model is characterized as follows. Consider the biologically active molecules, M_i, $1 \leq i \leq n$, and let F_k, $1 \leq k \leq m$ be the constituent functional groups responsible for recognition of these molecules and subsequent activation of the receptor. Further, let $\{P_{ij}\}$ collect the points j in the conformational hyperspace available to M_i and corresponding to the sterically allowed conformations. The 3m-6 pairwise distances between F_p and F_g are d_{pg} with $1 \leq p \leq g$ m in P_{ij} and $1 \leq i \leq n$. The pharmacophore model consists of the specification of $\{F_k\}$, $1 \leq k \leq m$, the correspondence between $\{F_k\}$ and M_i, $1 \leq i \leq n$, and the specification for any M_i of d_{pg} as $d_{pg} \in [d_{pg}^{min}, d_{pg}^{mx}]$, with d_{pg}^{min} and d_{pg}^{mx} as given constants. The pharmacophore model is valid if the set

$$\prod_{i=1}^{n} \{P_{ij}\} = \{P_{ik} | (d_{pg}^{\alpha} - d_{pg}^{\beta}) \leq \delta$$

$$1 \leq p < g \leq m, \quad 1 \leq \alpha \leq \beta \leq n\} \tag{47}$$

is not empty. Here δ is a given constant.

The relative location of the pharmacophore groups F_k in M_i can be represented by a mapping of $\{P_{ij}\}$ onto a (3m-6) dimensional vector space, the orientation space (O-space). The hyperspace defined by the (3m-6) pairwise distances, d_{pg}^{min}, is a hypercube of edge r. The pharmacophore model is unique at resolution r if the points in the O-space fall within a single hypercube.

A molecular mechanics program, Maximin, is hereby included.[97] With this program, energy can be minimized while maintaining specified geometric relationships among given sets of atoms. Two alternative methods for conformational comparisons are supported: a set of flexible molecules can be mapped onto a rigid reference structure or, by treating all molecules as flexible entities one can minimize the conformational variance of the set.

The Maximin program uses a special force field with usual energy functionals for bond stretching, angle bonding, torsional, van der Waals, electrostatic, and hydrogen bond terms (compare Section II.C, Equations 13 and 14a to 14e). ''Fixing'' of potential functions can also be introduced. A choice of dielectric functions to calculate the dielectric constant, $D(r_{ij})$ for the separation distance, r_{ij}, between the charge carriers i and j, which accommodate most of the literature suggestions, is also given.[97] Maximin requires characterization of molecular topography, molecular topology, and chemical constitution; a virtual atom-type Du (''dummy'') is also provided in order to facilitate marking geometrical features of molecules.

2. Inhibition of Angiotensin-Converting Enzyme and Other Applications

Angiotensin-converting enzyme, ACE, catalyzes the hydrolytic removal of the carboxy-terminal dipeptid residue of the decapeptid angiotensin I, with production of the octapeptid, angiotensin II — a vasoconstrictor. ACE also hydrolyzes bradykinin. The structure of ACE is not known yet, but chemical and biochemical considerations indicate that it resembles the structure of carboxypeptidase A.[87,98,99]

Inhibitors of ACE are potential drugs for control of high blood pressure. Extensive structure-activity studies of ACE inhibitors are available.[98] They clearly establish the importance of a terminal carboxylate group, an amide carbonyl, and a Zn binding functional group, X. The carboxylate group, $-COO^-$, interacts with an NH_3^+ group of the enzyme and the amide carbonyl, –CO–, is hydrogen bonded to an enzyme NH group. The presence of these functional groups is required for high levels of enzyme inhibitions and they are considered as pharmacophore groups F_1, F_2, and F_3. The vector basis of the O-spaces represent the pairwise distances (r_1, r_2, . . . ,r_5) between the probable location of the Zn atom, amide carbonyl oxygen, and carboxylate oxygen and are illustrated in Figure 12.

A number of ACE inhibitors and the corresponding activity values ($A = -\log IC_{50}$ mol/l, quoted from Reference 98) are listed in Figure 13.

To facilitate evaluation of the quality of geometrical fit the ''dummy'' atoms marking the probable location of the enzyme groups (X,Y, and Z in Figure 12) which interact with the three pharmacophoric groups of the inhibitors (A,B, and C in Figure 12) were chosen to match. The location of

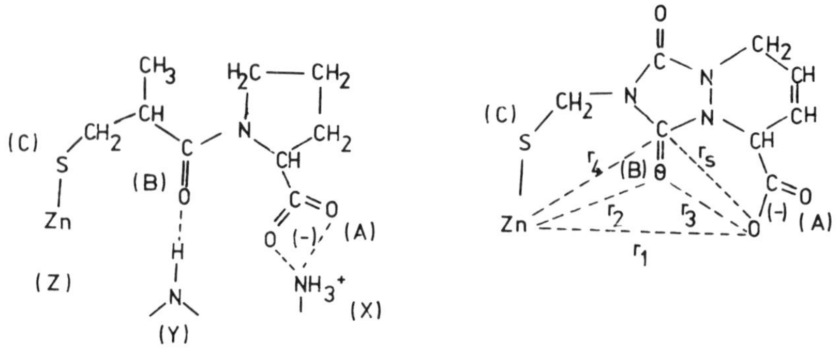

FIGURE 12. Pharmacophore hypothesis for ACE inhibition. Two different inhibitor molecules are depicted. Correspondence of the functional groups A, B, and C in the inhibitor molecules with the corresponding ACE binding sites X, Y, and Z (enzymatic NH_3^+ and NH groups, Zn^{2+} ion) and pairwise r_1, r_2, r_3, r_4, and r_5 distances of the 0 space.

Zn has been established using geometric parameters from data on appropriate Zn complexes; an NH enzyme group, hydrogen bonded to amide, and the protonated nitrogen of the $-NH_3^+$ enzyme group were all placed into a standard geometry. Then, 15 ACE inhibitors were mapped upon molecule 5, used as rigid reference, and other molecules fitted upon it.[97] For a quality fit of between 0.01 and 0.04 Å, the calculated bond-stretching and angle-bending energies were small and positive (0.34 to 4.54 and 1.44 to 18.03 kcal/mol), respectively, but the van der Waals energies were negative and somewhat higher (-3.36 to -12.25 kcal/mol).[97] The numbers of valid inhibitor conformations were also calculated, as well as the corresponding upper and lower limits of van der Waals energies.[94] There were some difficulties in computing electrostatic interaction energies because of the distance dependence of the dielectric constant, $D(r_{ij})$ and therefore this type of interaction was not included.[97]

The final results[98] can be expressed by the conformations of the 28 ACE inhibitors that fit the final OMAP and the corresponding five distances:

$$r_1 = 8.543\text{Å} \qquad r_2 = 4.938\text{Å} \qquad r_3 = 3.808\text{Å}$$
$$r_4 = 5.278\text{Å} \qquad r_5 = 4.095\text{Å} \tag{48}$$

The differences between the energy of the conformer of each inhibitor that fits the final OMAP and the energy of the same conformer after minimization were also calculated. They are rather small, ranging between 0.275 and 3.90 kcal/mol.

As a conclusion — first, the positions of two of the functional groups of the inhibitor are very restricted, while the third has a little more orientational

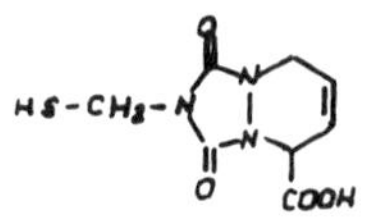

1 (6.12)

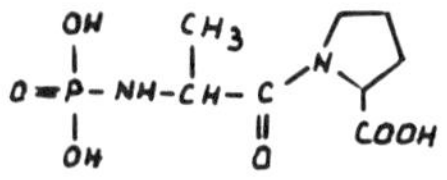

7 (8.85)

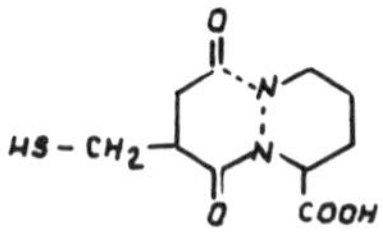

2 (7.42)

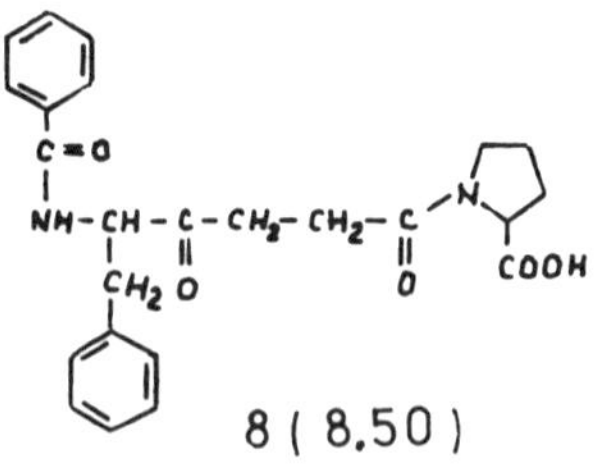

8 (8.50)

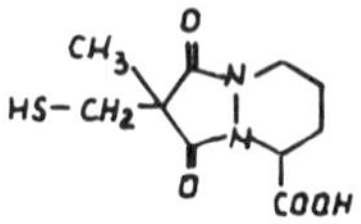

3 (6.00)

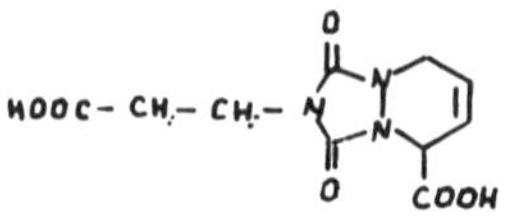

9 (9.00)

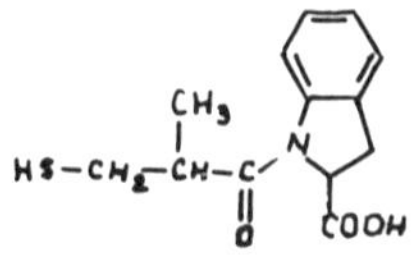

4 (8.43)

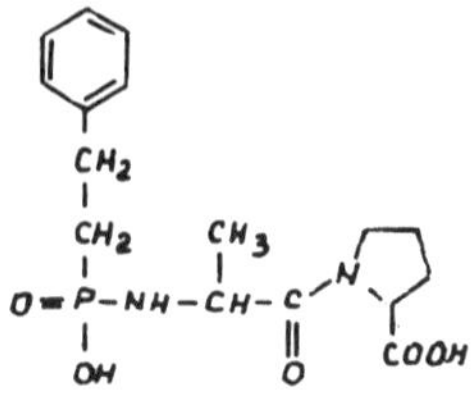

10 (8.15)

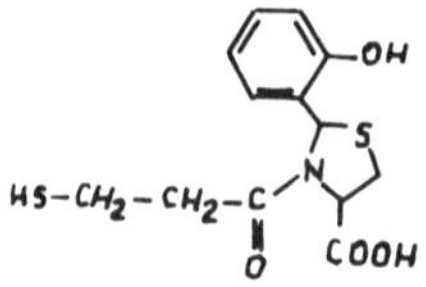

5 (8.22)

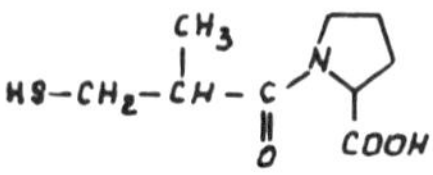

11 (7.64)

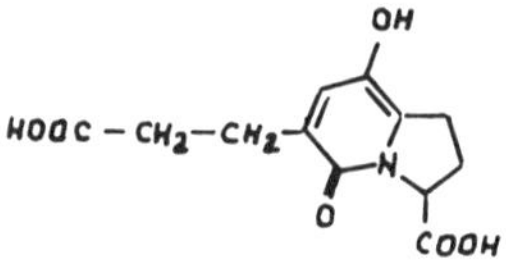

6 (8.00)

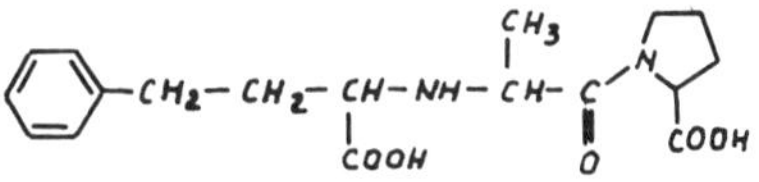

12 (8.92)

FIGURE 13. ACE inhibitor molecules and activity values (A = $-\lg IC_{50}$ mol/l)

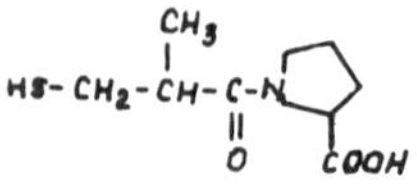

13 (8.05)

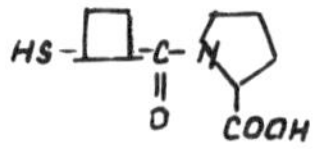

19 (7.30)

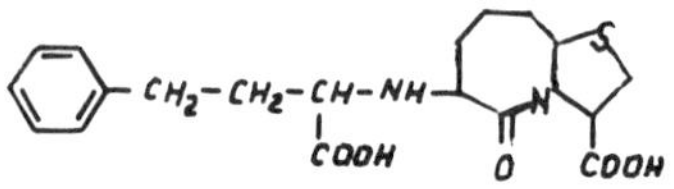

14 (9.22)

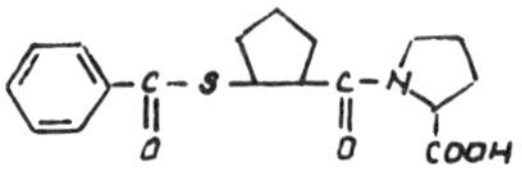

20 (7.70)

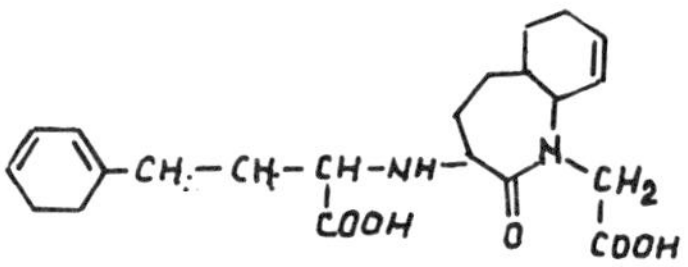

15 (8.55)

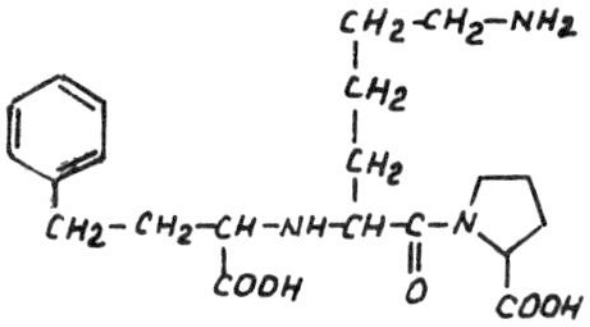

21 (8.92)

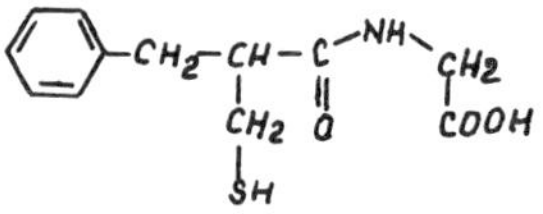

16 (7.00)

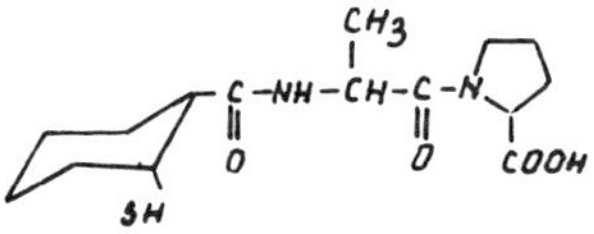

22 (8.52)

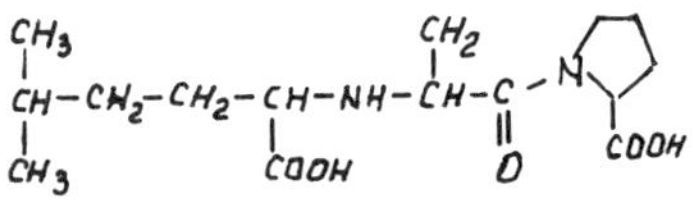

17 (8.59)

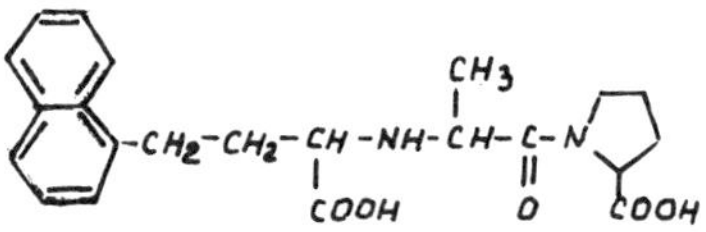

23 (8.96)

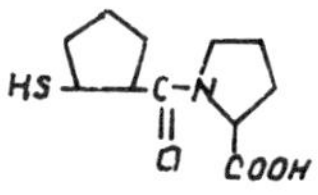

18 (7.40)

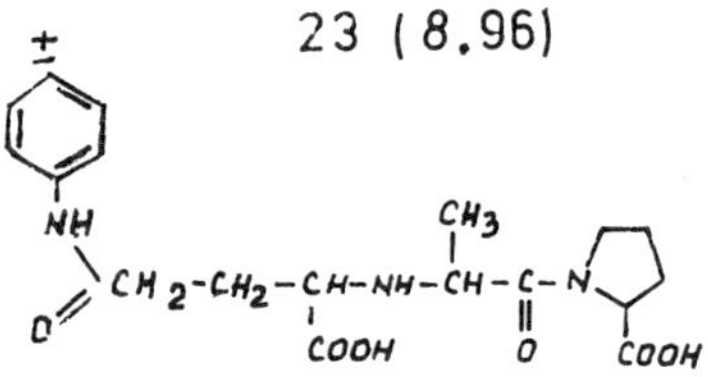

24 (9.66)

FIGURE 13. (continued)

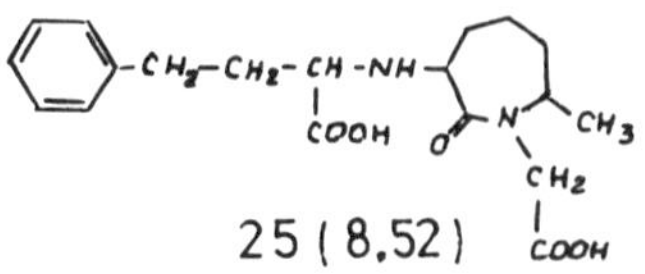

25 (8.52)

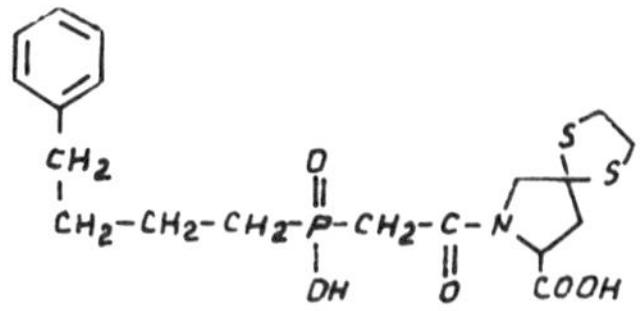

26 (8.40)

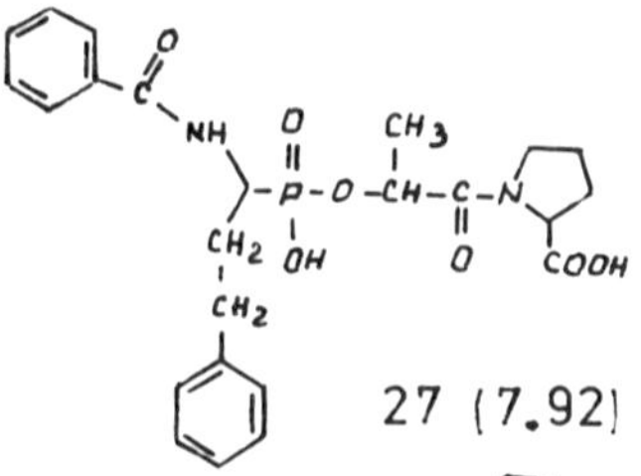

27 (7.92)

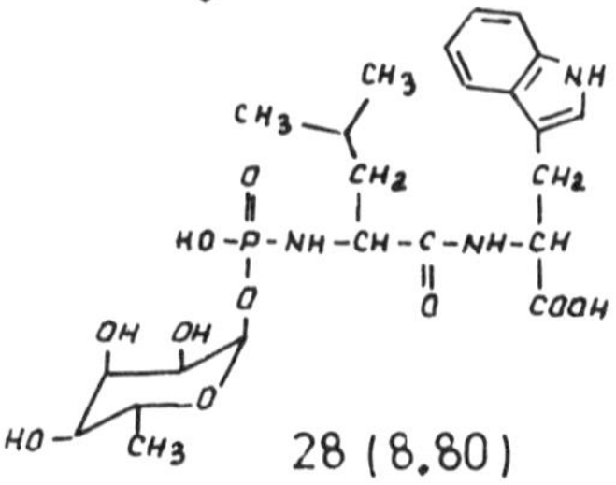

28 (8.80)

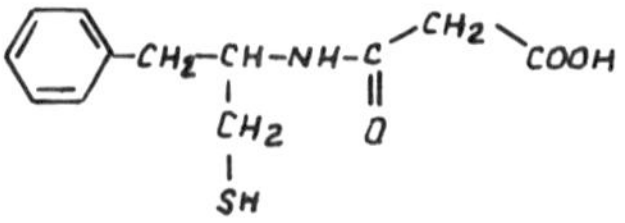

29 (inact)

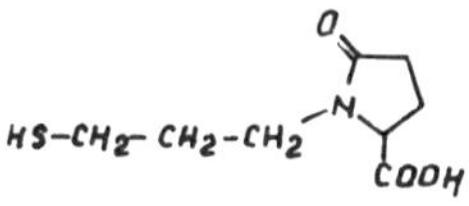

30 (inact)

FIGURE 13. (continued)

freedom, being connected with the position of the Zn ion in the enzyme structure. Second, the amide carbonyl group in many compounds may adopt two alternative orientations relative to the carboxyl group. Third, the position of the terminal carboxyl group was oriented in all cases in an axial position with respect to the ring system, avoiding steric interaction with the amino carbonyl group. Last, the orientation of Zn-binding functions differ among the various inhibitors, probably because ligands can form both tetracoordinate and pentacoordinate complexes with Zn over a similar range of approach angles. These conclusions agree with observations derived by X-ray crystallography and other methods, concerning structural features of various ACE inhibitors and can be used for prediction of new, more potent inhibitors.[98]

The inhibition site of ACE was also studied by the CoMFA method (quoted by Marshall et al.[92]), for an extended series of 67 molecules. A five-compound model within the CoMFA method yields a cross validated $r^2 = 0.76$. Within the method presented in this paragraph, a common geometric arrangement representing the active site geometry should be present for each compound. In a study of the binding of inhibitors to chymotrypsin, Naruto et al.[100] showed some movements of active site residue to accommodate different ligands. Nevertheless, the studies of ACE inhibitors found a unique active site geometry for the enzyme which compares favorably with that of carboxypeptidase A, an enzyme of analogous function.[92]

The three-dimensional structure-activity relationship approach was also applied successfully in designing inhibitors for 3-hydroxy-3-methyl-glutaryl coenzyme A reductase[85,86,94] (see also Section IV.A.3).

3. Comparative Molecular Field Analysis (CoMFA)

The CoMFA method[79,101] presents some similarities to the distance geometry approach of Ghore and Crippen,[102] concerning procedures to orient molecules in the active receptor site model and in the description of this site by points (atoms with a certain geometric position) presenting certain interaction characteristics. In the distance geometry approach these points are given according to heuristic considerations and interaction energy parameters are obtained by an optimization procedure. Only recently, the selection of the characteristic site points and their geometry is performed on a more systematic basis, with some similarities to the MTD method.[103]

In the CoMFA method, the receptor site model is a three-dimensional lattice, at the intersection of which the probe atoms are situated. In principle, for any molecule M_i of the studied series, one calculates the interaction energy of each point with each M_i molecule. Two terms, a coulumbian, U_{el}^{ik}, and a van der Waals, U_{vdW}^{ik}, are used for each grid point atom, k, molecules atom, li, interaction:

$$U_{el}^{ik} = \frac{1}{4\pi\epsilon_o D}\,\frac{e \cdot q_{li}}{r_{kli}}\ ;\ U_{vdW}^{ik} = -\frac{1}{r_{kli}^6} + \frac{1}{r_{kli}^{12}} \qquad (49)$$

The A and B parameters correspond to sp^3 carbon atoms of charge +1.0 e. The interaction energy of the lattice with molecule M_i, or the corresponding biological activity, A_i, is given by a linear equation:

$$\hat{A}_i = a_o + \sum_{k=1}^{M} b_k \cdot U_{el}^{ik} + \sum_{k=1}^{M} c_k \cdot U_{vdW}^{ik} \tag{50}$$

As the total number, M, of grid points can easily reach 1000, data analysis by partial least squares (PLS) is used in the optimization processes to determine the regressional coefficients a_o, b_k, c_k or, more exactly, a batch of them able to explain the biological activity (A_i) differences between the N molecules M_i. Also, docking techniques are used to optimize the orientation of M_i molecules with respect to the grid. Cross validation is used to maximize the likelihood that the results have predictive validity.

The CoMFA results are presented graphically, as contoured, colored three-dimensional coefficient plots. Unfortunately, without supplementary material (which can be obtained on request), the published results cannot be reproduced and verified.

In an application[79] for steroid binding potencies of a corticosteroid binding globulin (CBG) and a testosterone-binding globulin (TeBG), the CoMFA method yields much better results than various classical QSAR approaches. For N = 21 steroids (Table 12 and Figure 14) for CBG binding, the CoMFA-QSAR yields $r^2 = 0.897$, while for the cross validation procedure, $r^2 = 0.662$. With the same proteins, for TeBG binding for the CoMFA-QSAR, $r^2 = 0.873$, while for cross validation $r^2 = 0.555$ is obtained. For a test series of N = 10 steroids, the predicted vs. measured CBG binding yields $r^2 = 0.65$. The steric contour maps indicate, for CBG binding, regions where lower steric interactions that increase binding are in the vicinity of 2β, 4β, and 17β substituents, while regions where higher steric interactions increase binding are in the region of 11β substituents, a more extended region in the 17β direction and the region of the D ring. For TeBG binding, lower steric interactions are favorable in the region 4β, the 16-C atom, and 16-β substituents of the D ring, and an extended region in the 17β direction; higher steric interactions are favorable in the 11β region.

Other CoMFA results include the ACE inhibitors,[92] already discussed here in the previous section (IV.A.2), inhibition of dihydrofolate reductase by substituted benzylpyrimidines,[101] inhibition of prostaglandin synthetase by 3-benzyl-2-amino-phenylacetic acid substitution derivatives ($r^2 = 0.5$, classical QSAR unsuccessful),[104] and a recent study[105] concerning the affinity for the benzodiazepine receptor inverse agonist site of N = 37 substituted β-carbolines and pyridoindoles (cross validated $r^2 = 0.59$).

The principles of the CoMFA method, description of the active receptor site by a grid of atoms at about 2 Å interdistance, and direct use of van der Waals and electrostatic potentials for receptor-molecule interaction energies,

TABLE 12
Steroids for CoMFA Study of Binding to Testosterone Binding Globulins (TeBG) and Corticosterone Binding Globulins (CBG)

Compound	Ring type and substituents	A_{Te}	A_C
1. Aldosterone	II,3-O,11β,13βO-CH-OH,17β-$COCH_2OH$	−5.32	−6.28
2. Androstanediol	I,3β-OH,17β-OH	−9.11	−5.00
3. Androstanediol	III,3β-OH,17-β-OH	−9.18	−5.00
4. Androstanedione	II,3-O,17-O	−7.46	−5.76
5. Androsterone	I,3β-OH,3-O	−7.15	−5.61
6. Corticosterone	II,3-O,11β-OH,17β-$COCH_2OH$	−6.34	−7.88
7. Cortisol	II,3-O,11β-OH,17αOH,17-β$COCH_2OH$	−6.20	−7.88
8. Cortisone	II,3-O,11-O,17αOH,17-β$COCH_2OH$	−6.43	−6.89
9. Dehydroepiandrosterone	III,3β-OH,17-O	−7.82	−5.00
10. Deoxycorticosterone	II,3-O,17-β-$COCH_3OH$	−7.38	−7.35
11. Deoxycorticol	II,3-O,17αOH,17-β-$COCH_2OH$	−7.20	−7.88
12. Dihydrotestosterone	I,3-O,17-β-OH	−9.74	−5.92
13. Estradiol	IV,3-OH,17β-OH	−8.83	−5.00
14. Estriol	IV,3-OH,16α,17-β-diOH	−6.63	−5.00
15. Estrone	IV,3-OH,17-O	−8.18	−5.00
16. Etiocholanolone	I,3α-OH,3-O	−6.15	−5.20
17. Pregnenolone	III,3β-OH,17β-$COCH_2OH$	−7.15	−5.00
18. Hydroxypregnenolone	III,3β-OH,17-α-OH,17-β-$COCH_2OH$	−6.36	−5.00
19. Progesterone	II,3-O,17-β-$COCH_2OH$	−6.94	−7.30
20. Hydroxyprogesterone	II,3-O,17-α-H,17-β-$COCH_2OH$	−7.00	−7.70
21. Testosterone	II,3β-OH,17-β-OH	−9.20	−6.89

Note: A_{Te} and A_C: logarithms of dissociation constants. Steroid skeleton types (I, II, III, IV) — see legend of Figure 15.

are very appealing. Nevertheless, there are some debatable points.[106,107] The receptor site is considered rigid, without possible different deformations by different M_i molecules. Repulsive interactions-steric misfit is considered by 6–12 van der Waals potentials (U_{vdW}, Equation 49) with a rather high cutoff, +30 kcal/mol. For a grid atom, k, producing attractive van der Waals contributions for several M_i molecules but a high repulsive interaction with even one molecule (due to short r_{kli} interdistances), a very small $c_k = 0$ regressional coefficient will result. In the real case, as discussed in Section III.D, steric fit is avoided at a much lower energy cost than the +30 kcal/mol cutoff by local conformational changes within the receptor, or by deformation of torsional angles. This overestimation of repulsive effect in CoMFA could be avoided if the van der Waals, U_{vdW} (Equation 49) were used only for attractive van der Waals contributions (with a cutoff at +0 kcal/mol), while special repulsion terms:

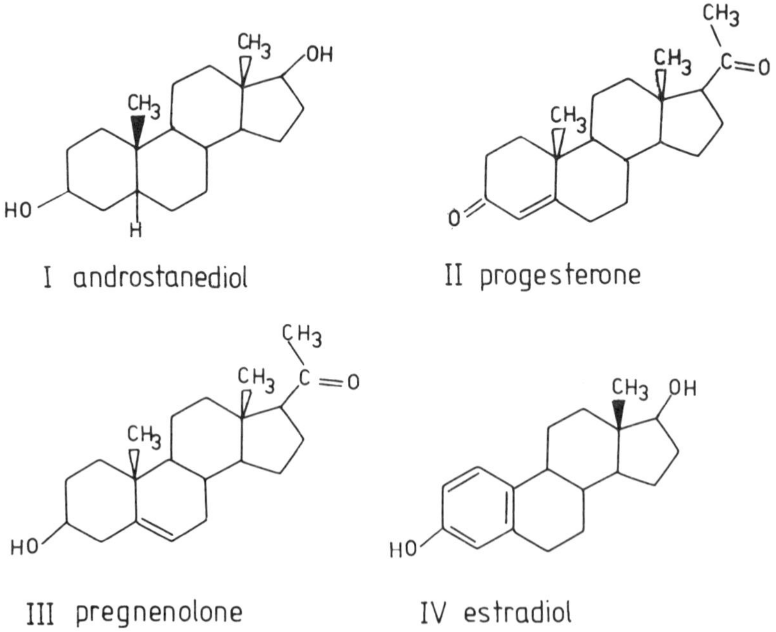

FIGURE 14. Type of steroid rings used in the CoMFA-QSAR for CBG and TeBG steroid binding. **I:** steroid cycle without double bonds; **II:** steroid cycle with one 4,5-Δ double bond; **III:** with one 6,7-Δ double bond; and **IV:** with a benzenoid A ring.

$$U_{rep}^{ik} = \frac{B}{r_{kli}^{12}} \quad \text{(cutoff at } +30 \text{ kcal/mol)} \tag{51}$$

would be considered as a third type of terms with separate d_k regressional coefficients in Equation 50 for the biological activity, A_i.

In a recent MTD study concerning the binding to CBG and TeBG of the steroids of Table 12, the cross-validation results are even better than those of the corresponding CoMFA study.[79,108]

V. GENERAL CONCLUSIONS

The calculation of the receptor-effector affinities also requires, supplementary to the calculation of the direct interaction energy ΔU_{R*E}, calculation of hydration effects (of the complex, ΔG_3, and of the receptor and effector separately, ΔG_1) and the energy, ΔU_{conf}, related to conformational changes within the complex.

Ab initio quantum chemistry methods require further developments of computational and programming techniques before it is possible to apply them currently to molecules of interest. The intermolecular force potentials of the molecular mechanics approach can be used with good results for calculating the direct interaction energy, ΔU_{R*E}, and for energetic effects in conformational changes of effector micromolecules. Hydration effects (ΔG_1,ΔG_3) can be calculated by molecular dynamics, by the free energy perturbation approach, but large amounts of supercomputer time are required for each special interaction-pair ($R*E_o$ compared to a single $R*E_1$). The bottleneck is caused by the lack of success in calculating conformational energy effects in proteins and in predicting steric structure from primary structure in proteins, due to the astronomical number of possible interactions between distant amino acid residues as primary structure. This is a serious drawback if one considers that there are large numbers of proteins with known primary structure, but of which only a relatively small number have a known steric structure.

One may speculate that receptors with a high specificity (e.g., antibody-combining sites) will be rather rigid, while receptors with low specificity (lytic enzymes, binding sites of transport proteins) will have a rather high conformational flexibility, and rather shallow binding sites, imposing few steric restrictions on the molecules to be bonded.

The older, more intuitive approach, concerning types of intermolecular forces and their strengths, is certainly very helpful in understanding mechanisms of receptor-effector interactions and variations in affinity produced by modifications in the effector molecule. No systematic predictions of receptor-effector affinities or of structure-activity relations are possible in this case.

The quantitative structure-activity approach, used for drug design, can also yield information concerning receptors whose structure was not deciphered by X-ray crystallography or NMR, or concerning structural changes induced by receptor-effector interactions. The three-dimensional or computational chemistry approach to structure activity relations has had interesting results in several cases, but does not seem to work in cases in which different rigid effector molecules induce different local conformational changes in the receptor. The classical Hansch-type approaches with simplified consideration of steric misfit (MTD and SIBIS) may be successful in certain cases. They can certainly be recommended as a preliminary approach, especially in establishing models for receptors without known structure.

APPENDIX 1

How exact must a steric fit be? — Let us start from Equations 22 and 22a of Section III.D. We obtain:

$$-A/r^6 + B/r^{12} = -A^2/8B$$

Multiplying both terms with $8B.r^{12}/A^2$ and substituting $r^6 = Z$ produces:

$$Z^2 - \frac{8B}{A} Z + \frac{8B^2}{A^2} = 0$$

with the solutions:

$$Z_{1,2} = \frac{2B}{A} (2 \pm \sqrt{2}) = r_o^6 \cdot (2 \pm \sqrt{2})$$

$$r = r_o + r = r_o \cdot (2 + \sqrt{2})^{1/6} = \begin{cases} 1.227\ r_o \\ 0.916\ r_o \end{cases}$$

APPENDIX 2

Calculation of the entropy of the mixing of reactions in solution — For a reaction in solution, if hydration of reactants and products are similar the entropy will be determined mainly by the entropy of mixing, ΔS_{mix}. Consider a reaction in solution, producing a supplementary product molecule:

$$A + \sigma S \rightarrow B + C + (\sigma - 1)S \ ; \ \Delta S'_{mix}$$

$$A + \sigma S \rightarrow B + C + \sigma S \ ; \ \Delta S''_{mix}$$

S is the solvent molecule, σ is the number of moles of solvent 1 l of solution, wherein the concentration of each of the solutes is 1 mol/l (activities equal to 1 for ideal solutions). For molar solutions in water, the number of moles of H_2O/l of solutions is $\sigma \cong 50$; thus $\sigma >> 1$. For the first reaction, in 1 l solutions there are N_A molecules A and σ N_A molecules of S before, and N_A molecules A, N_A molecules of B, and $(\sigma - 1)N_A$ molecules of S after the reaction. From the usual formula for entropy in physical statistics:

$$\Delta S'_{mix} = S_{final} - S_{initial}$$

$$= k \cdot \ln \frac{[N_A(\sigma + 1)]!}{N_A!N_A![N_A(\sigma - 1)]!} - k \cdot \ln \frac{[N_A(\sigma + 1)]!}{N_A!(N_A \cdot \sigma)!}$$

$$= k \cdot \ln \frac{(N_A\sigma)!}{N_A![N_A(\sigma - 1)]!}$$

With Stirling's approximation: $x! \cong (x/e)^N$

$$\Delta S'_{mix} = k \cdot \ln \frac{(N_A\sigma)^{N_A}}{N_A^{N_A}[N_A(\sigma - 1)]^{N_A(\sigma - 1)}} \cong R(1 + \ln\sigma)$$

Similarly, for the second reaction:

$$\Delta S''_{mix} = R[1 + \ln(\sigma + 2)]$$

For $\sigma \cong 50$, $\Delta S'_{mix} \cong \Delta S''_{mix} = +9.9$ cal/mol°K. Thus, at 300°K, the free enthalpy, ΔG, of the reaction is about 3.0 kcal/mol.

REFERENCES

1. **Moelwyn-Hughes, E. A.,** *Physical Chemistry,* Pergamon Press London, 1961, chap. 6.
2. **Bolis, L., Keynes, R. D., and Wilbrand, W., Eds.,** *Role of Membranes in Secretory Processes,* Elsevier/North-Holland, Amsterdam, 1972.

2a. **Marshall, G. R. and Cramer, R. D., III,** Three dimensional structure-activity relationships, *Trends Pharmacol. Sci.,* 9, 285, 1988.

3. **Clementi, E.,** Supercomputers and Supercomputing in Chemical Research and Development. The MOTTEC Initiative, Intelligent Instruments & Computers, 1989, 245; 1990, 14.
4. **Naray-Szabo, G., Surjan, P. R., and Angyan, J. G.,** *Applied Quantum Chemistry,* Reidel Publishing, Dordrecht, and Akademiai Kiado, Budapest, 1987.
5. **Cohen, N. C., Blaney, J. M., Humblet, Chr., Gund, P., and Barry, D. C.,** Molecular modelling software and methods for medicinal chemistry, *J. Med. Chem.,* 33, 88, 1990.
6. **Hirschfelder, J. O.,** *Intermolecular Forces,* McGraw-Hill, New York, 1970.
7. **Burkert, U. and Allinger, N. L.,** *Molecular Mechanics,* American Chemical Society, Washington, D.C., 1982.
8. **Clark, T.,** *A Handbook of Computational Chemistry,* John Wiley & Sons, New York, 1985.
9. **McCammon, J. A.,** Computers aided molecular design, *Science,* 238, 486, 1987.
10. **De Grado, W. F., Wassermann, Z. R., and Lear, J. D.,** Protein design, a minimalist approach, *Science,* 243, 622, 1989.
11. **Maelicke, A.,** Predictions of protein structures, *Nachr. Chem. Tech. Lab.,* 36, 1119, 1988.
12. **Howard, A. E. and Kollman, P. A.,** Analysis of current methodologies for conformational searching of complex molecules, *J. Med. Chem.,* 31, 1669, 1988.
13. **Dill, K. A.,** Calculation of conformer numbers for polypeptides, *Biochemistry,* 24, 1501, 1985.
14. **Wutrich, K.,** NMR — Techniques in protein structure, *Science,* 243, 45, 1988.
15. **Dean, M. P.,** *Molecular Foundation of Drug — Receptor Interactions,* Cambridge University Press, Cambridge, 1987.
16. **Tainer, J. A., Getzoff, E. D., Alexander, H., Houghton, R. A., Olson, A. J., Lerner, R. A., and Hendrickson, W. A.,** The reactivity of antipeptide antibody as function of atomic mobility of sites in a protein, *Nature,* 312, 127, 1984.
17. **van Gunsteren, W. F.,** Binding of thermolysine inhibitors to thermolysine by the free energy perturbation method, in *Computer Simulation of Biomolecular Systems: Theoretical and Experimental Applications,* van Gusteren, W. F. and Weiner, P. P. K., Eds., ESCOM Science Publishers, Leyden, 1989, 27.
18. **Webb, J. L.,** *Enzymes and Metabolic Inhibitors,* Academic Press, New York, 1963, chap. 6.
19. **Hinze, J. and Jaffe, H. H.,** Electronegativity. I. Orbital electronegativity of neutral atoms, *J. Am. Chem. Soc.,* 84, 540, 1962.

20. **Labanowski, J. K., Dammkoehler, R. A., and Motoc, I.,** Orbital electronegativity and analytical representation of atom valence state energy, *J. Computat. Chem.,* 10, 1026, 1989.
21. **Valceanu, R., Simon, Z., and Kurunczi, R.,** Quantum chemical calculations for dipole moments of phosphorus compounds, *Rev. Roum. Chim.,* 34, 377, 1989.
22. **Bacaloglu, R., Bacaloglu, I., and Simon, Z.,** MO calculations with specific parametrisation (MOSP). I. Description of the method. Dipole moment calculations, *Rev. Roum. Chim.,* 36, 1061, 1991.
23. **Orozco, M. and Luque, F. J.,** A practical procedure for determination of electrostatic charges of large molecules, *J. Comp. Aided Molec. Design,* 4, 411, 1990.
24. **Mullay, J.,** A simple method for calculating atomic charges in charged molecular systems of biochemical interest, *J. Computat. Chem.,* 12, 369, 1991.
25. **Simon, Z.,** *Quantum Biochemistry and Specific Interactions,* Abacus Press, Tunbridge Wells, Kent, U.K., 1976, chap. 3.

25a. **Raevsky, O. A., Grigoriev, V. Yu., Kireev, D. B., and Zefirov, N. S.,** Complete thermodynamic description of H-bonding in the framework of a multiplicative approach, *Quant. Struct. Act. Relat.,* 11, 49, 1992.

26. **von Hippel, P. H. and McGhee, D. A.,** DNA-protein interactions, *Annu. Rev. Biochem.,* 41, 490, 1972.
27. **Brenner, S. L. and McQuarrie, D. A.,** A self consistent calculation of the free energy and electrostatic potential for a cylindrical poly-ion, *J. Theoret. Biol.,* 39, 343, 1973.
28. **Katschalski, A.,** Polyelectrolytes, *Pure Appl. Chem.,* 26, 327, 1971.
29. **Pimentel, G. C. and McClellan, A. C.,** *The Hydrogen Bond,* W. H. Freeman, San Francisco, 1963.
30. **Zuika, I. V. and Bankovskii, Ju. A.,** Hydrogen bonds in sulphur containing compounds, *Usp. Himii(Moscow),* 42, 39, 1973.
31. **Nakano, N. J. and Igarashi, S. J.,** Molecular interactions of purines, pyrimidines and other heteroatomic compounds in aqueous solution, *Biochemistry,* 9, 577, 1970.
32. **Scheraga, H. A.,** Role of hydrophobic bonds in protein structure, *Ber. Bunsenges. Phys. Chem.,* 63, 838, 1964.
33. **Ray, A.,** Solvophobic interactions and micelle formation in structure forming non-aqueous solvents, *Nature,* 231, 313, 1971.
34. **Clothia, C.,** Hydrophobic bonding and accessible surface area in proteins, *Nature,* 248, 338, 1974.
35. **Reynolds, J. A., Gilbert, D. B., and Tanford, Ch.,** Empirical correlation between hydrophobic free energy and aqueous cavity surface area, *Proc. Natl. Acad. Sci. U.S.A.,* 71, 2925, 1974.
36. **Leo, A., Hansch, C., and Elkins, D.,** Partitions coefficients and their use, *Chem. Rev.,* 71, 525, 1971.
37. **Hansch, C. and Leo, A.,** *Substituent Constants for Correlation Analysis in Chemistry and Biology,* John Wiley & Sons, New York, 1979.
38. **Recker, R. F. and De Kort, H. M.,** Hydrophobic fragmental constants, *Eur. J. Med. Chem.,* 14, 479, 1979.
39. **Draskovits-Schuch, E., Holban, S., Ciubotariu, D., and Simon, Z.,** Partition coefficients and steric factors, *An. Univ. Timisoara, Ser. Stiinte Fiz. Chim.,* 17, 73, 1979.
40. **Mason, R. P., Rhodes, D. G., and Herbette, L. G.,** Reevaluating equilibrium and kinetic binding parameters for lipophylic drugs based on a structural model for drug interactions with biological membranes, *J. Med. Chem.,* 34, 869, 1991.
41. **Record, M. T. and Spolar, R. S.,** Some thermodynamic principles of non-specific and site specific protein-DNA interactions, in *The Biology of Non-Specific DNA-Protein Interactions,* Revzin, A., Ed., CRC Press, Boca Raton, FL, 1990, 33.
42. **Tanford, C.,** Contribution of hydrophobic interactions to the stability of globular conformations of proteins, *J. Am. Chem. Soc.,* 84, 4240, 1962.

43. **Le Pecq, J. B.,** Resonance energy transfer between ethidium bromide molecules to nucleic acid: Does intercalation wind or unwind the DNA-helix?, *J. Mol. Biol.,* 59, 43, 1971.
44. **Razska, M. and Mandel, M.,** Interaction of aromatic aminoacids with neutral poly-adenilic acid, *Proc. Natl. Acad. Sci. U.S.A.,* 68, 1190, 1970.
45. **Craig, A. M. and Isenberg, A.,** Intercalations of aromatic hydrocarbons in the poly-adenine double helix, *Proc. Natl. Acad. Sci. U.S.A.,* 67, 1337, 1970.
46. **Briegleb, B.,** *Electron Donating-Accepting Complexes,* Springer-Verlag, Berlin, 1961, chap. 3 and 9.

46a. **Burley, S. K. and Pestko, G. A.,** Aromatic-aromatic interactions: A mechanism for stabilization of protein structure, *Science,* 229, 23, 1985.

47. **Pullman, B. and Pullman, A.,** *Quantum Biochemistry,* John Wiley & Sons, New York, 1963.
48. **Fischer, E.,** Einfluss der Configuration auf die Wirkung der Enzyme, *Chem. Ber.,* 27, 2985, 1894.
49. **Williams, R. J. P.,** Flexible pharmacomolecules and dynamic receptors, *Angew. Chem.,* 89, 805, 1979.
50. **Simon, Z., Chiriac, A., Holban, S., Ciubotariu, D., and Mihalas, G. I.,** Minimum steric difference, *The MTD — Method for QSAR Studies,* Research Studies Press, Letchworth, U.K., 1984.
51. **Gergen, I., Bohl, M., Simon, H., and Simon, Z.,** Structure-activity relations for steroids by the MTD-method. Superposition procedure for molecules with different condensed cycles, *Rev. Roum. Chim.,* 34, 995, 1989.
52. **Lopez de Compadre, R. L., Pearlstein, R. A., Hopfinger, A. J., and Seydel, J. K.,** Extended QSAR analysis of 4-aminodiphenylsulphone antibacterial agents, *J. Med. Chem.,* 30, 900, 1987.
53. **Andrews, R. P., Craik, D. J., and Martin, J. L.,** Empirically derived energies for binding organic molecules to receptors, *J. Med. Chem.,* 27, 1648, 1984.

53a. **Dougherty, D. A. and Stauffer, D. A.,** Acetylcholine binding by a synthetic receptor. Implications for biological recognition, *Science,* 250, 1558, 1990.

54. **Martin, Y. C.,** *Quantitative Drug Design,* Marcel Dekker, New York, 1978.
55. **Franke, R.,** *Theoretical Drug Design Methods,* Elsevier, New York, 1984.
56. **Marsden, C. A.,** *Quantitative Drug Design in Comprehensive Medicinal Chemistry,* Vol. 4, Hansch, C., Sammes, P. G., and Taylor, J. B., Eds., Pergamon Press, Oxford, 1990.
57. **Rose, V. S., Croall, J. F., and MacFie, H. J. H.,** An application of unsupervised neural network methodology (Kohonen-Topology-Preserving Mapping) to QSAR analysis, *Quant. Struct. Act. Relat.,* 10, 6, 1991.
58. **Balaban, A. T.,** Topological correlations and cancer, in *Modeling of Cancer Genesis and Prevention,* Voiculetz, N., Balaban, A. T., Niculescu-Duvaz, I., and Simon, Z., Eds., CRC Press, Boca Raton, FL, 1990, 131.
59. **Berzuker, I. B., Dimoglo, A. S., Gorbachov, M. Yu., and Vlad, P. F.,** Origin of musk fragrance activity: the electron-topologic approach, *New J. Chem.,* 15, 307, 1991.
60. **Topliss, J. G. and Edwards, P. P.,** Chance factor studies of QSAR, *J. Med. Chem.,* 23, 1238, 1979.
61. **Cramer, R. D., III, Bunce, J. D., Patterson, D. E., and Frank, J.,** PLS and cross validation in QSAR, *Quant. Struct. Act. Relat.,* 7, 18, 1988.
62. **Darvas, F. and Darvas, Zs.,** Abstract of publications related to QSAR, *Quant. Struct. Act. Relat.,* 9, 40–82, 136–199, 234–293, 348–390, 1990.
63. **Blaney, J. M., Hansch, C., Silipo, C., and Vittoria, A.,** Structure-activity relations of dihydrofolate reductase inhibitors, *Chem. Rev.,* 84, 333, 1984.
64. **Verloop, A., Hoogenstraten, W., and Tipper, J.,** *Development and Application of New Steric Substituent Parameters in Drug Design,* Vol. 8, Ariens, E. J., Ed., Academic Press, New York, 1976, 165.

65. **Ferguson, L. N.,** *The Modern Structural Theory of Organic Compounds,* Prentice-Hall, Englewood Cliffs, NJ, 1964.
66. **Recanatini, M., Klein, T., Yang, C.-Z., McClurin, J., Langride, R., and Hansch, C.,** Quantitative structure-activity relationships and molecular graphics in ligand-receptor interactions: amidine inhibition of trypsin, *Molec. Pharmacol.,* 29, 433, 1986.
67. **Hansch, C. and Klein, T. E.,** Molecular graphics and QSAR in the study of enzyme-ligand interactions. On the definition of bioreceptors, *Acc. Chem. Res.,* 19, 392, 1986.
68. **Summers, J. B., Kim, K. H., Mazdiyasni, H., Holms, J. H., Ratajczyk, J. D., Stewart, A. Q., Dyer, R. D., and Carter, G. W.,** Hydroxamic acid inhibitors of 5-lipoxygenase quantitative structure-activity relationship, *J. Med. Chem.,* 33, 992, 1990.
69. **Gupta, S. P.,** QSAR studies on drugs acting at the central nervous system, *Chem. Rev.,* 89, 1765, 1989.
70. **Simon, Z. and Szabaday, Z.,** Minimal steric difference parameter and the importance of steric fit for structure-biological activity correlations, *Stud. Biophys. (Berlin),* 39, 123, 1973.
71. **Simon, Z.,** Specific interactions. Intermolecular forces, steric requirements and molecular size, *Angew. Chem. Int. Ed. Engl.,* 13, 719, 1974.
72. **Balaban, A. T., Chiriac, A., Motoc, I., and Simon, Z.,** *Steric Fit in Quantitative Structure-Activity Relations,* Springer-Verlag, Heidelberg, 1980, chap. 4.
73. **Motoc, I. and Dragomir-Filimonescu, O.,** Molecular interactions in biological systems. I. Steric interactions, the SIBIS algorithm, *Math. Chem. (Weinheim),* 12, 117, 1981.
74. **Motoc, I., Holban, S., Vancea, R., and Simon, Z.,** Minimal steric difference computed as nonoverlapping volumes. Correlation with enzymatic hydrolyses of ribonucleotides, *Stud. Biophys. (Berlin),* 66, 715, 1977.
75. **Bruns, W., Motoc, I., and O'Driscoll, K. F.,** *Monte-Carlo Applications in Polymer Science, Lecture Notes in Chemistry,* Vol. 27, Springer-Verlag, Berlin, 1981, chap. 1.
76. **Motoc, I.,** Biological receptor maps. I. Steric maps. The SIBIS method, *Quant. Struct. Act. Relat.,* 3, 13, 1984.
77. **Motoc, I.,** Quantitative comparison of shape of bioorganic molecules, *Z. Naturforsch.,* 38a, 1342, 1983.
78. **Stouch, T. R. and Jurs, P. C.,** Quantitative structure-activity relations for steroids, *J. Med. Chem.,* 29, 2125, 1986.
79. **Cramer, R. B., III, Patterson, D. E., and Bunce, J. D.,** Comparative molecular field analysis (CoMFA). I. Effect of shape on binding of steroids to carrier proteins, *J. Am. Chem. Soc.,* 110, 5959, 1988.
80. **Simon, Z., Dragomir, N., Plauchithiu, M. G., Holban, S., Glatt, H., and Kerek, F.,** Receptor sites mapping for cardiotoxic aglycones by the MTD-method, *Eur. J. Med. Chem.,* 15, 521, 1980.
81. **Bohl, M. and Simon, Z.,** MTD approach to QSAR for cardiotonic steroids, *Z. Naturforsch.,* 40C, 858, 1985.
82. **Bohl, M., Simon, Z., Vlad, A., Kaufmann, G., and Ponsold, K.,** MTD calculations on QSAR of steroid binding to the progesterone receptors, *Z. Naturforsch.,* 42C, 935, 1987.
83. **Bohl, M., Simon, Z., and Lochmann, J. R.,** Theoretical investigation on the role of steroids skeleton C4=C5 unsaturation in competitive aromatase inhibition, *Z. Naturforsch.,* 44C, 217, 1989.
84. **Simon, Z. and Bohl, M.,** QSAR by MTD for gestagenic steroids. The case of hard molecules and soft receptors, *Quant. Struct. Act. Relat.,* 11, 23, 1992.
85. **Motoc, I., Dammkoehler, R. A., Sit, S. Y., Balasubramanian, N., and Wright, J. J.,** 3-Hydroxy-3-methyl coenzyme A reductase: Molecular modelling, three dimensional structure-activity relationship, inhibitor design, *Quant. Struct. Act. Relat.,* 10, 30, 1990.

86. **Sit, S. Y., Parker, R. A., Motoc, I., Han, W., Balasubramanian, N., Catt, J. D., Brown, P. J., Harte, W. E., Thompson, M. D., and Wright, J. J.,** Synthesis biological profile and QSAR of series of novel 3-hydroxy-3-methyl-glutaryl-coenzyme A reductase inhibitors, *J. Med. Chem.,* 33, 2982, 1990.
87. **Motoc, I., Dammkoehler, R. A., Meyer, D., and Labanowsky, J.,** Three dimensional QSAR. General approach to the pharmacophore model validation, *Quant. Struct. Act. Relat.,* 5, 99, 1986.
88. **Horn, A. S. and De Ranter, C. J., Eds.,** *X-ray Crystallography and Drug Actions,* Clarendon Press, Oxford, 1984.
89. **Burgen, A. S. V., Roberts, G. C., and Tute, M. S., Eds.,** *Molecular Graphics and Drug Design,* Vol. 3, Elsevier, Amsterdam, 1986.
90. **Hopfinger, A. J.,** Molecular shape analysis and structure-activity relationships for dihydrofolatereductase inhibitors, *J. Med. Chem.,* 26, 990, 1983.
91. **Wold, S., Albano, O., Dunn, W. J., Edlund, V., Esbensen, K., Geladi, P., Hellberg, S., Johanson, E., Lindberg, W., and Sjostrom, M.,** Multivariate data analysis in chemistry, in *Chemometrics,* Kowalsky, R. B., Ed., Reidel Publishing, Dordrecht, 1984, 17.
92. **Marshall, G. R., Mayor, D., Naylor, C., Hodkin, E. E., and Cramer, H. D., III,** Mechanisms based analysis of enzyme inhibitors of amide bond hydrolysis, in *QSAR: Quantitative Structure-Activity Relationships in Drug Design,* Fauchere, C., Ed., Alan R. Liss, New York, 1989, 287.
93. **Marshall, G. R., Barry, C. D., Bosshard, H. E., Dammkoehler, R. A., and Dunn, D. A.,** The conformational parameter in drug design. The active analog approach, in *Computer Assisted Drug Design,* Olson, E. C. and Christoffersen, R. E., Eds., American Chemical Society, Washington, D.C., 1979, 205.
94. **Motoc, I., Harte, S. Y., Sit, N., Balasubramanian, N., and Wright, J. J.,** 3-Hydroxy-3-methyl glutaryl-coenzyme A-reductase: three dimensional structure-activity relationship and inhibitor design, *Math. Comput. Model.,* 14, 517, 1990.
95. **Motoc, I., Silverman, J. N., Polansky, O. E., and Olbrich, G.,** The effect of molecular topology on pi-molecular orbital energies, *Theor. Chim. Acta,* 67, 63, 1985.
96. **Motoc, I., Dammkoehler, R. A., and Marshall, G. R.,** Three dimensional structure-activity relationship and biological receptor mapping, in *Chemistry, Mathematics and Computational Concepts,* Trinajstic, N., Ed., Ellis Horwood, Chichester, U.K., 1986.
97. **Labanowski, J., Motoc, I., Naylor, C. B., and Dammkoehler, R. A.,** Three dimensional quantitative structure-activity relationship. 2. Conformational mimicry and topographical similarity of flexible molecules, *Quant. Struct. Act. Relat.* 5, 138, 1986.
98. **Mayer, D., Naylor, C. B., Motoc, I., and Marshall, G. R.,** A unique geometry of the active site of angiotensin-converting enzyme consistent with structure-activity studies, *J. Comput. Aided Molec. Design,* 1, 3, 1987.
99. **Hartsuch, J. A. and Lipscomb, N. W.,** *Enzymes,* Vol. 3. Boyer, P. D., Ed., Academic Press, New York, 1971.
100. **Naruto, S., Motoc, I., Marshall, G. D., Daniels, S. B., Sofia, M. J., and Katzenellenbogen, J. A.,** Analysis of the interaction of haloenol lactone suicide substrates with α-chymotrypsin using computer graphics and molecular mechanics, *J. Am. Chem. Soc.,* 107, 5262, 1985.
101. **Cramer, R. D., III, Patterson, D. E., and Bunce, J. D.,** Recent advances in comparative molecular field analysis, in *QSAR: Quantitative Structure-Activity Relationship in Drug Design,* Alan R. Liss, New York, 1989, 11, 161.
102. **Ghore, A. K. and Crippen, G. M.,** The distance geometry approach, *J. Comput. Chem.,* 6, 350, 1985.
103. **Ghore, A. K. and Crippen, G. M.,** Modeling the bensodiasepine receptor binding site by the general three dimensional structure directed quantitative structure-activity relationship method REMOTEDISC, *Mol. Pharmacol.,* 37, 725, 1990.

104. **Walsh, D. A., Moran, H. W., Shamblee, D. A., Uwaydah, I. M., Weisted, W. J., Sancilio, L. F., and Dannenburg, W. J.,** Quantitative structure-activity relations for inhibition of prostaglandin synthetase by substituted 3-benzyl-2-amino-phenylacetic acids, *J. Med. Chem.*, 27, 1379, 1984.
105. **Allen, M. S., Tan, Y.-C., Trudell, M. L., Narayanan, K., Schindler, L. R., Martin, M. J., Schultz, C., Hagen, T. J., Koehler, K. F., Codding, P. W., Skolnick, P., and Cook, J. M.,** Synthetic and computer-assisted analyses of the pharmacophore for the benzodiazepine receptor inverse agonist site, *J. Med. Chem.*, 33, 2343, 1990.
106. **Simon, Z.,** Computational chemistry and Hansch-type approaches in QSAR and receptor-effector interaction, *Roum. J. Biophys.*, 1, 13, 1991.
107. **Simon, Z.,** Comparative molecular field analysis. Critical comments, *Rev. Roum. Chimie,* 1992 (in press).
108. **Simon, Z., Cibotariu, D., Oprea, T., and Sulea, T.,** Minimal steric differences vs. comparative molecular field analysis for binding of steroids to carrier proteins, *Quant. Struct. Act. Relat.* (in press).

Chapter 3

SPECIFICITY IN ENZYMATIC REACTIONS

Adrian Chiriac and Zeno Simon

TABLE OF CONTENTS

0-8493-5398-X/93/$0.00 + $.50

I. INTRODUCTION

Specificity in enzymatic reactions can be considered from several points of view: first, substrate specificity, in the sense that the enzyme catalyzes the transformation of only one molecule — the natural substrate and at sensibly lower rates — of other similar molecules. Specificity with respect to inhibitors can also be important — differences, with respect to inhibitors of the same enzyme (such as functionality) in bacteria and man are important for drug design. Finally, specificity also can be considered for substances modulating enzymatic activity.

The degree of this specificity is variable for different enzymes. Lytic enzymes of digestive processes, pepsin, trypsin, etc., are expected to have a rather reduced substrate specificity — they are expected to produce hydrolysis of all sorts of peptides and proteins. Enzymes of the DNA replication complex, on the other hand, must have a very high specificity. From the natural deoxyribonucleotide triphosphates, the probability of false base incorporation into the growing DNA chain is of only 10^{-8} to 10^{-11} per base pair[1] (probability of point mutations). These enzymes can easily be tricked by bromouracil which is incorporated instead of thymidine and does readily pair, at the next round of replication, with guanosine triphosphate instead of adenosine triphosphate.[2] The specificity toward foreign, natural substances is usually low. Iodine acetate (ICH_2COO^-), a specific inhibitor of glycerol aldehyde dehydrogenase, sensibly decreases the activity of at least 55 other enzymes and As(III) compounds, is considered as a specific keto-acid oxidation inhibitor and can produce the inhibition of at least 52 other enzymes (at 10^{-3} *M* concentration).[3]

Let us analyze, according to Section II.B.3. of Chapter 1, the specificity requirements imposed on the formation of enzyme substrate-activated complex (i.e., the reaction rate).[4] The overall number of micromolecular species within the living cell is not excessively large. Starting with the building bricks — 20 amino acids, 5 bases, several kind of monosaccharides, fatty acids, etc., and also counting their biosynthesis intermediates — the groups upon macromolecular surfaces which come into contact with the enzymes — a total of $N = 10^3$ micromolecular species will not be exceeded. A maximum of $p = 0.01$ for catalysis of reaction with a false partner would probably suffice for precision. Let us apply Equation 23 of the above-quoted paragraph, and calculate the size n of the combining site in number of atoms (not monomeric units!). The number of types of various atoms to be considered is about $\alpha = 10$. The average decrease Δg of the enzyme site-correct substrate affinity, for one substitution of a substrate atom, should be in the order of a fraction of 1 kcal/mol. Thus, $r = \exp(\Delta g/RT) = 0.5$. Considering similar concentrations, $[B_i]$, for various micromolecular species, one has $[E_{tot}]/p[E_o] = N/p$ and

$$n > \frac{\lg([E_{tot}]/p_f[E_o])}{\lg\alpha - \lg[1 + r(\alpha - 1)]} \approx \frac{\lg 10^3/0.01}{\lg 10 - \lg 5.5} \approx 12 \qquad (1)$$

A ''touching'' of about 12 atoms from the substrate molecule by the enzyme site is probably sufficient for most enzymes to recognize the correct reaction partner among the micromolecular species within the cell.

Concerning the general mechanism and the efficiency of enzymatic reactions, the so-called efficiency function of Albery and Knowles[5] (roughly, the quotient of the enzymatic reaction rate and the rate of diffusion of the reacting species to the catalytic center) gives some interesting insights. The efficiency function is not far from the maximal value 1 for several enzymes. For the enzyme triozophosphate isomerase this function is 0.6 as compared to $2.5 \cdot 10^{-11}$ for the catalysis by the acetate ion of this isomerization. This high efficiency is generally considered to be produced by a cooperation of several effects, usual but singular in organic catalysis performed by micromolecules (for reviews see References 6 to 8). A first effect consists in achieving a high ''local'' concentration of reactants, as well as a mutual orientation favorable to the reaction. For a usual bimolecular reaction, the number of collisions experienced by one molecule (for 1 mol/l concentrations) is about 10^9 to 10^{10} s^{-1}, which must be multiplied by a steric factor $p \approx 10^{-4}$ to 10^{-5}, corresponding to a negative entropy of activation, ΔS. For two reacting molecules, fixed into a suitable mutual orientation in the reaction site of the enzyme, similar to the orientation within the activated complex, $p = 1$, and the number of collisions undergone by a molecule corresponds to some weak vibration with $\nu \approx 10^{12}$ to 10^{13} s^{-1}. Another effect is the lowering of activation energy by simultaneous intervention of intermolecular bonds which facilitate the reaction.

Finally, one should also mention more intricate mechanisms proposed decades ago for enzymatic reactions,[9-11] which go beyond the usual Born-Oppenheimer image of separation of electronic and nuclear motions (conformon, coupling of resonance energies, etc.). In most cases, such mechanisms do not seem to be required in order to explain enzymatic catalysis.

II. DIHYDROFOLATE REDUCTASE

Dihydrofolate reductase. NADP oxidoreductase, EC 1.5.1.3. catalyzes the NADPH-dependent reduction of 7,8-dihydrofolate (H_2F) to 5,6,7,8-tetrahydrofolate (H_4F). This enzyme is essential for cell division processes and for maintaining intracellular pools of H_4F and its derivatives, which are essential cofactors in the biosynthesis of purines, thymidylate, and several amino acids. The structure and function of this enzyme is widely known — for a review see Reference 12. A study of the enzyme is a good approach for designing drugs based upon species-specific enzyme inhibition. It is the target

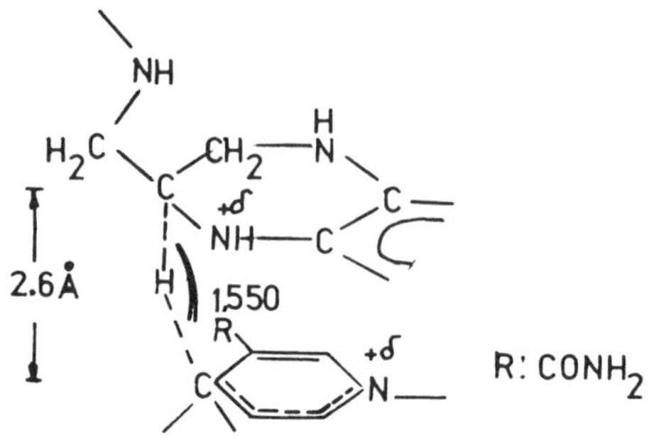

FIGURE 1. Transition state geometry for H_2F reduction by NADPH in the enzyme · H_2F · NADPH complex.

enzyme of a group of antifolate drugs, such as methotrexate (MTX), trimethoprim, and pyrimethamine — drugs widely used as antitumor and antimicrobial agents. Structure-activity relationships of dihydrofolate reductase inhibitors have been thoroughly studied; for a review of these QSAR studies see Reference 13.

A. STRUCTURE AND FUNCTION

The visualization of active-site amino acids was provided by X-ray diffraction studies on crystals of the *E. coli* DHFR · MTX complex and of the *Lactobacillus casei* DHFR · MTX · NADPH complex; the importance of individual amino acids in ligand binding was assessed by kinetic analysis and protein mutagenesis.[12]

The backbone structure of the *E. coli* and the *L. casei* enzymes are similar, despite the presence of less than 30% homology for the two sequences. The active site for both enzymes is a cavity about 15 Å deep that is lined by hydrophobic side chains (Phe^{31}, Ile^{50}, Leu^{54}, and Thr^{113} for the *E. coli* enzyme; Phe^{49} instead of Ile^{50} for the *L. casei* enzyme), with a single polar side chain, that of Asp^{27}. In the enzyme NADPH · H_2F complex, the same interactions appear as in the ternary complex with MTX, but with the pteridine ring turned over with respect to the side chain by a 180° rotation about the C6-C9 bond, as demanded by the stereochemical course of the reduction. Molecular mechanism calculations suggest that the energies of these alternate forms are nearly equal (within a few kcal/mol) for both H_2F and MTX bound to DHFR. Theoretical calculations also indicate, for the transition state, an optimal carbon-carbon bond distance of 2.6 Å (Figure 1.).[12]

Other contacts in the ternary complex of *L. casei* DHFR-NADP-MTX, according to molecular graphics studies,[13] are as described below. The unit cell of *L. casei* DHFR contains about 1200 water molecules. The nicotinyl moiety of NADPH is hydrogen bonded to the carbonyl and imino groups of Ala^6. The nicotinamide interacts via the ring H-atoms 2,4, and 6, with three roughly coplanar O-atoms from amino acid residues 13, 97, and 145, which places the functional part of NADPH in a very polar environment. The MTX molecule is hydrogen bonded with N-1 and 2-NH_2 to the carboxylate group of Asp^{26}, the second H atom of 2-NH_2 to Thr^{116}. The 4-NH_2 group is hydrogen bonded to carbonyl groups of Leu^4 and Ala^{97}.

The kinetic sequence for *E. coli* DHFR, that predicts steady state kinetic parameters and full-time course kinetics under a variety of conditions, was established by Fierke, Johnson, and Benkovic[14] (see Reference 14) by measuring ligand association and dissociation rate constants and pre-steady state reaction transients with the use of stopped-flow fluorescence and absorbance spectroscopy. The key features, at 25°C, of this scheme are (1) steady state turnover is permitted at pH 6 to 7 by H_4F release; (2) dissociation of H_4F is fastest from the mixed ternary E · NADPH · H_4F complex, consequently the kinetic pathway for steady state turnover at saturating substrate concentration follows the pathway in which H_4F dissociation from DHFR occurs after NADPH replaces NADP so that the catalytic cycles are interlocked; and (3) the reaction strongly favors H_4F formation with an overall equilibrium constant $K_{ov} \approx 10^4$ at pH 7. The last process is partially reflected in the internal equilibrium E · NADPH · $H_2F \rightleftharpoons$ E · NADP · H_4F, for the reactive ternary complexes, with an internal equilibrium constant $K_{int} \approx 10^3$ at pH 7. The observed rate for reduction, $k_H \approx 950\ s^{-1}$ is not rate limiting.

A comparison of *E. coli* and *L. casei* DHFRs, including all amino acidic residues at 5 Å within MTX, shows an active site amino acid homology of 55% from an ensemble of 40 amino acids. The total solvent-accessible surfaces for the two proteins (within 5 Å of MTX) are similar (an observed difference of only 7% in favor of the *L. casei* enzyme). The kinetic schemes of the two enzymes are also nearly identical. The major difference observed is that the free energy of binding NADPH to the *L. casei* enzyme is 2 kcal/mol more favorable than to the *E. coli* enzyme, whereas the remainder of free energy differences are less than 1 kcal/mol.[12]

The effect of some mutations on kinetic and thermodynamic parameters of *E. coli* DHFR are described in Table 1. Substitutions of Phe^{31} to Tyr or Val decrease hydrophobic interaction and that of Thr^{113} to Val produces the loss of a hydrogen bond. In terms of binding ΔG, Leu^{54} contributes 3.9 kcal/mol to the binding of H_2F and 5.0 kcal/mol to that of MTX. The ΔG for the Phe^{31} to Val substitution include also the loss of edge-to-face weakly polar interaction between the aromatic side chain and the ligand. The study of Asn^{27} and Ser^{27} mutants estimates about 1.8 kcal/mol for the salt bridge formation between the 2-amino group of MTX and the carboxylate group of Asp.[27]

B. CONCLUSIONS CONCERNING SUBSTRATE AND INHIBITOR BINDING

The combination of site-specific mutagenesis and complete kinetic analysis of the enzyme-catalyzed transformations offers insights into the importance of various active-site residues. Hydrophobic interactions appear to dom-

TABLE 1
Effect of Mutations on Kinetic and Thermodynamic Parameters of *E. coli* DHFR at 25°C[12]

Mutation	k_H (s^{-1})	V_{max} (s^{-1})	$K_D(H_4F)$ (M)	$K_D(H_2F)$ (M)	$K_D(MTX)$ (M)	$K_D(H_2F)$ G (kcal/mol)	$K_D(MTX)$ G (kcal/mol)
Wild type	950	12	$6 \cdot 10^{-8}$	$2.1 \cdot 10^{-7}$	$2.0 \cdot 10^{-11}$	—	—
Phe^{31} to Val	400	26	$3 \cdot 10^{-7}$	$6.6 \cdot 10^{-6}$	$3.2 \cdot 10^{-11}$	2.0	3.0
Phe^{31} to Tyr	—	—	—	—	—	1.5	2.7
Leu^{54} to Gly	12	14	$>1.5 \cdot 10^{-5}$	$1.4 \cdot 10^{-4}$	$1.0 \cdot 10^{-7}$	3.9	5.0
Val^{31} + Gly^{54}	0.9	0.9	—	$2.0 \cdot 10^{-3}$	$9.0 \cdot 10^{-7}$	5.4	6.3
Thr^{113} to Val	120	32	$4 \cdot 10^{-6}$	$3.0 \cdot 10^{-3}$	$5.0 \cdot 10^{-10}$	2.9	1.9

Note: k_H — reduction rate, V_{max} — overall rate of substrate saturation. The error for free energy changes is ±0.5 kcal/mol.

inate the binding of substrates or inhibitors as well as catalysis. The rate of hybrid transfer relates linearly to the binding of H_2F, which is consistent with the hypothesis that the active side residues act as an ensemble to create a particular surface for binding and catalyzing the reaction. The same surface can be constructed by different combinations of amino acids of the *E. coli* and, the *L. casei* enzyme; consequently there must be compensating changes. This surface can be unfavorably perturbed by single changes at key residues. As to the inhibitory activity of MTX, this is based on similarity to the substrate, H_2F. The inhibitor has a higher affinity to the enzyme site-dissociation constant 3 to 4 orders of magnitude lower. Reduction of the –N(5)=C(6) double bond of MTX does not take place, possibly because of the more pronounced aromatic character of the pteridyl moiety in MTX, as compared to H_2F. MTX is protonated in the bound state and this probably explains its higher affinity as compared to H_2F.[13]

C. QUANTITATIVE STRUCTURE-ACTIVITY RELATIONS AND SPECIFICITY IN DHFR INHIBITION

The first structure-activity studies on DHFR inhibitors were performed by the group of Baker.[14] Methothrexate was synthesized in 1949. Several other classes of substances with good inhibitory efficiency have been synthesized and tested since then, all based upon some structural similarity to folic acid. Several such classes are illustrated in Figure 2.

Most QSAR studies concerning DHFR inhibitors were performed by the school of Hansch.[13] The largest series described by one correlational equation is a series of N = 244 triazine derivatives (**III,** Figure 2), which relates the concentration, C, producing 50% inhibition of L1210 leukemia cell and Walker 256 rat tumor DHFR, with parameters of the substituted X in the benzenic cycle:

$$\lg 1/C = 0.68\pi_3 - 0.12\pi_3^2 + 0.23MR_4 - 0.024MR_4^2 + 0.24I_1 - 2.53I_2 - 1.99I_3 + 0.88I_4 + 0.69I_5 + 0.70I_6 + 6.49$$

$$N = 244 \; ; \; r = 0.923 \qquad (2)$$

where π_3 is the hydrophobicity parameter of meta substituents; MR_4 is the molecular refractivity of para substituents, and I_1, I_2, . . . ,I_6 are indicator variables. Optimal hydrophobicity and molecular refractivity values are

$$\pi_3^o = 2.9 \quad \text{and} \quad MR_4 = 4.7 \qquad (2a)$$

Structure-activity studies also were performed in order to discover differences between the enzymes of different tissues. As examples, we list the results obtained for meta-substituted triazine derivative (3-X-, **III,** Figure 2). In most cases, apparent inhibitor binding constants, K, were correlated.

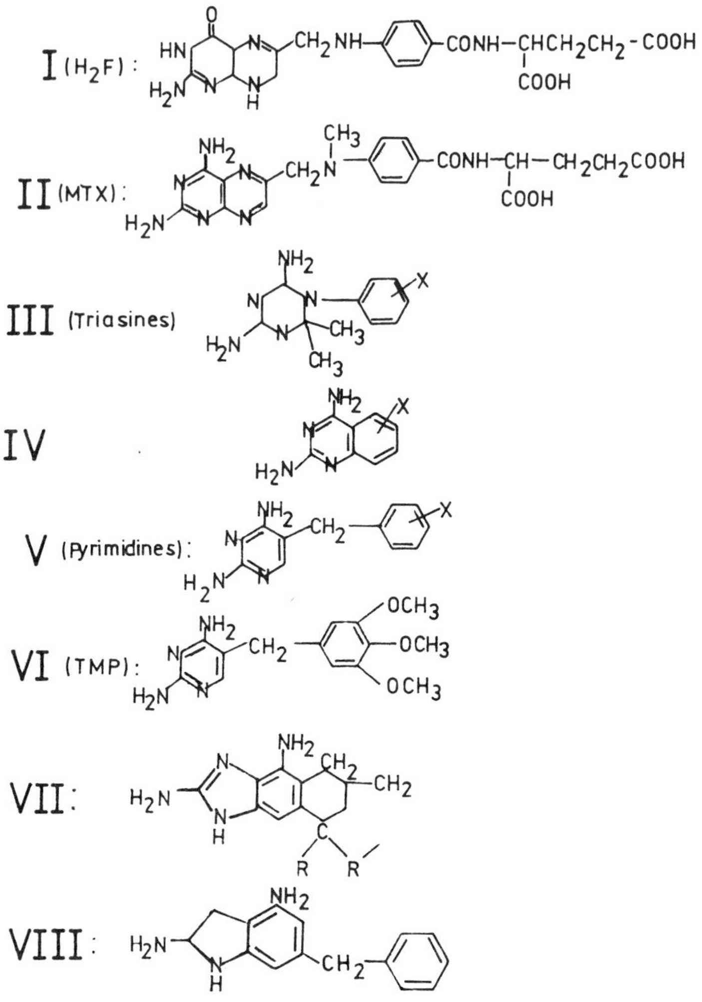

FIGURE 2. Typical structures of DHFR inhibitors.

Human DHFR:

$$\lg 1/K = 1.07\pi_3' - 1.10 \lg(\beta \cdot 10^{\pi'} + 1) + 0.50I + 0.82\sigma + 6.04$$

$$N = 60 \; ; \; r = 0.890 \; ; \; \pi_3^0 = 1.84 \qquad (3)$$

Chicken DHFR:

$$\lg 1/K = 1.01\pi_3' - 1.16 \lg(\beta \cdot 10^{\pi'} + 1) + 0.86I + 6.33$$

$$N = 59 \; ; \; r = 0.906 \; ; \; \pi_3^0 = 1.80 \qquad (4)$$

Bovine DHFR:

$$\lg 1/K = 1.10\pi_3' - 1.23\lg(\beta \cdot 10^{\pi'} + 1) + 0.61I + 7.08$$

$$N = 38 \; ; \; r = 0.914 \; ; \; \pi_3^0 = 1.72 \qquad (5)$$

Murine tumor DHFR:

$$\lg 1/K = 1.19\pi_3' - 1.38\lg(\beta \cdot 10^{\pi'} + 1) + 0.50I + 0.90\sigma + 6.20$$

$$N = 38 \; ; \; r = 0.935 \; ; \; \pi_3^0 = 1.56 \qquad (6)$$

Rat liver DHFR:

$$\lg 1/K = 1.12\pi_3' - 1.34\lg(\beta \cdot 10^{\pi'} + 1) + 6.80$$

$$N = 18 \; ; \; r = 0.877 \; ; \; \pi_3^0 = 1.68 \qquad (7)$$

***L. casei* DHFR:**

$$\lg 1/K = 0.83\pi_3' - 0.91\lg(\beta \cdot 10^{\pi'} + 1) + 0.71I + 4.60$$

$$N = 38 \; ; \; r = 0.961 \; ; \; \pi_3^0 = 2.69 \qquad (8)$$

DHFR of L5178YS, leukemia cells sensitive to MTX:

$$\lg 1/c = 1.32\pi_3' - 1.70\lg(\beta \cdot 10^{\pi'} + 1) + 0.44I + 8.10$$

$$N = 37 \; ; \; r = 0.929 \; ; \; \pi_3^0 = 0.76 \qquad (9)$$

DHFR of L5178YR, leukemia cells resistant to MTX:

$$\lg 1/\epsilon = 0.57\pi_3' - 0.35\lg(\beta \cdot 10^{\pi'} + 1) - 0.15MR_3 + 5.12$$

$$N = 42 \; ; \; r = 0.932 \; ; \; \pi_3^0 = 6 \qquad (10)$$

Regressional coefficients and optimal hydrophobicity, π_3^0, values are rather similar for different organisms; *L. casei* DHFR and the L-5178-enzymes have more different values. The Kubinyi $\lg(\beta \cdot 10^{\pi'} + 1)$ dependence is related to membrane permeability of the drug.[13] The π_3' parameters are modified hydrophobicities. Hydrophobicity is important, but electronic effects (σ Hammett parameter) and more special structural effects (as expressed by indicator variables, I) are also important.

The importance of such differences for drug design is obvious. The mechanism of drug resistance (Equations 9 and 10) is not very clear. Possibly, it is due to the high rate of production of DHFR in MTX-resistant cells,[a] result of increased levels of DHFR-mRNA, brought about by gene amplification.[15,16]

Regarding other QSARs developed by Hansch et al.,[13] we mention the one for quinazoline derivatives (**IV,** Figure 4), rat liver DHFR inhibitors

$$\lg 1/c = 0.81MR_6 - 0.064MR_6^2 + 0.77\pi_5 - 0.73I_1 - 2.15I_2 - 0.54I_3 - 1.40I_4 + 0.78I_6 - 0.20MR_6I_1 + 4.92$$

$$N = 101 \; ; \; r = 0.961 \; ; \; MR_6^o = 6.4 \qquad (11)$$

and the one for pyrimidine inhibitors (**V,** Figure 4.), bovine DHFR inhibitors:

$$\lg 1/K = 0.55\pi_{3,4,5} - 0.43\lg(\beta \cdot 10^{\pi'} + 1) - 0.20MR_4 + 0.32\sigma + 4.46$$

$$N = 39 \; ; \; r = 0.900 \qquad (12)$$

The marked dependence of molecular refractivity for the 6-X substituent in the quinazoline derivative suggests interaction with an enzyme pocket, with both apolar and polar groups, able to accommodate both hydrophobic and hydrogen bonding substituents.

MTD, distance geometry, and molecular shape analysis were also applied to DHFR inhibition, especially in attempts to ascertain steric characteristics. Unfortunately, the bioreceptor mapping (which such methods aim at), is not easy because of possible flexibility of the receptor site and ability to bind similar ligands in more than one way.[13]

Molecular graphics were also used in the testing of DHFR inhibition. In a beautiful example of drug design, trimethoprim (**VI,** Figure 2) was modified by substituting one of the meta OCH_3 side chains by $O(CH_2)_nCOOH$ groups, because the study of crystallographic analyses indicates that, for n = 4, the carboxylate group will interact with Arg^{57} of the *E. coli* enzyme. In reality, for n = 3 to 6, the derivatives were up to 30 times more potent inhibitors than TMP.[13]

A combination of classical QSAR and molecular graphics models constructed from X-ray crystallography of DHFR inhibitor complexes was used for inhibition of chicken liver DHFR by substituted pyrimidine derivatives (5-X-substituted, Figure 2).[17] The regressional equation is

$$\lg 1/K = 0.43\pi_3 - 1.3\lg(\beta \cdot 10^{\pi'} + 1) - 0.66MR_5 + 0.59\pi_4 - 0.63 \cdot \lg(\beta_4 \cdot 10^{\pi} + 1) + 0.48\pi_5 + 0.14MR_3 + 4.64$$

$$N = 53 \; ; \; r = 0.921 \; ; \; \pi_3^o = 2.00 \; ; \; \pi_4^o = 1.53 \qquad (13)$$

The nature of the various contacts, as suggested by the regressional equation, is compared with molecular graphics models constructed from the X-ray crystallographic coordinates of TMP and 5-(3,4-dimethoxy-4-isopropylbenzyl)-2,4-diaminopyridine bound to the enzyme. There is good correspondence between the two types of models.

A receptor-based design of benzimidazole and indole derivatives (**VII** and **VIII,** Figure 2) as inhibitors of DHFR was attempted based on molecular modeling with respect to X-ray structure of DHFR.[18] Hydrogen bond distances and pK_a values of these derivatives seem reasonable as compared to 5-benzyl-2,4-diaminopyrimidine, although electronic densities and bond angles are rather different. Although some compounds (**VII,** Figure 2, with R, R′: $-SCH_2CH_2$-S–) presented a good fit to the enzyme site in modeling studies, the most active compound is 2 orders of magnitude less potent than diaminopyrimidine with respect to bacterial DHFR.

III. ALPHA-CHYMOTRYPSIN

The hydrolytic enzyme, α-chymotrypsin, is a characteristic hydrolytic enzyme of a rather low specificity, since it is present in the digestive process, implying the cleavage of all sorts of proteins. As to the so-called primary specificity,[19] the minimal structural requirement, this enzyme hydrolyzes peptide bonds whose acylic terminal is an aromatic amino acid or methionine. Of the next residues toward the carboxyl end, Lys, Val, and Ile favor, and Asp, Asn, His, Ser, and Pro obstruct hydrolysis. A marked influence of some more-remote amino acid residues is also observed.[20] Not only peptides are hydrolyzed, but also nitrophenolic esters of various acids or amino acids[21] or phenyl hippurate derivatives.[22]

The scheme of the hydrolysis reaction and the hydrolysis rate can be described as:

$$S + H_2O \xrightarrow{E} P_1 + P_2$$

$$E + S \underset{k_{-1}}{\overset{k_1}{\rightleftarrows}} ES \xrightarrow{k_2} ES' \xrightarrow[+H_2O]{k_3} E + P_2$$

$$v = \frac{k_o[E][S]}{K_m + [S]} \qquad (14)$$

Here [S] and [E] are the substrate and the total enzyme concentrations, and K_m the Michaelis constant. The relations to the rate constants are[25]

$$K_m = K_S \cdot \frac{k_3}{k_2 + k_3} \; ; \; K_S = \frac{k_{-1} + k_3}{k_1} \approx \frac{k_{-1}}{k_1} \; ; \; k_{cat} = \frac{k_2 \cdot k_3}{k_2 + k_3} \qquad (15)$$

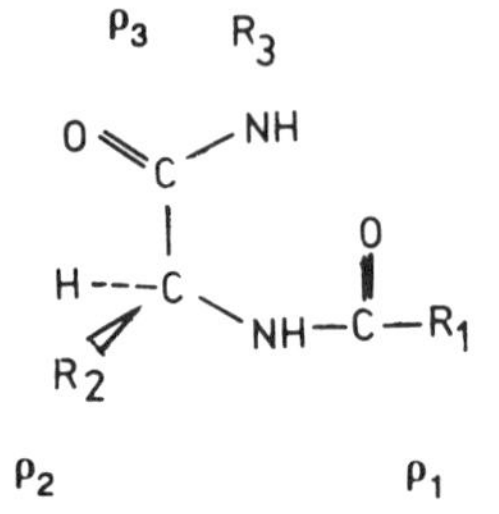

FIGURE 3. The Hein and Niemann structure for α-chymotrypsin substrate molecules; ρ_1, ρ_2, ρ_3 — the three substituents attaching to the corresponding enzyme sites.

A first comprehensive view on the specificity of this enzymes is given in a review of Hein and Niemann.[24] With respect to binding of substrates of type $R_1CONHCHR_2COR_3$, three regions are distinguished ρ_1, ρ_2, and ρ_3 to which the three substituents (attached to the C-α of the central amino acidic residue) should attach. The ρ_2 site should be a rather hydrophobic one, so as to accommodate the R_2 site chain for the central L-amino acid residue (Figure 3.).

Chymotrypsin is an excellent enzyme for QSAR studies. The mechanism of action is rather well understood. Its X-ray crystallographic structure has been determined[25] and models have been suggested for binding of substrates to this enzyme. A comprehensive study of this enzyme based upon QSAR and molecular graphic analysis was also published.[20]

The active site of α-chymotrypsin includes residues 41–43, 55–59, 102, 142–143, 189–196, 213–222, and 225–258, among them the sequences[26] Gly^{184}AlaSerGlyValSerCysMet^{192} and Ile^{212}ValSerTrpGlySerSerThrCysSer-ThrSerThrProGlyVal^{227}.

In the QSAR study of the hydrolysis of X-phenylhippurates[22]

$$XC_6H_4OCOCH_2NHCOC_6H_5 + H_2O \rightarrow$$
$$XC_6H_4OH + HOOCCH_2NHCOC_6H_5 \qquad (16)$$

Of the N = 28 quoted experimental hydrolysis parameters, there is a very reduced interval for the $\log k_{cat}$ (min^{-1}) values (from 1.13 for X:3,5-di-CF_3 to 1.67 for X:3-F or 1.66 for X:4I), but there is a more substantial interval for $\log 1/K_m$ values (from 3.73 for X:4-NH_2 to 4.71 for 3,5-di-CF_3). For the unsubstituted derivative, X:H, $\log k_{cat}$ (min^{-1}) = 1.26 and $\log 1/K_m$ = 3.89 — all values corresponding to 25°C and pH 6.9.

The attempt to correlate Michaelis constants, $\log 1/K_m$ values, was successful:

$$\log 1/K_m = 0.42\sigma^- + 0.28\pi'_3 + 3.87 \; ; \; N = 28 \; ; \; r = 0.945 \qquad (17)$$

Attempts to correlate $\log k_{cat}$ with substituent constants were unsuccessful. The correlation with $\log(k_{cat}/K_m)$ was also rather poor, with $r = 0.787$. Dropping the most poorly fitted points (4-I, 3-F, 3-CN, and 3-CF_3), one obtains:

$$\log(k_{cat}/K_m) = 0.45\sigma^- + 0.24\pi'_3 + 5.21 \; ; \; N = 24 \; ; \; r = 0.891 \qquad (18)$$

The correlational results are not ascertained by a crossvalidation procedure, but there is a fair constancy of regressional coefficients for σ^- and π_3' in going from Equation 18 to 17 and in comparing the correlation equations for $\lg 1/K_m$, obtained separately for only the meta- and para-substituted derivatives with Equation 17 for all $N = 28$ derivatives.

As to the substituent parameters, π_3' is the hydrophobic constant for hydrophobic meta X-substituents. Hydrophylic substituents with negative π values are all assigned $\pi_3' = 0$. Inclusion of π' for para X-substituents does not improve the correlation and neither does the inclusion of MR values. The σ^- Hammett-type constant and its positive regressional coefficient attest to the importance of electron withdrawing effects, especially by through-resonance. Hydrophobic meta substituents should make contact with the hydrophobic surface of the enzyme, but hydrophylic substituents should orient themselves into the exterior aqueous space. For substituents bound to the hydrophobic surface of an enzyme one finds a coefficient of about 0.5, and when binding occurs in a deep cleft or hole, the coefficient for π is found to be near 1, as suggested by a survey of QSAR data.[27] The hydrophobic substituents should thus not make full contact with a hydrophobic surface patch on the enzyme. A comparison of $\lg 1/K_m$ values for the X:3-$NHCONH_2$ and X:3-$NHCOCH_3$ derivatives, which are all almost equal to the one for X:4 ($\log 1/K_m = 3.89 \pm 0.01$), taking into account the small electronic effects ($\sigma = 0.03$ and 0.05, respectively), indicates that large polar groups cannot produce a steric effect. This confirms the hypothesis of their reorientation toward the aqueous solution. The electronic effects, as revealed by the $\rho = 0.42$ regressional coefficients, are somewhat lower than for hippurate hydrolyses by similar enzymes (papain, $\rho = 0.57$; ficin, $\rho = 0.57$; actidin, $\rho = 0.74$; bromelain B, $\rho = 0.70$; bromelain D $\rho = 0.63$; and subtilisin Carlsberg $\rho = 0.49$), and sensibly lower than the basic (OH^-) catalysis ($\rho = 1.66$) or for the uncatalyzed reaction ($\rho = 191$).

Using X-ray crystallographic coordinates for chymotrypsin, and computer graphics, a model was constructed and used to interpret the QSAR.[22] The model for the α-chymotrypsin-hippurate complex was based on the lowest energy covalent structure of L-N-acetyltryptophanamide. In this model, one hydrogen atom of the CH_2 group of the hippurate is directed toward Met[192], the carboxyl oxygen of the leaving group (O_2CR) in the oxyanion hole formed by the NH groups of Gly[192] and Ser[195], and the NH of the acylamide group was hydrogen bonded to the carbonyl moiety of Ser[214]. As a whole, the substrate in the ES complex must be near the transition state for the enzymatic acylation, which is suggested also by the rather low electronic effects ($\rho = 0.42$, only) and by the necessity to consider the hippurate molecule torsioned to 20° in the complex, in order to satisfy the requirements of Equation 17. In this conformation, the 4-SO_2NH_2 derivative is situated with the benzoyl moiety near the bottom of the so-called hydrophobic hole, whereas the phenyl group makes contact with the largely hydrophobic enzyme surface. For 3,5-dimethylphenyl hippurate, the molecular graphics results show how the two CH_3O groups make only limited contact with the hydrophobic surface. A slight tilt of the phenyl ring projects hydrophilic meta substituents into the exterior aqueous space.

The enzyme substrate models were optimized by several methods, such as rigid body optimization and bond rotations. Electrostatic charges for the inhibitors were also calculated (STO-3G level using Gaussian 80) but molecular mechanics calculations for the enzyme-substrate interaction were not performed because the small range of $\log 1/K_m$ values implied ΔG changes too small to be modeled by current techniques. As a conclusion, this report provides further evidence that the application of QSAR and molecular graphics methods to the study of enzymes can yield a self-consistent picture.

Another class of substrates studied both from the point of view of the hydrolysis kinetics and QSAR are D- and L-N-acyl-α-aminoamides,[28]

$$\underset{\displaystyle R}{R'CONH\overset{}{CH}}\text{–}CHNHX + H_2O \rightarrow R'COOH + \underset{\displaystyle R}{H_2N\overset{}{CH}}\text{–}CONHX \quad (19)$$

While some derivatives are hydrolyzed, others are competitive inhibitors, but enzyme binding can be treated within a single series for QSAR. The lowest measured binding constant $\lg 1/K = 0.46$ is the one for $CH_3COGlyNH_2$, for L-isomers $\lg 1/K = 1.21$, for L-CH_3CO-Tyr-NHOH, the highest $\lg 1/K = 2.66$ for L-C_6H_5CO-Tyr-$NHNH_2$. For D-isomers, the lowest $\lg 1/K = 1.40$ is for D-Trp-NHOH, the highest, $\lg 1/K = 3.74$, for D-4-$NO_2C_6H_4CO$-Trp-NH_2. From the $N = 49$ data available, a good QSAR was obtained for a series of $N = 45$ derivatives:

$$\lg 1/K = 0.720\text{MR–L} + 0.230\text{MR–S} + 0.323I_1 + 0.311I_2 - 1.062 \quad (20)$$

MR-L is the molecular refractivity of the larger of the R′ or R substituent, MR-S, represents the smaller, $I_1 = 1$ for hydrazides, 0 for all other cases; $I_2 = 1$ for D-congeners and glycine amides which are inhibitors and 0 for L-congeners. Four derivatives (L-tBu-CO-Tyr-$NHNH_2$ and some nicotinyl derivatives) were excluded. The regressional coefficients for MR-L, MR-S, and I-1, for separate equations for D-isomers, L-isomers, and for Equation 20 for

TABLE 2
Hydrolysis Rates For $R_2R_1CHCOOC_6H_4NO_2$ Derivatives

R_1CH	R_2	$lg(k_{cat}/K_m)$
CH_2	H	4.54
ClCH	H	5.76
H	H	5.29
$(CH_3)_2C$	H	2.07
L-CH_3CH	$C_6H_5CH_2OCONH$	5.18
D-CH_3CH	$C_6H_5CH_2OCONH$	4.03
L-$CH_3CH(CH_3)CH$	$C_6H_5CH_2OCONH$	3.82
D-$CH_3CH(CH_3)CH$	$C_6H_5CH_2OCONH$	3.03
L-$C_6H_5CH_2CH$	$C_6H_5CH_2OCONH$	7.91
D-$C_6H_5CH_2CH$	$C_6H_5CH_2OCONH$	5.68

the common series are similar, which attests that the success is not due to chance correlation.

The QSAR suggests that the two substituents R and R′ are bound to the ρ_1 and ρ_2 enzyme pockets, while the ρ_3 pocket accommodates the $-CONHX$ moiety for L-congeners and the H atom of C_α for D-congeners, which are inhibitors. It is interesting that MR and π are not significantly intercorrelated for the large and small substituents, and QSARs with π yield a sensibly lower correlation. The fact that L-$C_6H_5CO(4\text{-}CH_3C_6H_4SO_2NH)$Phe-$NH_2$ has a much lower observed $\log 1/K = 1.58$ value than the one calculated by Equation 20, $\log 1/K = 3.14$, suggests that the so-called ''hydrophobic'' pocket has limited dimensions and does not accommodate the large substituents.

A minimal steric difference study of α-chymotrypsin-catalyzed hydrolyses based upon hydrolysis rates of Dupaix et al.[21] was performed by Simon et al.[29] The general formula and hydrolysis reaction for these compounds is

$$R_1R_2CHCOO\text{–}C_6H_4\text{–}NO_2\text{–}4 + H_2O \rightarrow R_1R_2CHCOOH + 4\text{–}HO\text{–}C_6H_4\text{–}NO_2$$

The series consists of $N = 42$ derivatives of various aliphatic and aromatic acids and of D- and L-amino acids. Table 2 lists some log (k_{cat}/K_m) values, including minimal and maximal values.

The following correlational equations were obtained:[29]

$$lg(k_{cat}/K_m) = 3.207 + 2.296\sigma + 0.442\pi + 0.284ES$$
$$r = 0.676 \; ; \; N = 42 \qquad (21)$$

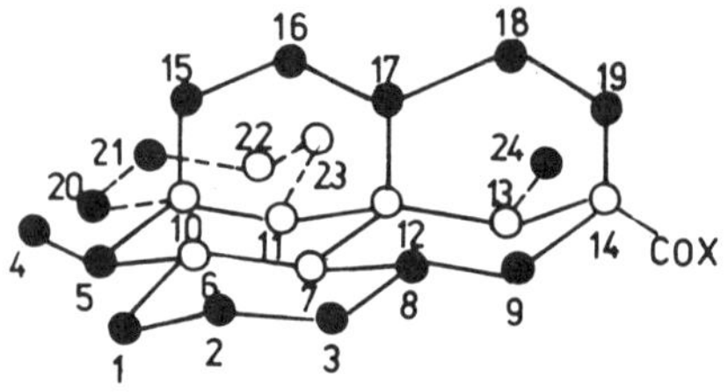

FIGURE 4. Hypermolecule and receptor map for α-chymotrypsin-catalyzed hydrolysis of nitrophenol esters of amino acid and carboxylic acids. Open circles — cavity vertices; black circles — wall vertices.

$$\lg(k_{cat}/K_m) = 5.076 + 2.537\sigma + 0.342\pi + 0.289\text{MTD}$$

$$r = 0.852 \; ; \; N = 42 \qquad (22)$$

The hypermolecule and optimized receptor map are depicted in Figure 4.

The receptor map was obtained from two different series of α-chymotrypsin-catalyzed hydrolyses of nitrophenolic esters of aromatic carboxylic acids and methyl esters of L-amino acids.[29,30] The parameters σ and π refer to the sum for R_1 and R_2. The Taft steric constant, ES, also refers to R_1. The inclusion of MTD instead of ES for the steric effects produces a sensible increase in the correlation coefficient, r. In the receptor map of Figure 4 the R_1 substituent of the L-amino acid derivative mostly occupies the beneficial vertices j:5–7, 10–12, and 20–24, while D-amino acidic derivatives mostly occupy detrimental vertices j:1–3 and 6–8; the R_2 substituent takes the way of vertices j = 19 and from there, probably in the exterior aqueous solution. The σ-coefficient is much higher than for the hydrolyses of hippurate derivatives, the coefficient for π being somewhat higher. For the most active L-phenylalanine derivative of Table 2, $\log(k_{cat}/K_m) = 7.91$, while for the hydrolyses of the most active hippurate derivative ($3,5\text{-di-}CF_3\text{-}C_6H_3\text{-}OOCCH_2NHCOC_6H_5$)[22] $\log(k_{cat}/K_m) = 5.84$ which is lower, but comparable with the rate for several other derivatives of Table 2.

The series studied by Dupaix et al.[21] has a much higher range of rate values than the hippurate series[22] and the range of binding constants of N-acyl-α-amino acid amides.[28] The structural variations within the Dupaix series are also important. As a conclusion, the intimate mechanism of α-chymotrypsin-catalyzed hydrolysis rates presents some differences for various classes of derivatives. In particular, the high ρ values for electronic effects for the Dupaix derivatives (Equation 22) near the noncatalyzed reactions suggests a mechanism based more upon stabilization of products than an ES complex with a conformation near the transition state.

IV. ACETYLCHOLINESTERASE

Acetylcholinesterase is implied in the transmission of nervous influx through neuromuscular synapses in both higher and lower animal organisms. Its inhibition is the mechanism of action of several pesticides. The enzyme hydrolyzes acetylcholine to choline and acetic acid by a push and pull mechanism[31] depicted in Figure 5. This mechanism corresponds to the usual three-step reaction scheme (Equation 14) for hydrolysis reactions.

Of the inhibitors, the mechanism of action of phosphororganic pesticides is depicted in Figure 6.

The three-dimensional structure of acetylcholinesterase (from the electric organ of *Torpedo californica*) was recently determined by X-ray crystallography for a glycolipid homodimer.[32] The monomer of the enzyme is a protein with 536 amino acid residues and consist of a 12-stranded mixed β-sheet surrounded by 14 α-helixes. This structure bears a striking resemblance with several other hydrolyses, but the active site contains a Ser-His-Glu (and not Ser-His-Asp) catalytic triad. This active site is situated near the bottom of a narrow and deep gorge that reaches halfway into the protein. Modeling of acetylcholinesterase binding to the enzyme suggests that the quaternary ammonium cation binds to some of 14 aromatic residues that line the gorge (cation-π-electron interaction) and not with an anionic site. This hypothesis is strengthened by binding of acetylcholine to a completely synthetic receptor with aromatic nuclei and without a negative charge,[33] corresponding to a dissociation constant of $K = 2 \cdot 10^{-4} M$.

Concerning substrate specificity, the uncharged analogue of acetylcholine $CH_3COOCH_2CH_2C(CH_3)_3$ is hydrolyzed 20 times slower, while the phenylated derivative $CH_3COOCH_2CH_2N^+(CH_3)_2CH_2C_6H_5$ is hydrolyzed 6 to 15 times slower. The cholinesterase is obtained from horse blood plasma and from cow erythrocytes, respectively, at 25°C and pH 7.5.[34]

Various acetic acid esters CH_3COOR are also hydrolyzed. In a series of $N = 25$ esters, the Michaelis constants, K_S, vary between 0.1 M (R:C_2H_5) and $1.4 \cdot 10^{-3}$ M (R:$CH_2CH_2C(CH_3)_3$) in similar conditions.[33] The corresponding hydrolysis rates (k_{cat}/K_S) are situated between values lower than $2 \cdot 10^2\ M^{-1}s^{-1}$ (R:$C(CH_3)_3$) and $2 \cdot 10^6\ M^{-1}s^{-1}$ (R: $(CH_3)_3CCH_2CH_2$) or even $5.3 \cdot 10^6\ M^{-1}S^{-1}$ (R:C_6H_5).

An attempt to establish a QSAR yields:[35]

$$\log(k_{cat}/K_S) = 1.84 + 300\sigma + 1.61\pi + 1.07E_S$$

$$N = 19 \; ; \; r = 0.999 \qquad (23)$$

but only if 6 of the 25 esters are eliminated. A more consistent QSAR can be obtained if the MTD method is used. The esters and the required structural parameters are listed in Table 3, the hypermolecule in Figure 7.

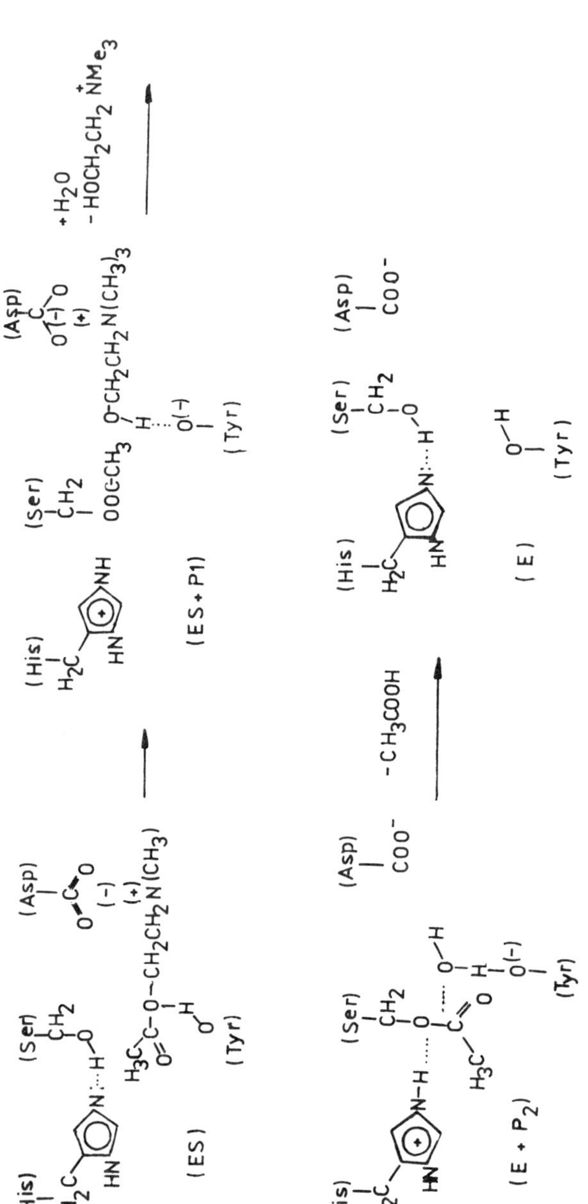

FIGURE 5. The three-step mechanism for acetylcholinesterase-catalyzed hydrolysis.

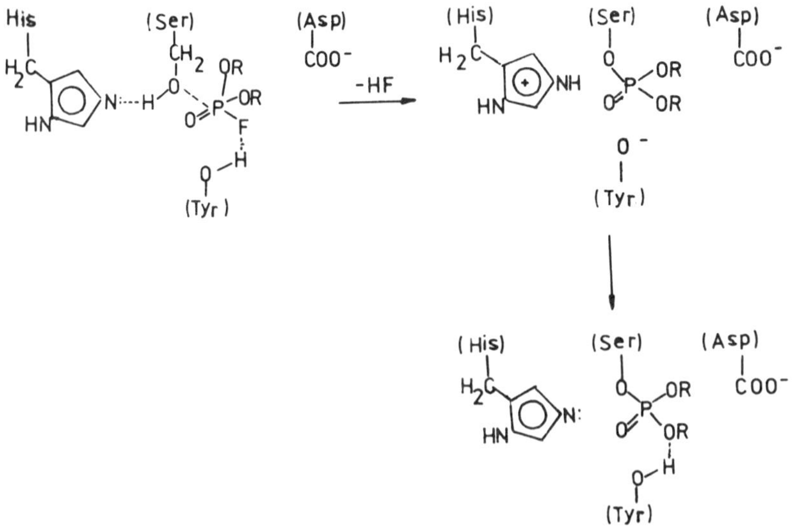

FIGURE 6. Mechanism of acetylcholinesterase inhibition by phosphororganic pesticides.

A recent QSAR study with the multiconformation MTD method[36] yields the following optimized receptor map and correlational equation:

$$S^* \begin{cases} j(\epsilon = -1): 3\text{–}8, 10, 11 \\ j(\epsilon = +1): 1, 2, 14, 15 \\ j(\epsilon = 0): \quad 9, 12, 13 \end{cases}$$

$$\hat{A} = 8.245 + 1.824\sigma - 0.729\text{MTD} \; ; \; r = 0.926 \qquad (24)$$

In a cross-validation procedure (half of the series for setting up the QSAR, the other half as test series, and vice versa) $r_{cv} = 0.828$ is obtained. Conformational energy computations by molecular mechanics were in this case also used in constructing the hypermolecule and in selecting the low energy conformations.

The strong favoring effect of the electron-withdrawing esteric group is confirmed. The cavity vertices ($\epsilon_j = -1$) suggest a cavity somewhat larger than the one accommodating strictly the esteric group of the natural substrate, acetylcholine, which occupies vertices j:4–8. The optimized receptor map, S*, is also depicted in Figure 7.

A. ACETYLCHOLINESTERASE INHIBITION

Acetylcholinesterase inhibition is the basis of a series of pesticides.[37,38] A large series of substances, including carbamates and various classes of organophosphorus compounds, act as inhibitors and one may expect various

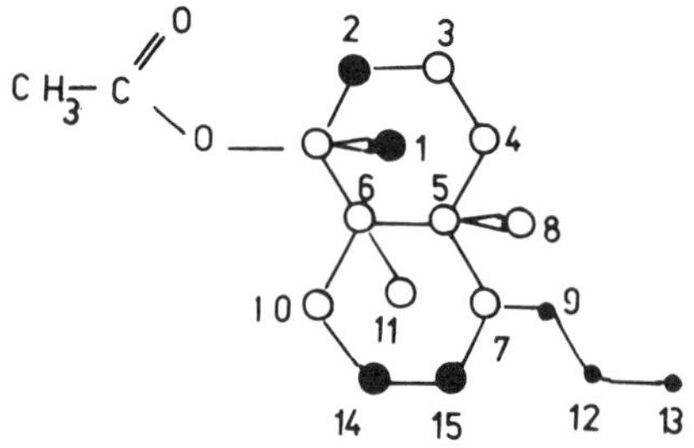

FIGURE 7. Hypermolecule for acetylcholinesterase-catalyzed hydrolysis of acetic acid esters.

mechanisms for this inhibition; carbamylation or phosphorylation of the enzyme (Figure 6) seem to be the final result of the inhibition mechanism. It is therefore of interest to see the results of QSAR attempts for the inhibition of this enzyme in order to obtain some information about the mechanism of inhibition of various classes of substances.

1. Carbamates As Acetylcholinesterase Inhibitors

According to O'Brien[37] and Metcalf and Fukuta,[38] these compounds act by combining with the enzyme, EOH:

$$CH_3HNCOOC_6H_4X + EOH \rightarrow CH_3NHCOOE + HOC_6H_4X$$

Correlation of inhibitory activity, $A = -\lg C_{50}$ (molar concentration producing >50% inhibition of enzyme activity) were performed by the QSAR group at the University of Timişoara.[39-42] The experimental data, for o series of $N = 97$ carbamates with different X-substituents, are taken from a paper by Metcalf et al.[43] The maximal activity is presented by some compounds with a trimethylammonium substituent: $A = 8.30$ for X:4-N^+Me_3, 3-iso-Pr; $A = 8.15$ for X:3-N^+Me_3, 5-iso-Pr; but only $A = 5.46$ for X:4-N^+Me_3. Of the noncharged compounds, the highest activity, $A = 7.48$ is observed for the X:3,5-di-iso-Pr and $A = 7.11$ for the X:3,5-di-*t*-Bu-derivative. Of the compounds with lowest activity: $A = 2.30$ for X:2-NO_2 and $A = 2.93$ for X:2,6-di-Cl derivatives.

A correlation with only two parameters, σ-sum of Hammett constants and π-sum of hydrophobicity constants for the X-substituents, yields only a poor correlation:[39]

$$\hat{A}_1 = 3.990 - 0.899\sigma + 0.813\pi \; ; \; r = 0.629 \qquad (25)$$

With a set of five structural parameters characterizing the X substituents (σ; MR, molecular refractivity; HB, hydrogen bonding character; EC, electric charge; and CT, charge transfer character) one obtains:

TABLE 3
Acetylcholine Catalyzed Hydrolysis Rates of Acetic Acid Esters

i	R	A	σ	π	$-E_S$	$j(x_{ij}=1)$	MTD°
1	C_6H_5	6.72	0.60	2.13	0.38	2–6	5
2	$Me_3CCH_2CH_2$	6.30	0	2.98	0.40	4–8	3
3	$EtSCH_2CH_2$	5.40	0.22	1.95	0.44	5–7,9	5
4	Me_2CHCH_21CH	5.32	0	2.30	0.43	4–6;5–7;5,6,8	5
5	$EtSCH_2$	5.35	0.56	1.45	0.44	4–6;4–6,8;5–8	4
6	$O_2NCH_2CH_2$	5.20	0.62	1.31	0.40	4–7;4–6,8;5–8	4
7	$ClCH_2CH_2$	5.02	0.39	1.39	0.48	5,6;6,10	6
8	$HC{\equiv}CH{-}CH_2$	4.81	0.55	0.94	0.60	6,11	6
9	nC_7H_{15}	4.75	0	3.50	0.40	5–7,9,12,13	5
10	$EtS(CH_2)4$	4.73	0.03	2.95	0.40	5–7,9,12,13	5
11	nC_5H_{11}	4.74	0	2.50	0.40	5–7,9	5
12	nC_4H_9	4.72	0	2.00	0.40	4–7;4–6,8;5–8	4
13	$cycloC_6H_{11}$	4.71	0	2.51	0.98	2–6	5
14	nC_6H_{13}	4.75	0	3.00	0.40	5–7,9,12	5
15	$C_2H_5S(CH_2)_3$	4.73	0.08	2.45	0.40	5–7,9,12	5
16	$C_6H_5CH_2$	4.66	0.25	2.26	0.38	5–7,10,14,15	6
17	Me_2CHCH_2	4.32	0	1.80	0.35	5,6,10	5
18	$CH_2{=}CHCH_2$	4.10	0.23	1.23	0.20	5,6;6,10	6
19	nC_3H_7	3.91	0	1.50	0.39	5,6;6,10	6
20	EtMeCH	3.69	0	1.80	0.93	1,5,6;2,5,6	7
21	C_2H_5	3.36	0.27	1.00	0.36	6	7
22	CH_3	3.00	0.20	0.50	0.07	—	8
23	Me_2CH	2.72	0	1.30	0.93	1,6;2,6	8
24	Me_3C	2.30	0	1.98	1.74	1,2,6	9
25	$EtMe_2C$	2.30	0	2.48	1.74	1–3,6;1,2,5,6	8

Note: A: biological activities — A = lg (k_{cat}/K_S), according to kinetic data published by Järv et al.[35]; σ: Hammett constants; π: hydrophobicities; E_S: steric Taft constants; $j(x_{ij} = 1)$: vertices occupied by different conformations in the hypermolecule of Figure 9; MTD: minimal steric differences according to the optimized receptor map, S*, of Equation 23.

$$\hat{A}_2 = 4.200 - 0.768\sigma + 0.058MR + 0.440HB + 1.082EC + 0.320CT \; ; \; r = 0.764 \tag{26}$$

Inclusion of minimal steric difference (MSD) with respect to the standard, S_o, yields:[40]

$$\hat{A}_3 = 5.161 - 0.540\sigma + 0.057MR + 0.476HB + 0.638EC + 0.058CT - 0.160MSD$$

$$r = 0.800 \tag{27}$$

The partial correlation coefficients are $r(\sigma) = -0.150$; $r(MR) = 0.630$; $r(\pi) = -0.045$; $r(HB) = 0.418$; $r(EC) = 0.424$; $r(CT) = -0.285$; and $r(MSD) = -0.312$.
Finally, the use of MTD instead of MSD yields:[41]

$$\hat{A}_4 = 7.192 + 0.058MR + 0.047\pi + 0.669HB + 0.301EC - 0.397MTD$$

$$r = 0.876 \tag{28}$$

An increased volume (MR) of the X substituents seems to favor inhibition, as well as the electron attracting character (negative regressional coefficient for σ), but hydrophobicity does not seem important. A hydrogen bonding character of the X substituents also seems to favor inhibition, as well as a positive electric charge. As to the steric requirements, although the partial correlation coefficient for MTD is significant, $r(MTD) = -0.510$, the receptor map presents an interspersed distribution of cavity, wall, and exterior vertices, which could be the results of the optimization process only. The standard for MSD computations, S_o, is the CH_3-NHCOO-C_6H_4-3-N^+Me_3-derivative, with three C atoms between the O and N^+ atoms, while only two C atoms are present in acetylcholine, the natural substrate of the enzyme. If the ortho derivative $CH_3NHCOOC_6H_4$-2-N^+Me_3 is used as standard (which allows a good superposition upon $CH_3COOCH_2CH_2N^+Me_3$) no improvement with respect to $\hat{A}2$ is obtained.

2. Schrader-Type Organophosphorus Derivatives

Organic phosphorus compounds of the Schrader type:[44] $R_1R_2POR_3$ (R_3 group with acidic character; an S atom instead of the O atom is present for some derivatives) are used as pesticides. Several QSAR studies were performed by the QSAR group at the University of Timişoara.[45-49]

A first QSAR was performed for $N = 71$ derivative of the type $R_1R_2POR_3$ with $A = -\lg(LD_{50}/M)$, 50% lethal dose per body weight, according to experimental data from the treatise of Schrader.[44] R_1 and R_2 are various alkoxy groups (MeO, EtO, PrO, iPrO, and BuO); R_3 is a more acidic group, such as $-CH{=}CCl_2$, $-C_6H_4NO_2$, $-C(Me){=}CHCOOEt$, etc. Some of the experimental data were adopted form Selepeva et al.[50] The QSAR results are[45,47]

$$\hat{A}_5 = 0.555 + 0.476HA_{1,2} - 0.017MR_{1,2} + 0.344\sigma_3 - 0.402AR_3 + 0.611CT_3 - 0.063MSD$$

$$r = 0.760 \tag{29}$$

$HA_{1,2}$ is the number of atoms with lone electron pairs in the R_1O and R_2O substituents, $MR_{1,2}$ is the molecular refractivity, σ_3 the electron attracting

(Kabachnik) constants for R_3, AR_3 and presence of an aromatic ring, and CT_3 is charge transfer character. Partial correlation coefficients are

$$r(HA) = 0.508,\ r(MR) = -0.240,\ r(\sigma) = 0.289,$$
$$r(AR) = -0.078,\ r(CT) = 0.285,\ r(MSD) = 0.048 \qquad (30)$$

The highest activity is A = 3.20 for the derivatives with R_1, R_2: MeO or EtO, R_3: $-CH{=}CCl_2$; the lowest activity is A = 0.63, for $R_1 > R_2$: i-BuO, R_3: $-CH_2O{=}CH$.

The standard, S_o, for calculation of MSD values was acetylcholine, $CH_3COOCH_2CH_2N^+(CH_3)_3$, with the carbonyl group superposed upon the P = O and the trimethylethanolamine residue upon R_3.

In comparison with structural requirements for carbamate inhibitors, there are rather significant differences. Hydrogen bonding character for the R_1O and R_2O substituents favors the inhibiting character, as well as charge transfer interactions within R_3; other structural requirements are different.

Another QSAR study was performed for phosphorylation rates of acetylcholinesterase for N = 60 Schrader-type compounds, according to data of Godovikov,[51] of the general type $R_1R_2POR_3$ with A = lgk ($M^{-1} \cdot s^{-1}$). The highest phosphorylation rate, A = 7.28, is for R_1:CH_3, R_2:C_2H_5O, R_3: $SCH_2CH_2S^+(CH_2)_RC_6H_{13}$; and the lowest, A = 0.63, is for R_1:CH_3, R_2:CH_3O; and R_3:S-CH_3. The correlation results are[46]

$$\hat{A}_6 = 2.941 + 0.136\pi_{1,2} - 0.060\sigma_3 - 0.272MSD$$
$$- 0.028C_3 - 3.003EC_3\ ;\ r = 0.880 \qquad (31)$$

with partial correlation coefficients $r(\pi) = 0.205$, $r(\sigma) = 0.296$, $r(MSD) = -0.717$, $r(CT) = 0.189$, and $r(EC) = 0.852$. MSD was calculated with acetylcholine as standard, S_o; MSD and EC values are largely intercorrelated, r(MSD, EC) = −0.645. Thus, the presence of a positive electric charge in R_3 (as in the natural substrate, acetylcholine), is the main structural feature favoring enzyme phosphorylation.

For a series of N = 15 vinyl enolphosphates (R_1, R_2: Me, Et, MeO, EtO; R_3: various vinyl derivatives such as $-C(Me){=}CHCOOMe$, $-C(Phe){=}CHCl$, etc.) a good correlation with electronic parameters, calculated by the simple HMO method, was obtained:[48]

$$\hat{A}_7 = -8.155 + 7.477\epsilon_{HOMO} + 0.126\epsilon_{\pi}\ ;\ r = 0.990 \qquad (32)$$

ϵ_{HOMO} represents the energy of the highest occupied molecular orbital, ϵ_{π} — the π electron energy of the R_3 substituent. The number of only N = 15 derivatives is rather small, but the results suggest interaction of the highly polarizable π electron system of the vinylic substituents with the enzyme site.

Inhibition by organophosphorus derivatives of butyrylcholinesterase, an enzyme similar to acetylcholinesterase, was reported by Russian authors[52] which determined phosphorylation rates and various rate and equilibrium constants for reversible and irreversible inhibition, for a total of N = 148 compounds of the type $R_1R_2P(O)SR_3$. A QSAR study was performed on the basis of mono- and biparametric correlations.[49] The structural parameters, tested in this work, were σ Hammett constants, hydrophobicity, molecular refractivity, MR, and electric charge, EC, for the groups R_1+R_2 and R_3, respectively. The experimental data are k_{II}, experimental rates constants for bimolecular phosphorylation; k_a, real rate constants (independent of inhibitor concentration); and K_r, equilibrium constant for reversible hydrolyses. For a number of N = 60 derivatives A = $\lg k_{II}$ data were available and the best biparametric QSAR obtained[49] is

$$A_8 = 4.813 + 2.256EC_3 - 3.205\sigma_3 \ ; \ r = 0.87 \qquad (33)$$

while for monoparametric correlation with $\lg k_{II}$, for the N = 60 derivatives: $r(\sigma_3)=0.323$, $r(\pi_3)=0.386$, $r(MR_3)=0.770$, and $r(EC_3)=0.630$.

For biparametric correlations with other kinetic parametrics the highest correlation coefficient obtained was r = 0.71, while for monoparametric correlation r = 0.55 (N = 114) was the highest value. The most pertinent structural features favoring butyrylcholinesterase inhibition are positive electric charge, electron attractive tendency, and overall size of the R_3 substituents.

3. Some Conclusions Concerning Acetylcholinesterase Inhibitors

QSARs for different classes of acetylcholinesterase inhibitors give rather different results for structural features of the inhibitors, although inhibition is considered to be the result of phosphorylation or carbamylation of the enzyme. One common feature is the favoring effect of electron-attracting power (σ Hammett constant) of the leaving group (R_3 or $-OC_6H_4X$). A positive electric charge in the leaving group, as in the substrate acetylcholine, also favors inhibition. The effects of hydrophobicity, molecular refractivity (size), and other structural parameters are less clear and seem to differ from one class of inhibitors to another. Structural requirements seem of rather reduced importance. Possibly, different inhibitors may adopt different spatial orientations with respect to the enzyme site in the activated complex for the enzyme phosphorylation or carbamylation reaction.

REFERENCES

1. **Drake, D. W.,** Rate of point mutation, *Nature,* 221, 1132, 1969.
2. **Bresch, C. and Hannsman, R.,** *Classical and Molecular Genetics,* Springer-Verlag, Berlin, 1970, chap. 6.
3. **Webb, J. L.,** *Enzymes and Metabolic Inhibitors,* Academic Press, New York, 1963, chap. 16.
4. **Simon, Z.,** *Quantum Biochemistry and Specific Interactions,* Abacus Press, Tunbridge Wells, Kent, U.K., 1976, chap. 7.
5. **Albery, D. J. and Knowles, J. R.,** Specifiency and evolution of enzymatic catalysis, *Angew. Chemie,* 89, 395, 1977.
6. **Page, M. J.,** Entropie. Bindungsenergie und enzymatische katalyze, *Angew. Chemie,* 89, 457, 1977.
7. **Page, M. J. and Jencks, W. P.,** Entropic contributions to rate acceleration in enzymic and intramolecular reactions and the chelate effect, *Proc. Natl. Acad. Sci. U.S.A.,* 68, 678, 1971.
8. **Kosower, E. M.,** *Molecular Biochemistry, Part III,* McGraw-Hill, New York, 1962.
9. **Volkenstein, M. V.,** The conformon, *J. Theoret. Biol.,* 34, 193, 1972.
10. **Volkenstein, M. V.,** K theory fermentativnovo kataliza, *Molek. Biol. (Moskow),* 6, 431, 1972.
11. **Ressler, N.,** Control of cellular processes by coupling of resonance energy in hydrogen transfer reactions. I. Applications to enzymatic reactions, *J. Theoret. Biol.,* 23, 425, 1969.
12. **Benkovic, S. J., Fierke, C. A., and Naylor, A. M.,** Insight into enzyme function from studies on mutants of dihydrofolate reductase, *Science,* 239, 1105, 1988.
13. **Blaney, J. M., Hansch, C., Slipo, C., and Vittoria, A.,** Structure, activity, relationship for dihydrofolate reductase inhibitors, *Chem. Rev.,* 81, 333, 1984.
14. **Baker, B. R.,** *Design of Active Site Directed Irreversible Enzyme Inhibitors,* Wiley, New York, 1967, 192.
15. **Kellems, R. E., Alt, F. W., and Schimke, R. T.,** Mechanism of resistance towards DHFR inhibitors, *J. Biol. Chem.,* 251, 6987, 1976.
16. **Numberg, J. H., Kaufman, R. J., Schimke, R. T., Urlanb, G., and Chasom, L. A.,** Gene amplification in resistance towards DFHR inhibitors, *Proc. Natl. Acad. Sci. U.S.A.,* 75, 5553, 1978.
17. **Dias Selasie, C., Fong, Z. X., Li, R. I., Hansch, C., Klein, T., Langridge, R., and Kaufman, B. T.,** Inhibition of chicken liver 1-dihydrofolate reductase by 5-(substituted benzyl)-2,4-diaminopyrimidines. A quantitative structure-activity relationship and molecular graphics analyses, *J. Med. Chem.,* 29, 621, 1986.
18. **Ochmy, K. A. and Roth, B.,** Receptor based design of novel dihydrofolate reductase inhibitors: benzymidasole and indole derivatives, *J. Med. Chem.,* 34, 1283, 1991.
19. **Neil, G. L., Niemann, C., and Hein, G. E.,** Primary specificity in alpha-chymotrypsin catalysed hydrolises, *Nature,* 210, 903, 1966.
20. **Vernick, L. M. and Fruton, J. S.,** Specificity in alpha-chymotrypsin catalysed hydrolyses, *Proc. Natl. Acad. Sci. U.S.A.,* 68, 257, 1971.
21. **Dupaix, A., Bechet, J. J., and Roucous, C.,** Alpha chymotrypsin catalized hydrolysed rates of nitrophenol esters of various carboxylic acids and amino acid derivatives, *Biochemistry,* 12, 2559, 1973.
22. **Morgensten, L., Ricanatini, M., Klein, T. E., Steinmeyr, W., Yang, C. Z., Langridge, R., and Hansch, C.,** Chymotrypsin hydrolyses of K-phenyl hippurates. A QSAR and molecular graphics analyses, *J. Biol. Chem.,* 262, 10767, 1987.
23. **Silver, M. S., Stoddard, M., Sone, T., and Matta, M. S.,** Alpha chymotrypsin. The use of substrates of restricted geometry to define the reactive conformation of methyl N-acetyl-L-phenylalanine, *J. Am. Chem. Soc.,* 92, 3151, 1970.

24. **Hein, G. and Niemann, C.**, Specificity in alpha chymotrypsin catalyzed reactions, *Proc. Natl. Acad. Sci. U.S.A.*, 55, 664, 1966.
25. Bruchaven Protein Data Banck, File:4CNA.
26. **Dickerson, R. E. and Geri, I.**, *Proteins*, Harper & Row, New York, 1969, 84.
27. **Hansch, C., Klein, T., McClarin, J., Langrige, R., and Cornell, W.**, QSAR and molecular modeling studies in alpha chymotrypsin catatysis, *J. Med. Chem.*, 26, 615, 1986.
28. **Yoshimoto, M. and Hansch, C.**, Quantitative and relationships of D- and L-acil, amino acid ligands binding to chymotrypsin. On the problem of combined treatment of stereoisomers, *J. Org. Chem.*, 41, 2269, 1976.
29. **Simon, Z., Holban, S., Motoc, I., Mracec, M., Chiriac, A., Cherek, F., Ciubotariu, D., Szabadai, Z., Pop, R. D., and Schwartz, I.**, Minimal steric difference in structure biological activity correlation for alpha chymotrypsin catalyzed hydrolysis, *Stud. Biophys. (Berlin)*, 59, 181, 1976.
30. **Chiriac, A. and Simon, Z.**, Connectivity restrictions in the MTD optimization procedure. Substrates for alpha chymotrypsin catalyzed hydrolysis, *University Timişoara, Ser. Chimia*, No. 5, 1985, Preprint.
31. **Kabachnik, M. I.**, Fosfoorganaicheskie fisiologicheskie aktivnye veshchestva, *Vestnik, A.N.S.S.S.R.*, 61, 1267, 1966.
32. **Sussman, J. L., Hurel, M., Frolov, F., Oefner, C. H. R., Goldman, A., Toker, L., and Sillman I.**, Atomic Structure of acetylcholinesterase from *Torpedo californica:* A prototypic acetylcholine binding protein, *Science*, 253, 872, 1991.
33. **Dougherty, D. A. and Staufer, D. A.**, Acetylcholine binding by a synthetic receptor. Implications for biological recognition, *Science*, 250, 1558, 1990.
34. **Brestnik, A. P., Bric, J. L., and Teplov, N. E.**, Kinetica hidroliza fenilnovo analoga acetilcolina pod deystviem choliesterazy loshadnoi syvorotki i acetilcholinesterazy bychnih eritrotsitov, *Biochimia (Moscow)*, 33, 1059, 1968.
35. **Jarv, J., Kesvalatela, T., and Aaviksaar, A.**, Structure activity relationships in acetylcholinesterase reactions. Hydrolysis of non-ionic acetic esters, *Eur. J. Biochem.*, 67, 315, 1976.
36. **Ciiubotariu, D., Deretey, F., Chiriac, A., Sulea, T., Oprea, T., Kurunczi, L., and Simon, Z.**, Multiconformational MTD methods. Acetylcholinesterase hydrolysis rates (in preparation).
37. **O'Brien, R. D.**, *Insecticide Action and Metabolism*, Academic Press, New York, 1967.
38. **Metcalfe, R. L. and Fukuta, T. R.**, Carbamates as acetylcholinesterase inhibitors, *J. Agric. Food Chem.*, 15, 1022, 1967.
39. **Chiriac, A., Vilceanu, R., Ciubotariu, D., Szabadai, Z., and Simon, Z.**, Structure-anticolinesterasic activity correlations in phenyl methyl carbamates, *University Timişoara, Ser. Chemie, No.* 3, 1974, Preprint.
40. **Chiriac, A., Ciubotariu, D., Szabadai, Z., Vilceanu, R., and Simon, Z.**, Structure-activity correlation in phenyl, methyl carbamates. *In vitro* anticholinesterase activity, *Rev. Roum. Biochim.*, 42, 143, 1975.
41. **Bădilescu, I. I., Stepan, A., Chiriac, A., and Simon, Z.**, Mapping of acetylcholinesterase inhibition site by correlation with minimal topologic (steric differences), *An. Universitătii Timişoara, Ser. Fiz-Chim.*, 16, 29, 1976.
42. **Chiriac, A., Motoc, I., Holban, S. and Vancea, R.**, Minimal steric difference, MSD (simple version) and MCD, *Universitatii Timişoara, Ser. Chimie, No.* 3, 1980, Preprint.
43. **Metcalf, R. L., Fukuta, T. R., Wilkinson, C. F., Fahm, M. H., El-Alziz E. A., and Metcalf, E. R.**, Phosphoric acids esters as pesticides, *J. Agric. Food Chem.*, 14, 555, 1966.
44. **Schrader, G.**, Die Entwicklung neuer insektizider Phosphorsäure Ester, *Verlag Chemie*, Weinheim, 1963.

45. **Vilceanu, R., Chiriac, A., Szabadai, Z., and Simon, Z.,** Multiple correlations in the anticholinesterasic activity of organic phosphorus compounds, *Rev. Roum. Biochem.*, 10, 239, 1973.
46. **Chiriac, A., Simon, Z., and Vilceanu, R.,** Structure-biological activity correlations in phosphororganic compounds. Role of electric charge and other parameters for acetylcholinesterase inhibition, *Stud. Biophys. (Berlin)*, 51, 183, 1975.
47. **Vilceanu, R., Chiriac, A., Szabadai, Z., and Simon, Z.,** Structure-activity correlations for the toxicity of organic phosphorus compounds of the Schrader type, *An. Universitätii Timişoara, Ser. Fiz.-Chim.*, 15, 62, 1977.
48. **Chiriac, A., Simon, Z., Chiriac, V., and Vilceanu, R.,** Correlations entre structure chimique et activite biologique des composes organo-phophorique des type vinyl-(enol)-phosphates. 1. Correlations a parametres quantochimiques, *An. University Timişoara, Seria Fiz-Chim.*, 12, 51, 1974.
49. **Chiriac, A., Chiriac, V., and Simon, Z.,** Mono and biparametric structure-activity correlations for large series of butylcholinesterase inhibiting organophosphorus compounds, *University Timişoara, Ser. Chimie,* No. 2, 1979, Preprint.
50. **Selepeva, E. C., Savin, I. I., Kagan, Y. S., Borodac, M. S., Gelfer, A. P., Paikyn, D. M., and Gamper, N. M.,** Khimia i primenenie phosphororganicheskih soedinenyi, *Izdat. Nauka, Moscow,* 1972, 484.
51. **Godovicov, N. N.,** Khimia i primenenie phosphororganicheskih soedinenyi, *Izdat. Nauka, Moscow,* 1972, 423.
52. **Brestkin, A. P., Brik, I. L., Godovicov, N. N., Kabachnik, M. I., Kirceva, E. G., and Trifunova, S. A.,** Phosphororganichesckie soedinenyia inhibitory butiryl cholinesteraza, *Izvest. Acad. Nauk SSSR, Otd. Khim. Nauk,* 429, 1976.

Chapter 4

SPECIFICITY IN REGULATION OF GENE EXPRESSION

Nicolae Voiculetz

TABLE OF CONTENTS

0-8493-5398-X/93/$0.00 + $.50

I. INTRODUCTION

Regulation of gene expression is one of the most intensely studied topics in molecular biology. The goal of this chapter is to review the evidence for recognition processes as the basis of specificity in gene regulation.

Major biological events such as development, growth, and differentiation are based on the mechanisms of both function and regulation of genes. This regulation takes place not only with high topographical (and also topological) specificity but also according to precise timing.

In our analysis we started from the following elements:[1]

1. The genome is an assembly of virtually functional and nonfunctional structural units (genes);

2. The genes are interrelated by feedback mechanisms (the whole system being considered as a biochemical oscillator);
3. The activation of genes is selective, the number of active genes being much smaller than the total number of genes forming the genome;
4. The selective activation of genes induces cell differentiation — in maintaining the same degree of cell differentiation, regulatory mechanisms of genes expression are involved;
5. The gene expression is a highly regulated process generally controlled at the transcriptional level during initiation or elongation of the nascent transcript — transcriptional regulatory proteins (transcriptional regulators) are responsible for such processes; and
6. In eukaryotes, the control systems are of a combined type (positive and negative regulation).[2,3]

The regulatory processes are organized as a network and are multistep in development. Thus, the regulation mechanisms of gene activity depend on genes encoding regulatory proteins which in turn activate other genes. These transcriptional regulators are DNA binding proteins that specifically interact with some target sites of DNA to form a variety of complexes.[4] What is generally agreed is that before RNA synthesis the transcriptional apparatus is often in contact with DNA binding regulatory proteins.[5]

To simplify the problem of gene regulation, two main aspects have to be considered: (1) the DNA-protein interaction, and (2) the competition of regulatory proteins for nonspecific and specific sites of the DNA sequence; the balance between these two mechanisms will determine the sequence of the transcriptional events.

The binding of proteins involved in the organization of the genome configuration and the regulation of its function (replication, transcription, repair, and recombination) is based on the recognition of specific sequences of DNA base pairs, which represent the control elements involved in gene expression. Some of these regulatory proteins have a high specificity for the target sites of DNA sequences (e.g., the competitive binding of a *lac repressor* to a single *lac operator* site on the *E. coli* genome), but others may be nonspecific (e.g., the DNA-histone complexes in eukaryotic nucleosomes). The amazing thing is how proteins, such as enzymes, repressors, activators, and other regulatory proteins, can recognize their specific sequences embedded in the enormous amount of competing sites of the genome, bind to the right site, and act in controlling, regulating, or initiating gene expression. The phenomenon, not yet fully understood, is very complex, and besides, the sequence recognition must be treated from a thermodynamic and kinetic point of view, too.[6]

Since there is a major difference in topographical organization of the prokaryote genomes, the gene regulatory mechanisms of both systems will be discussed.

II. COLINEARITY OF GENES AND PROTEINS

A. GENES AND PROTEINS ARE COLINEAR IN PROKARYOTIC GENOME

The chromosome from either the bacterial or eukaryotic cell, contains a long uninterrupted frame of DNA sequences represented by thousands of genes. In bacteria, the chromosome consists of a highly organized circular DNA molecule which forms a nucleoid body. In these organisms the double-stranded DNA is supercoiled, exhibiting a large number of loops (up to 50), each of them forming an independent region. This kind of organization may play an important role in gene expression. The active genes lie on the external part of the loops, this location being favorable to the DNA binding proteins involved in regulation.[7]

In prokaryotic cells and their viruses, the genes and proteins are colinear, which means that the sequence of each gene exactly corresponds to the amino acid sequence found in the translated protein. In a prokaryotic cell, transcription and translation proceed simultaneously. The reason for this behavior derives from the lack of compartments specific to these cells. The ribosomes are directly associated with the mRNA transcript (forming the so-called polysomes) and remain physically bound to the bacterial chromosome. Transcription takes place only on one strand of the DNA double helix. Thus, the mRNA sequence is complementary to the sequence of one DNA strand of the gene (in RNA, T corresponds to a U moiety), and it is exactly identical to the other strand of the DNA gene.

The gene that encodes a protein is generally thought to contain the full sequence transcribed in the mRNA. On each side of the region containing the information regarding the protein structure, the gene usually possesses several sequences that are not translated in the final product. This means that the transcriptional unit and the mRNA product are longer than the encoded protein (Figure 1).

As a response to evolutionary pressure, the bacterial structural genes tend to organize in clusters. This means that all enzymes required for a metabolic pathway are encoded by adjacent genes. In such a case, the transcript appears as a single polycistronic mRNA, which is further sequentially translated into each enzyme. Under these conditions, gene regulation may be accomplished by a single regulatory protein which specifically binds to the promoter of the cluster genes.[8,9]

B. INTERRUPTED GENES APPEAR IN EUKARYOTIC GENOME

In contrast, the genes of eukaryotic DNA are interrupted because they are formed by alternate transcriptional and nontranscriptional sequences called *exons* and *introns*. The transcription of the gene gives rise to a RNA molecule (primary transcript) which exactly corresponds to the sequence of the gene. This precursor must be processed to mRNA in order to translate its own

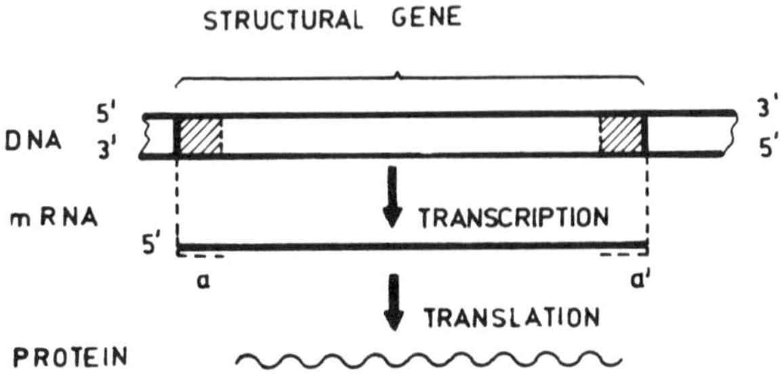

FIGURE 1. The gene and the RNA transcribed may be longer than the sequence of protein translated; *a* and *a'* = short sequences not translated in the protein.

protein. During this process (called RNA splicing) the introns are removed and the resulting mRNA contains only exons which, in fact, are the coding sequence of the gene.

During the splicing process, the exons are usually joined together in the mRNA molecule in the same order as they are located in the gene sequence. Therefore, we can say that the gene and the protein appear to be colinear, because the individual exons are translated in the corresponding part of the protein chain (Figure 2).

It should be mentioned that not all eukaryotic genes are compulsorily interrupted. The existence of alternative splicing patterns can be observed. In this case, the exons are not connected in the same order as they lie in the DNA sequence.

The majority of genes contain a DNA encoding sequence for only one protein, but there is the case (e.g., overlapping genes) in which a part of the gene sequence may be transcribed independently (Figure 3). Thus, the encoding region of the gene specifies a protein which represents, in fact, either the first half region or the second half of the entire gene. This situation appears when two genes share the same sequence of DNA and when transcription initiation occurs in more than one reading frame. This phenomenon is repeated during the transcription of some viral and mitochondrial genes.

III. THE RECOGNITION PROBLEM IN GENE REGULATION

The binding of regulatory proteins to specific sequences of DNA base pairs and especially the recognition of a particular functional site represents the molecular basis of specificity in the regulation of gene expression.

The recognition processes range from extremely selective to fairly permissive, depending on the requirement imposed by their functions.[10] In the first category, we can mention the restriction endonuclease which recognizes (and cleaves) small sites of 4 to 6 bp on foreign DNA; on the other hand, regulatory proteins have larger target sequences, ranging from 12 to 30 bp.

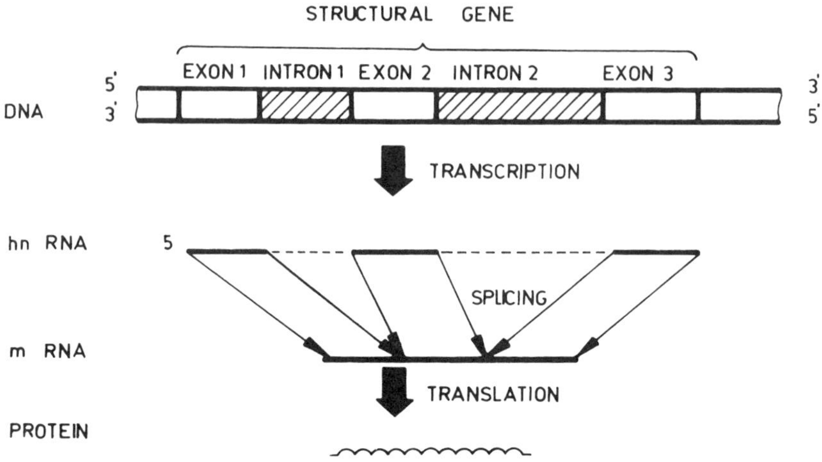

FIGURE 2. Structural genes are interrupted by introns. The gene is expressed via a hnRNA (precursor), from which the introns are removed; and the exons are spliced together to give the mRNA, which is translated into the protein.

The general question is how does a regulatory protein recognize the functional target in DNA and interact with it in a specific manner? Before the interaction of two macromolecules to form a complex, they must come into close contact. To clarify this process, we have to discuss the types of bonds encountered during this interaction, the modality by which molecules involved in interaction come in close contact, and the conditions which could influence this interaction.

A. BONDS INVOLVED IN DNA-PROTEIN INTERACTION

First of all, we must consider that DNA-protein binding can be divided evenly in a convenient approximation into two classes: specific and nonspecific binding. Binding is based on the electrostatic interaction between the positively charged groups of the protein and the negatively charged phosphates from the DNA backbone. The bonds encountered in biological macromolecules interactions are usually weak, noncovalent, and may be classified into four types:[3,11] ionic bonds, hydrogen bonds, van der Waals attractions, and hydrophobic forces created by the three-dimensional structure of the water. In aqueous solution, each of the noncovalent bonds mentioned above are 30 to 3000 times weaker than the classical covalent bonds. In comparison with the average energy of thermal motions at 37°C, the noncovalent bonds are only slightly stronger (by about 6 to 10 times). For this reason, a DNA complex which has to resist the thermal collision (that could separate the partners) requires a large number of noncovalent bonds. As in the case of enzyme-substrate interaction, the number of noncovalent bonds realized between two surfaces depends on the number of suitable atoms on the two surfaces that

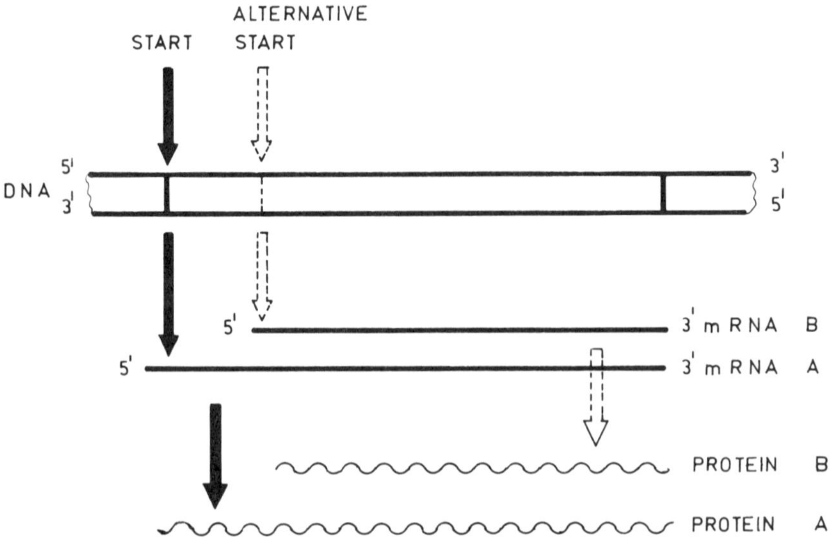

FIGURE 3. A single gene may code more than one protein by starting (or terminating) transcription at different points.

are able to fit exactly, one to another. This close fitting of the two domains of macromolecules accounts for the specificity of biological recognition.

The weak noncovalent bonds determine how different domains of the same macromolecules fit together. In a biological environment, specificity and stability may be under either thermodynamic or kinetic control.[11] The modifications which appear during the hydrophobic effect, polyelectrolysis and hydrogen binding, could have several implications of great importance in the stability of DNA-protein complexes. The stability of the complex formed in an aqueous solution may be critically dependent on the bulk concentrations of the surrounding and exchanged small molecules, such as H_2O and electrolyte ions. Thus, it will be difficult or even impossible to predict the stability of such a complex based only on structural configuration.

In addition, since the individual ions may play a stoichiometric role in DNA-protein complex formation, the type and concentration of the electrolyte ions could be used as a parameter, both in the thermodynamic and kinetic computations of these interactions. The equilibrium binding constant can give information about the relative contributions of ion interactions and also of nonelectrostatic interactions involved in the DNA-protein complex formation. The general feature of these interactions is that they are highly ion dependent and, usually, the observed affinity decreases sharply with an increase in ion concentration. Experimental data have shown that this constant depends not only on the DNA and protein molecules, but also on the charge-neutralizing ion concentration.[12] In the case of a *lac* repressor association with the DNA *lac* operator *in vitro,* modification of the NaCl concentration from 0.1 to 0.2

M induces a change in the affinity by a factor of 20. The high electrostatic potential of the negatively charged phosphate residues of the DNA backbone leads to the accumulation of counterions (e.g., K^+) in the neighborhood of the DNA which partially neutralize the closely spaced backbone phosphates.

The interaction of proteins (positively charged) with DNA causes a perturbation on the electrostatic potential around the DNA region which can lead to the removal of the surrounding counterions into the bulk solution. The appearance of these counterions in a low ionic concentration environment can produce an increase in the entropy of the system which, as a result, makes an important favorable contribution to the free energy of the interaction. This represents a major contribution to the binding free energy associated with DNA-protein complex formation.[6,13]

B. RECOGNITION AND DNA BINDING INTERACTION

The main problem in the complex process of regulation of gene expression is how gene regulatory proteins recognize and bind to specific control DNA sequences within a functionally useful time period.

The specificity for DNA is determined partially by the contact of proteins with DNA bases (direct or read out). The predominant form of DNA *in vivo* is the B form, in which the base pairs are displaced from the helical axis in an asymmetric location, which creates the minor and major grooves in the DNA structure.

The first step in DNA-protein interaction is a nonspecific charge interaction between positively charged groups of the protein and negatively charged phosphates of the DNA backbone. Then, a more specific interaction with the base pair sequences located mostly in the DNA grooves (major) takes place. The electrostatic potential helps to fit the protein into correct binding position in which it establishes a specific interaction with a particular target DNA sequence. If the specific binding does not occur, the protein remains in a nonspecific binding situation. In any case, the nonspecific binding of the protein has important consequences for the kinetics of the association and, thus, it may be considered as an integral part of the molecular design which causes specific site interaction.[7] Taking into consideration this assumption, it appears that the main determinant for the specific binding probability *in vivo* is the competition of regulatory proteins for nonspecific and specific sites of DNA sequences.

By simple diffusion, caused by thermal motions, the macromolecules come into close contact and form a complex through multiple, weak noncovalent bonds. Generally, the strength of the binding between protein and DNA macromolecules could be considered as an index for the specificity of the recognition process.[3]

Taking into account all experimental and theoretical data, it was proposed that DNA-protein interaction occurs in two steps: a first nonspecific interaction which is mainly electrostatic (protein is bound to the DNA phosphate backbone), and a second specific-site interaction by other weak noncovalent bonds formed between the amino acids of the protein and the base pairs located in the major groove of the B form of DNA, without disruption of the double-stranded structure.[12] These specific interaction mechanisms take place by at least two pathways: by the protein sliding till it reaches the specific target of DNA sequences, or by intersegment transfer, by which the proteins can be transferred from a nonspecific site to a specific one, without becoming free in solution. However, under *in vivo* conditions, both pathways could work together.

1. Sliding

In the sliding pathway the protein is associated with DNA in a nonspecific binding due to the electrostatic charge of both molecules.

The nonspecific complex formed as an isopotential surface favors the protein to slide along the DNA strand, being propelled under physiological conditions by the thermal fluctuations. The rate of binding and the complex lifetime determine the duration of the binding to a certain target site. Under these conditions, the nonspecific interaction can help specific association by properly aligning the functional groups of the two macromolecules. In addition, the isopotential electrostatic surface of the DNA backbone may be a geometry that favors the protein to run in two directions in order to find its specific target sequences.

The specific binding can be achieved between amino acid and base interactions (by weak noncovalent binding) in the major groove of the DNA. The preferred base is A or T. The T methyl group provides an additional sequence recognition element that can be easily identified in the major groove. The major groove in the DNA is more suitable for the α helix structure of proteins.

The sliding pathway has been revealed in the *E. coli lac* repressor-operator interaction,[14] in a promoter search by *E. coli* RNA polymerase, and in the EcoRl endonuclease-DNA interaction.[10] The rate enhancement determined in the above-mentioned cases was found to correspond to a one-dimensional diffusion (sliding) rate of about 10^{-9} cm^2 s^{-1}. Berg[7] established that in the case in which sliding is a major pathway in DNA-protein interaction, the mean time required by a protein molecule to find a certain specific site can be approximately expressed by the equation:

$$T_{s1} = (1 + 1/D_T K_{RD}) L/\sqrt{4 k_D D_1} = (1 + 1/D_T K_{RD})(L/s)(1/k_D) \qquad (1)$$

where D_T is the concentration of nonspecific DNA sites, $K_{RD} = k_A/k_D$ is the nonspecific binding constant, L is the length (cm) of the DNA region, D_1 is the one-dimensional diffusion (sliding) rate, and k_D is the nonspecific dissociation rate constant. The sliding distance (s) or effective target size is approximated by:

$$s = 2\sqrt{D_1/k_D} \qquad (2)$$

where s represents the distance on which the protein can slide by one-dimensional diffusion to a nonspecific DNA site during its residence time, $1/k_D$; s/2 is a more specific parameter which represents the distance from which the protein can run into a specific target with a probability of exp(− 1) during the residence time.

In Equation 2, Berg attributed to *s* the sliding distance or effective target size. Equation 1 represents the product between the number of nonspecific binding events, L/s, realized by the protein exploring the whole DNA chain, and the effective time required to find another target site, $1/k_D + 1/k_A D_T$.

Through the use of a proper mathematical model, it was deduced that when many nonspecific DNA events are presented, the minimum association time increases and the optimal sliding decreases.[7]

2. Intersegment Transfer

It was demonstrated that most regulatory proteins contain two DNA binding sites. Binding to the operator takes place simultaneously at both sites of the specific DNA sequence, which appears to be divided into two parts: left and right.[16] This nonspecific mode of binding can help in the search process for the specific target by permitting the protein to pass from a certain segment of the DNA to another, without becoming free in the environment and without any risk of being lost from the DNA segment. In this pathway the protein is bound transiently and nonspecifically and can be transferred between two different segments of DNA during the search for an operator. The intersegment transfer is favored by the loop domains which can appear between the two specific DNA sites, and therefore is advantageous because the search volume becomes much smaller for the regulatory proteins.

A relationship between the formation of the loop domains and the intersegment transfer has been identified in *lac, gal,* and *ara* repressors.[17] In eukaryotes, the function of many enhancers located at long distances from the promoter is consistent with the looping hypothesis.

Under certain conditions, the most efficient search pathway seems to be the hybrid process in which the regulatory proteins undertake both the sliding and intersegment transfer. This hybrid process could be considered a third possible mechanism of DNA-protein interaction.

The result for the association time calculated by Berg,[7] taking into account these pathways, appears to be as follows: in the sliding process it is 5 s, in the intersegment process it is 0.2 to 2 s, and in the hybrid process it is 0.1 to 0.3 s. From these data one can see that the hybrid process is faster than the previous two.

C. DNA SITE SEQUENCES ARE SYMMETRICAL OR REPEATED

We saw that DNA binding proteins are associated with great selectivity to some particular sites of DNA and binding occurs with a high affinity to the correct site and with a relatively low affinity to other DNA site sequences. The selectivity and affinity are explained by the fact that a protein which binds to DNA has to be able to form a fixed number of hydrogen bonds and other weak noncovalent bonds which finally provide a precise extent of binding energy (or enthalpy). In the binding process the center of protein volume (specific domain) must be exactly positioned against the DNA site. Protein binding increases if the configuration state of the protein and the DNA macromolecules, and particularly the conformational flexibility of these molecules, are convenient. Under these conditions the protein is held in the correct site and position.

D. HELIX-TURN-HELIX STRUCTURE

The helix-turn-helix structure as a model for DNA-protein interaction is supported by NMR, X-ray crystallography, and genetic studies[12] for many regulatory proteins (CRP of *E. coli,* bacteriophage λ, regulatory protein *cro,* repressor of tryptophan operator, 434 repressor, etc.). The proteins were purified and crystallized in the absence of DNA. X-ray crystallography established the structures of these proteins and their DNA contacting domains were recognized. Their surfaces have shown some protrusions, separated by approximately 35 Å, which consist of short α helix motifs oriented across the major groove of DNA. A second helix was predicted to be located partly within the groove and to be involved in specific amino acid residue-DNA interactions. The second α helix is considered to be the recognition helix. For the 434 repressor protein, the helix-turn-helix region was established to contact the DNA by the NH_2-terminal end of the recognition DNA backbone in the respective region, and it was shown that the protein has no contacts with the bases within the minor groove. The two subunits of the protein contact the edges of the bases from the two major grooves of DNA, A and T being the preferred bases. In the case of the 434 repressor, binding is accomplished by three or four hydrogen bonds between phosphate residues of the DNA backbone and the peptide amino groups. This interaction, between the bases from the major groove and the recognition helix seems to occur with three Gln residues, which form the following bonds: one Gln residue forms two hydrogen bonds, another forms a hydrogen bond and a van der Waals interaction with the methyl group from the T base, and the last Gln provides a single hydrogen bond to each of two consecutive bases.

Biochemical analysis has established that the amino acid sequences in the helix-turn-helix of *Cro,* λ repressor, and CRP are similar; this type of sequence has also been found in other regulatory proteins. This finding could be an additional argument that the helix-turn-helix structure is the major contacting region of DNA binding proteins.[12]

E. MOLECULAR ASPECTS OF PROTEIN-DNA RECOGNITION

The DNA double helix (considered here only in the B form) presents three elements for recognition: the major groove, the minor groove, and the phosphate backbone. The two grooves provide specific hydrogen binding patterns given by the sequences of the four base pairs (AT, TA, CG, and GC), while the phosphate backbone is identical for all base pairs.[18] These types of patterns, as well as the interatomic distances involved, are presented in Figure 4; the distances and orientation required for hydrogen double bonds between amino acid side chains and the proton-donating and -accepting groups are depicted in Figure 5.[19]

In the major groove, the dAT pairs present four recognition sites: the thymine O^4 (hydrogen acceptor), the thymine CH_3 group (van der Waals interaction), the adenine position 6 NH_2 group (hydrogen donor), and the adenine N^7 (hydrogen acceptor). The dTA pairs present the same recognition site, but in inverse sequence. The recognition sites available at the dGC pairs are guanine N^7, adenine O^6 (two hydrogen acceptors), and cytosine NH_2 (one hydrogen donor). For dCG base pairs the inverse sequence of these sites has to be considered. In any case, the dAT pairs vs. dTA pairs (and dGC vs. dCG) are easily distinguishable by the DNA binding proteins.

In the minor groove, a similar arrangement is also valid for the dAT (dTA) base pairs (two proton acceptor hydrogen bonding groups: adenine N^3 and thymine O^2). The dGC base pairs present two hydrogen acceptor groups (guanine N^3 and cytosine O^2) and one hydrogen donor group (guanine C^2 and NH_2). It should be mentioned that considerably less sequence information is presented when the reading is done from the minor groove.

Regarding the phosphate backbone, the DNA conformation (the twist roll, angles of the base pairs) and the electronegative potential could be considered among the biological recognition parameters.[20]

On the basis of such recognition elements, the theoretical possibility of preferential hydrogen bonding between amino acids and base pairs was investigated.[21] The formation of two hydrogen bonds seems possible between (a) the amidic group of Asn or Glu and $(NH_2)^2$ and the N^3 -guanine positions from the dGC pair (minor groove); (b) the ammonium cation of Lys and O^2(T) and N^3(A) from the dAT pair (minor groove); and (c) the guaninic Arg cation with O^6 and N^7 of guanine from dAT (major groove).[22] The spatial configurations of the amino acids will also be important, but some specificities in oligopeptide binding may be explained hereby. The preferential binding of poly-L-Lys to dAT-rich DNA[20] may be explained by the (b)-type binding, and the weak specificity[22] of poly-L-Arg to dGC-rich DNA by the (c)-type binding. Other mechanisms for specific DNA-protein interactions could be hydrophobic interactions with the $(CH_3)^5$ group of the dAT pair (major groove), aromatic side chain intercalation in the DNA double helix (similar to acridine or ethidium bromide),[23] and juxtaposition of charge distributions of bases and the intercalated amino acid. The comparatively weak preference of Trp for guanine with adenine and the marked preference for purine over pyrimidine base (see Chapter 2, Section III.C.2) due to a maximal overlapping area rule, could contribute to specific interactions between bases and intercalated aro-

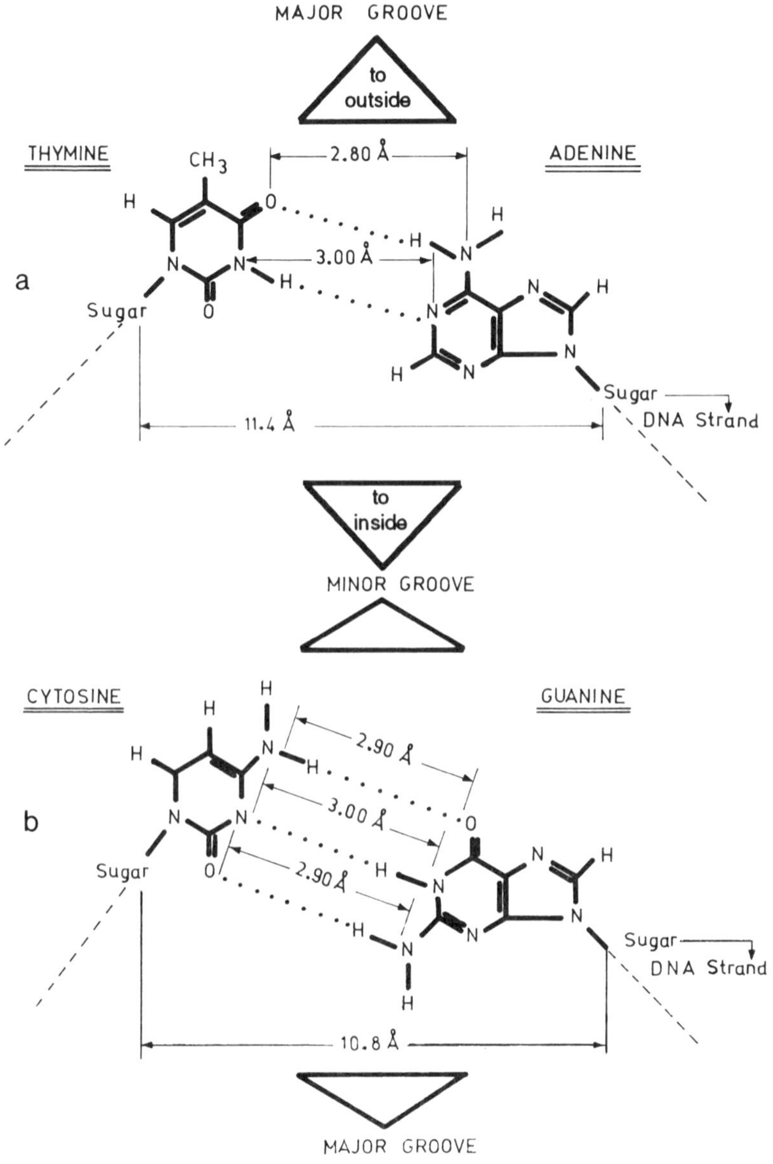

FIGURE 4. Hydrogen bonding groups and interatomic distances in the DNA base pairs: (a) thymine usually binds to adenine by two hydrogen bonds, and (b) cytosine to guanine by three hydrogen bonds. In every base pair all hydrogen bonds are strong because every hydrogen atom is directly oriented toward the correspondent acceptor atom (nitrogen or oxygen).

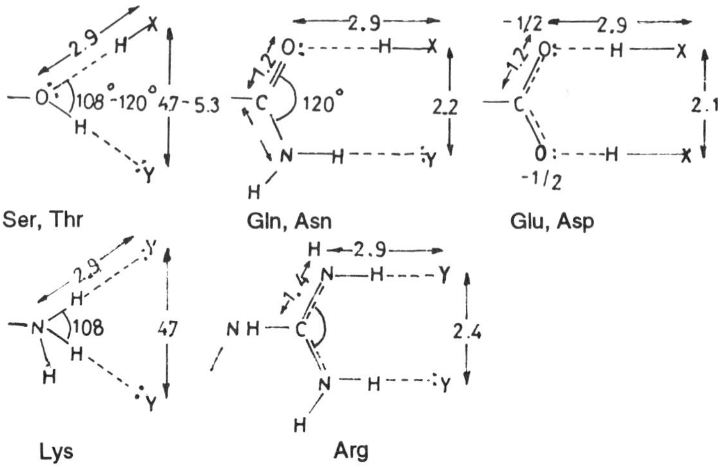

FIGURE 5. Hydrogen bonds between amino acids from a polypeptidic chain. Distances (Å) and orientations required for hydrogen double bonds between amino acid side chains and proton donating (X-H) and accepting (-:Y) groups.

matic amino acid chains. Finally, the electronegative potential of contact zones, shielding of polar interactions with the phosphate backbone by sufficient ionic strengths, or Mg^{2+} and Ca^{2+} cations, steric fit conditions, and DNA conformations may also play a role. Nevertheless, to date, an example demonstrating where the DNA conformation is determinant for protein recognition is still lacking, except perhaps for the nuclear HMG_1 protein, which specifically recognizes cruciform DNA.[24,25]

Several oligomeric drugs bind to DNA with remarkable sequence specificity,[26] for example, echinomycin to GC-rich DNA sequences and lexitropins to AT-rich DNA sequences, but the bonding constant ratios between high- and low-affinity sites is of the order of 10 to 20. In contrast, this ratio for the *lac* repressor binding to the operator sequence and to random DNA sequences ranges around 10.[5,27,28]

More recent data on the binding of peptides and oligomeric drugs, such as lexitropin and dystamicin to specific DNA double helix sequences have strongly contributed to our understanding of DNA-protein binding.[18] Most sequence-selective DNA binding proteins recognize the DNA from the major groove and the α-helical motifs are those which contact the floor of the major groove.[12]

Several oligomeric drugs are bound to the minor groove sequences, such as dystamicin, netropsin, DAP 1(4′,6-diamidino-2-phenylindole). These types of compounds express a significant preference for AT-rich sequences. This preference arises from the large electronegative potential of the AT-rich sequences located in the minor groove, by which the positively charged drug is attracted.[29]

F. SIZE-SPECIFICITY REQUIREMENTS

How large must the interaction segment (combining site) be in order to achieve recognition between the regulatory protein and the right DNA sequence?

According to Gilbert and Maxam,[30] in the *lac* repressor of *E. coli* a 27-bp DNA segment is protected from deoxyribonuclease digestion while the size of the λ bacteriophage DNA fragment that binds a repressor dimer is 35 nucleotide pairs.[31]

According to Equation 23 from Chapter 1, Section II.B.3,[32]

$$n > \frac{\lg([E_{tot}]/p_{er}[E_o])}{\lg\alpha - \lg[1 + r(\alpha - 1)]} \qquad (3)$$

for $N = 4.5 \times 10^6$ base pair length of the *E. coli* genome, $\alpha = 4$ base pairs, and $r = 0.027$, one obtains $n > 20$. A similar calculation by Sadler and Smith[33] yields $n > 18$. Both results are in satisfactory agreement with the experimental figures.

In eukaryotes, the genome length is about $N = 6 \times 10^9$ base pairs. According to Equation 3, this would increase the minimal required size up to $n > 25$ base pairs. This typical size of regulatory binding sites in eukaryotic genomes is about three-time repeats of 8 to 12 base pairs each,[34] which would yield the required size.

Equation 3 is approximate and yields rather an upper limit to the size. Thus, due to the chromatin structure, the number of available chromosomal sites in eukaryotes would be greatly reduced. Nevertheless, the size of the drug molecules designed to interact specifically with gene control elements should be large enough to interact with at least 20 to 25 base pairs. Repressor-operator binding usually involves about 15 base pairs of the operator DNA sequence. The restriction enzyme, EcoR1, binds to only 6 base pairs of DNA, producing a distortion of the DNA which increases the interstrand distance by 4 Å.[25]

Theoretical considerations concerning the interaction of regulatory proteins or other large ligands with polynucleotidic double helixes and their significance in specific protein-DNA recognition were studied by Gurskii and Zasedatelev[35] and by Bruskov et al.[36]

G. CHROMATIN STRUCTURE AND NONSPECIFIC DNA-PROTEIN INTERACTIONS

The accessibility of regulatory proteins to specific DNA segments in replication, recombination, and transcription represents one of the major features of the chromatin structure. The chromatin structure plays a key role in the repression of gene transcription and suggests how such repression can be overcome *in vivo*. Nevertheless, the mechanism by which the multiproteic pre-initiation complexes of other nucleoproteic complexes are accommodated into a 300-Å-thick chromosomal fiber, or even into a nucleosomal array, is scarcely known.[37]

In eukaryotic cells, the DNA is organized in highly ordered structures[38] forming the chromatin. Nucleosomes are the first ordered structures,[39] consisting of 146 to 166 base pairs wrapped around the octameric histone complex, which contain two molecules of each histone: H2A, H2B, H3, and H4. These cores are connected by DNA linker regions of 20 to 100 base pairs, which are particularly sensitive to staphylococcal nuclease. The Hl histone is associated with both cores and linker regions stabilizing the nucleosomes.[40] The second-order "pearl string" structure (formed from nucleosomes and linkers) appears in a more complex superstructure — the "thick fiber" — a spiralized superfiber with a diameter of 250 to 300 Å and containing about 45 kbp. This thick fiber consists of a "solenoid", a continuous coiling of core and linker regions. Each solenoidal turn has a helix with 6 to 9 nucleosomes which include about 1200 bp of DNA.

In interphasic nuclei and metaphasic chromosomes, the thick fiber appears to be additionally folded into even more complex structures termed loops or domains, comprising 35 to 90 kbp of DNA. The chromatin structure is considered to be firmly anchored at a support structure in the nucleus — the so-called matrix or scaffold. The sequences of DNA that are tightly associated with the nuclear matrix *in vitro* have been called matrix or scaffold attachment regions (MAR_s or SAR_s). MAR_s often are close to enhancer-like regulator sequences and were revealed to be common at the boundaries of transcription units, where they may delimit the end of the active chromatin region. This structure is related to the active and inactive transcriptional DNA regions, the euchromatin and heterochromatin, respectively.[41] More specific structures, like the Balbiani ring genes are loops of active genes of about 37 kbp, which begin and end in tightly packed chromatin.[42] The nontranscribed upstream region is a thin, extended flexible fiber, without nucleosomes, of about 500 bp of DNA and the untranscribed downstream region appears as a nucleofilament of a 2000-Å length, loosely coiled into a thick chromatin fiber containing about 3 kbp of DNA.

The H1 histone participates in repression in the fundamental genome state and various activating proteins selectively induce the expression of genes either by true activation or by releasing the basal repression.[43,44]

The chromatin structure depends largely on nonspecific DNA-protein interactions.[45] The characterization of such nonspecific complexes appears to be more difficult than expected. Several new binding modes and unexpected cooperational structures were discovered. Proteins interacting with nucleic acids often have multiple functions. For instance, the *CAP* protein modulates both the transcription of operons and its own gene repression. The Phage T4 protein 82 interacts with other proteins in complexes which replicate DNA, but it can prevent the transcription of its own DNA sequence, too.

In the formation of DNA-protein complexes, the releasing of ions and water, accompanied by an increase of entropy, often appears to be a crucial factor.[45] The binding of anions to proteins also can contribute to the ΔG of complexation. Electrostatic effects are important for both specific and non-specific interactions and explain the negative effect of an increase in ionic strength on binding. The apparent contributions of hydrophobic effects are large, $\Delta G_{hf} = -70$ kcal/mol, as compared to $\Delta G = -15 \pm 1$ kcal/mol for complex formation between the *lac* repressor and the corresponding *E. coli* operator gene. These great values for ΔG_{hf} are too large to result from "burial" of nonpolar surfaces only in the protein-DNA binding site and must be dependent on global conformational changes in the protein as well.

IV. GENE REGULATION IN PROKARYOTES

A. INDUCTION AND REPRESSION

In bacteria and lower types of eukaryotes (e.g., yeasts), the most common type of gene regulation occurs by induction and repression.

Induction represents the phenomena that trigger the transcription of a gene which specifies an enzyme(s) when a special substrate appears in the environment. A substrate that induces the increased synthesis of the enzyme(s) is called an inductor. Contrarily, a substance which decreases or prevents the synthesis of the enzyme is called a co-repressor.[9] They are represented by small molecules which determine the rate of enzyme synthesis at the transcriptional level. Induction and repression represent elements of an exogenous control. The classic example is represented by the response of *E. coli* growing in the presence of lactose. This experiment underlies the induction and repression concept of gene regulation.

The inductor or the co-repressor acts by binding one of the proteins encoded by the regulatory genes. The regulatory proteins control the expression of structural genes, determining the timing and rate of the bio-synthesis processes and the amount of the enzymes thus formed. The transcription of the genes is controlled by proteins binding to specific sites of the DNA called *operators*. These are generally situated within, or in the vicinity of, the promoters.

The control of gene expression is a highly specific process and may be considered to be positive when the regulatory protein is an *activator* and negative when it is a *repressor*. From the great number of repressors discovered, the two best known are the transcriptional repressors: *lac* (the lactose operon repressor) and the λ repressor (the repressor of bacteriophage).[47,48]

The expression of *lac* genes (*lac Z,* encoding β-galactosidase; *lac Y,* encoding lactose permease; and *lac A,* encoding lactose transacetylase) of *E. coli* is controlled by negative regulation. The repressor encoded by the *lac I* gene prevents transcription of *lac* structural genes by binding to a DNA sequence that overlaps the RNA polymerase binding site from which RNA synthesis begins. This means that structural gene clusters are coordinately regulated.

Here we use the terms "induction" and "repression" in reference to the action of a chemical that can exert an exogenous control of gene activity (e.g., amino acid, lactose). The terms "repressor" and "activator" are used for the regulatory proteins responsible for these activities (repression and activation).

B. REPRESSOR IS A DNA BINDING PROTEIN

The repressor is a tetrameric protein of identical 38-kDa subunits containing two types of binding sites: the inducer and the DNA binding site.

The DNA binding site of the repressor recognizes the sequence of the operator (for the lactose operon, *lac O*), showing some preferences for AT-rich DNA. The repressor binds very quickly and tightly to the operator, indicating that the repressor finds the operator sequence by diffusing along the DNA strand. The dissociation constant of the complex formed is about 10^{-13} *M*.

Using dominant negative mutation analysis, it was clearly demonstrated that the helix-turn-helix portion of the repressor protein recognizes and binds to a specific DNA sequence.[12,49] *Lac O* lies near the promoter (*lac P*) and when the *lac* repressor is bound to it, the initiation of the structural gene transcription (*lac* operon Z, Y, A, genes) by RNA polymerase is prevented. This phenomenon takes place when the inducer is absent from the environment. But when the inducer is present, it binds to the inducer binding site of the repressor, which loses or reduces its affinity for the operator sequence and dissociates from the DNA template. During the following step, RNA polymerase interacts with the promoter and initiates the transcription. Such a mechanism is described as an *allosteric control.* When a regulatory protein undergoes a modification in its function subsequent to a small-molecule interaction, the cell achieves an economical way for control of a large number of gene expressions as a response to an environmental stimulus.[50] Thus, the activities of key proteins in bacteria are controlled by allosteric change and reversible covalent modifications.

Because the *lac* genes are under a single operator control, when the inducer is added to the growth medium, all these genes are expressed. This means that the control is cis-acting.

C. STRUCTURE OF THE OPERATOR AND ITS INTERACTION WITH THE REPRESSOR

The *lac* operator has a symmetrical sequence extending from -8 to $+28$ which lies between the promoter (*lac P*) and the cluster of structural genes (*lac* Z, Y, A) just as the subunits of the *lac* repressor. The symmetry in the operator structure is due to the presence of inverted repeats (or almost-repeated base pairs). This configuration permits the operator sequence to divide into two sides: the left and the right.[16]

The repressor is bound to the operator by two subunits in $+1$ to $+6$ and $+21$ to $+16$ positions. These two sequences appear to be symmetrical and they seem to be relevant to repressor-operator interaction. The interaction takes place between each repressor dimer subunit with each half-side of the operator.[51,52]

This model of symmetry (operator-repressor) appears to be a required motif for gene regulation. The symmetry of the protein binding sites and its dimeric character are justified by the inverted repeat structure of the DNA binding target achieving a more tightly bound interaction. This mechanism avoids the synthesis of an excessively long transcript and also decreases the probability of errors.[12] From a thermodynamic point of view, a similar entropy is required for positioning both dimer and typical monomeric proteins.

When the repressor is bound to its operator, the association between RNA polymerase and the promoter is physically hindered. The inducer binding decreases the repressor affinity for the operator sequence by about 10^3 times. This phenomenon is based on changes in the repressor conformation in the DNA binding region and leads to the dissociation of the repressor from the operator.[53,54] Consequently, RNA polymerase can bind the promoter and initiate the transcription.

This kind of regulatory mechanism occurs under negative control. However, for a normal function the *lac* operator requires a positive control signal, too.[55,56] This positive control signal was revealed when glucose was used in the experiment. In this case, the largest number of *lac* operators were blocked. A protein which controls the catabolism of β-galactosides, such as lactose, arabinose, and maltose,[53] was discovered and was called the catabolite activator protein (CAP). When the CAP is associated with cyclic AMP (cAMP) the complex gains the ability to bind to a specific DNA site, thus enhancing the transcriptional level of the adjacent genes.[57] The CAP binds to a 22-bp twofold symmetric DNA target site with extremely high affinity ($K_{obs} = 4 \times 10^{10}\ M^{-1}$).[49] Thus, CAP represents the positive control factor for all genes which are sensitive to glucose because its presence is necessary to switch on gene transcription. CAP binding favors the association of RNA polymerase to the promoter,[56] the process being dependent on the cAMP concentration. When the level of cAMP is low in the medium, the transcriptional level of *lac* genes decreases and vice-versa. However, the catabolite repression, in this case, is based on the ability of glucose to reduce and maintain a low intracellular concentration of cAMP.

D. MOLECULAR ASPECTS OF REPRESSOR-OPERATOR INTERACTIONS

In prokaryotes as well as in eukaryotes, the control of gene expression is exerted primarily at the transcriptional level. A central role in this complex process is the repressor-operator interaction.[57] Control systems have been extensively studied in *E. coli*. Several activator proteins and repressors have been isolated and characterized, as well as their corresponding DNA binding sequences.[58,59] Both types of proteins are oligomeric proteins free of RNA, which could be found in quantities of about 10 to 40 molecules per bacterial cell. For a cell volume of 10^{-12} cm^3, the corresponding concentrations are 10^{-7} to 10^{-8} *M*. Structural studies of three such proteins, the CAP protein of *E. coli* (involved in activation and repression of several genes), the Cro protein, and the λ repressor from the λ phage (involved in regulation of lytic and lysogenic states) exhibit some common features which were also found in other regulatory proteins.[59] Specifically, these proteins act as dimers with a pair of α helixes exposed to interactions. Each of these α helical regions belongs to highly conservative regions of these proteins. The specific binding to DNA is based upon the common binary symmetry of protein dimers and of the recognized DNA sequences. Hydrogen binding between side chains of the α helixes and bases located in the DNA major groove is essential in the recognition processes. The physiological transduction signal in the case of the cAMP binding to CAP protein takes place via conformational changes in the regulatory protein. The DNA binding sequences are about 17 bp long.

One of the best-studied regulator proteins is the λ repressor,[33,51] which is a tetrameric negatively charged protein with a 150-kDa molecular weight. Crystallographic studies of several prokaryotic DNA regulatory proteins indicate a general similarity in the three-dimensional structure, although there are also a few similarities in the general amino acid sequence. Each protein contains a two-α-helix motif, connected by a short nonhelical fragment. The two helical fragments fit well into two sites on the major groove of the β-type DNA double helix,[25] although similar arguments apply to other forms of DNA, such as the A or Z form.[18]

1. Structure of Regulator Protein-DNA Complexes

The *lac* operator has a certain symmetry, being formed of two equivalent segments with several sites arranged as VIa — Va — IIa — IIIb — IVb — IIIa — IIb — Vb — VIb.[33,60] There are 50 amino acids in the *lac* repressor which seem to be involved in the binding to the operator gene, for which the following sequences have been proposed:

$$\begin{matrix} 5' \\ 3' \end{matrix} \begin{bmatrix} \text{-C,A -T-A-G, T-C, A-}(X)_{\overline{n}} \\ \text{-G,T -A-T-C, A-G, T-}(X^{-1})_{\overline{n}} \end{bmatrix}$$

X stands for an unkown base and C,A refers to either a cytosine or an adenine found in the respective position from these sites. Sites IIa, IIIa, and IVb are formed from one AT pair, IIb from three GC pairs, and IIIb and IVa from one GC pair; there are also two supplementary sites formed from two AT pairs. As already mentioned, the length of the DNA sequence protected by combination with the *lac* repressor is 27 bp.[30] If the AT pair in the 13th position of the operon is substituted with TA or GC, the repressor binding is greatly decreased, while substitution with G $(CH_3)^5$-C does not affect the binding; this demonstrates the importance of the CH_3 group of C in the binding process.

The crystalline structure of the λ repressor-operator complex was determined by X-ray crystallography at a 2.5 Å resolution.[61] This complex is stabilized by an extended network of hydrogen bonds between the protein and the pentoso-phosphatic backbone. Several amino acid side chains form hydrogen bonds with specific sites from the major groove, while hydrophobic contacts also contribute to the specificity of bonding. The general arrangement of the complex is similar to that predicted by modeling studies which fit the proteic dimer on the linear B form of the DNA. The X-ray study also revealed important side chain/side chain interactions which were not predicted by previous modeling or by genetic and biochemical studies.

High resolution X-ray crystallography also was used to study[62] the recognition of a DNA operator by the phage 434 repressor. Direct interaction between the major groove base pair groups and the regulatory protein determines if the DNA will adopt the conformation required by binding. Three centered hydrogen bonds, bifurcated between pairs of bases, explain the binding of the phage 434 repressor to certain sites.

Mutant Trp repressors with new DNA binding specificities were obtained by oligonucleotide-directed mutagenesis of the codons for Gln^{68}, Lys^{72}, Ile^{79}, Ala^{80}, and Thr^{81} of the *E. coli trp R* (tryptophane aporepressor) gene and used in a study of binding requirements for this repressor-operator interaction.[63] The new binding specificities, thereby obtained, indicate that the HO group of Thr^{81} makes a specific contact with one of the four critical base pairs in a *trp* operator halfsite, while the CH_3 group of Thr^{81} determines a specific contact at a second critical base pair. The first two amino acids of its "recognition α helix", Ile^{79} and Ala^{80}, are not used by the Trp repressor to make specific DNA contacts. The Trp repressor-operator interaction takes place in a fundamentally different way from the way that the phage lambda repressor, λ *Cro* protein, and coliphage 434 repressor bind to their specific sites. The left half-site of the *trp* operator is

$$\begin{matrix} (11) & & & & & (6) & & & (3) & & (1) & \\ \left[\begin{matrix} T & C & G & A & A & C & T & A & G & C & T \\ A & G & C & T & T & G & A & T & C & G & A \end{matrix}\right] \text{Dyad} \end{matrix}$$

and the four critical base pairs extend from position (3) to (6).

Single and double loop formation takes place when the *deoR* repressor binds to its natural operator site, as demonstrated by electron-microscopy studies.[64] There are three operator genes in this DNA, and by the binding of the *deoR* repressor oligomer to two of the three operators, single loops of 280, 600, and 880 bp, respectively, are formed. When the protein binds simultaneously to two operators, a double loop of 280 and 600 bp is also formed. This type of binding agrees with the cooperational effect of the three operator sites involved in repression (Figure 6). While the early studies of repressor-operator interactions have indicated a possibly similar molecular

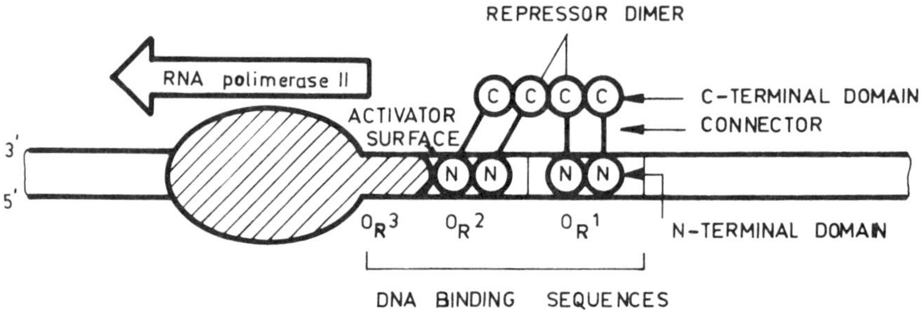

FIGURE 6. Interactions between a suppressor dimer with the O_R1 and O_R2 site of the promoter and with RNA polymerase in the lambda right control region. Under this condition the RNA polymerase starts transcription of the genes which lie at the left side of the O_R3 site.

mechanism,[25,59] further studies have demonstrated a wider diversity than was initially assumed.

2. Kinetic and Equilibrium Studies: Effects of Mutations

The *lac* repressor-operator interaction was studied both from the kinetic and thermodynamic point of view by Riggs and co-workers, using a highly sensitive membrane filter technique and tracer compounds.[4,65] The association-dissociation kinetics obey the second-order rule:

$$\frac{d[RO]}{dt} = k_a([O] - [RO])([R] - [RO]) - k_d[RO] \tag{4}$$

where k_a and k_d are association and dissociation rate constants; and O, R, and RO represent the concentration of the three-operator gene, free repressor, and repressor-operator complex,[66] respectively.

The association rate, $k_a = 7 \times 10^9\ M^{-1}s^{-1}$ is higher than the one calculated from diffusion data ($10^8\ M^{-1}s^{-1}$) and decreases with an increase in ionic strength (corresponding to a product of charges $Z_o \cdot Z_r = -6$). The high k_a value, superior to the diffusion rate, is due to the electrostatic attraction and not to a nonspecific repressor to DNA binding followed by sliding to the operator. This explanation is supported by the fact that the cleavage of DNA does not influence the association kinetics. In contrast, the dissociation equilibrium constant decreases strongly when the operator is cleaved out of *E. coli* DNA.

The dissociation equilibrium constant of the repressor under intracellular conditions is $K = k_a/k_d = 1 \times 10^{-13}\ M$; and the dissociation rate is $k_d = 8 \times 10^{-4}\ s^{-1}$. For dissociation from *E. coli* DNA without an operator, K is in the range of $10^{-3}\ M$. Thus, at a 10^{-7} to $10^{-8}\ M$ intracellular repressor concentration the repressor binds only to the operator. Thermodynamic data for *lac* repressor-operator equilibriums are listed in Table 1, according to Record and Spolar.[45]

TABLE 1
Thermodynamic Data for *lac* Repressor-Operator Interactions

Site	K(*M*)	G (kcal/mol)	H (kcal/mol)	T S (kcal/mol)	Z_R, Z_o
0^+	5×10^{-12}	-15 ± 1	14 ± 5	29 ± 6	-7 ± 1
0^c666	5×10^{-9}	-11 ± 1	4 ± 2	15 ± 3	-7 ± 1
0^c640	11×10^{-8}	-11 ± 1	16 ± 5	27 ± 6	-11 ± 1
0^c630	$>5 \times 10^{-4}$	>-10	—	—	—
DNA	1×10^{-4}	-5 ± 1	-5 ± 2	0 ± 3	-12 ± 2

Note: 0^+, 0^c, DNA — *lac* operator, different mutants, and *E. coli* DNA without operator, respectively; Z_R, Z_o — electric charge product for repressor and operator. Conditions: 23°C, pH = 7.4, ionic strength = 0.2 *M*.

The complexation process is endothermic, the increase of entropy being the driving force, due to the release of water molecules. The affinity difference between the repressor and operator, respectively and the nonoperator DNA is $\Delta\Delta G = 10$ kcal/mol.

The *lac* repressor also binds to renatured DNA, which indicates that the operator is very likely thermodynamically stable in the double helix state. This fact and the high association rate, k_a, higher than the diffusion rate, precludes a conformational transition during the repressor-operator interaction and makes unlikely the mechanism proposed by Gierer.[67] According to this mechanism, the specificity of the DNA-protein interaction is due to a particular structure (nondouble helical) of the binding DNA sequence.

The *lac* repressor has a rather high affinity for poly-dAT, $K = 10^{-9}$ *M;* its affinity for polynucleotides decreases according to the sequence: dAT > dA.dT > dT > dG.dC. In these interactions, the affinity also decreases with the increasing ionic strength, because of the electrostatic interactions.[66] Point mutations in the *lac* operator increase 8 to 100 times[33,60] the dissociation equilibrium constant, K; a fact that corresponds to a mean affinity decrease of $\Delta g = 2.1$ kcal/mol, or $r = 0.027$, per base pair substitution (see Chapter 1, Section II.B.3 for Δg and r). The wild type of the *lac* repressor is not the one that has the highest known affinity for the *lac* operator gene. The constitutive *lac E. coli* mutant produces a repressor whose dissociation constant, from the complex with the *lac* operator, is four times lower than for the wild-type repressor.[6]

V. GENE CONTROL IN EUKARYOTES

Eukaryotic cells have a different and much more complex cellular organization than do prokaryotes. The cell contains a nucleus in which DNA exists in the form of chromatin,[38] well separated from cytoplasm (see Section III.G).

Most of the DNA sequences are structurally inaccessible and functionally inactive, but the type of organization of the eukaryotic genome enables some sequences to come close to each other in chromatin, thus forming accessible recognition sites for large protein or environmental factors.

Generally, eukaryotic genes display a wide range of lengths of up to 500 kbp, but most often are in the 200-kbp domain. Recently a gigantic gene was discovered and cloned: the NF-1 gene has more than 2000 kbp and is inactivated in neurofibromatosis or Von Recklinghausen disease.[69]

Contrary to bacterial mRNA, the eukaryotic mRNA is monocistronic, encoding a single polypeptidic chain. Nevertheless, various splicing patterns of the mRNA are possible, so that more than one protein could be synthesized.

Active genes in transcription are found in euchromatin, while the inactive ones are located in heterochromatin. Active genes contain a short DNA sequence of 50 to 400 bp which are usually — but not always — located before the transcription initiation start point and have up to 10 times increased sensitivity to DN-ase I digestion. These sensitivity and hypersensitivity sites, free of nucleosomes, seem to be correlated with gene activation.[70] The exclusion of nucleosomes from these short DNA regions is accomplished by an unknown protein that probably binds to the initiation transcription sites and to the DNA short sequences, modifying the local conformation.

Eukaryotic chromatin also contains tissue-specific nonhistonic proteins. As demonstrated with reconstituted chromatin (nuclear DNA to which nonhistonic and histonic proteins are added), the mRNA transcribed by *in vitro* systems has a similar sequence to the mRNA transcribed *in vivo*. The nonhistonic protein must be added first and histones afterwards. If this order is reversed, the constituted chromatin will be transcriptionally inactive. Such nonhistonic proteins are DNA and RNA polymerase and several ancillary proteins involved in the regulation of growth, differentiation, replication, transcription, etc.

A. EUKARYOTIC GENE CONTROL ELEMENTS

The regulation of specific gene expression involves the synthesis of RNA representing one of the DNA strands. The transcription process occurs by nucleotide incorporation into RNA molecules catalyzed by RNA polymerase, and it is regulated by cis-acting DNA sequences. A gene or transcript unit (which may include one or several genes), also includes the control sequences by which the gene expression is regulated. Taking into account only the main stages of transcription (initiation, elongation, and termination), we can see how many points are available for regulation of gene activity.[71]

Let us consider two types of control sequences for the positive and the negative control of gene expression. The control sequences involved in positive control include promoters and enhancers. In contrast, the control sequences for negative regulation are known as de-enhancers, attenuators, or silencers.

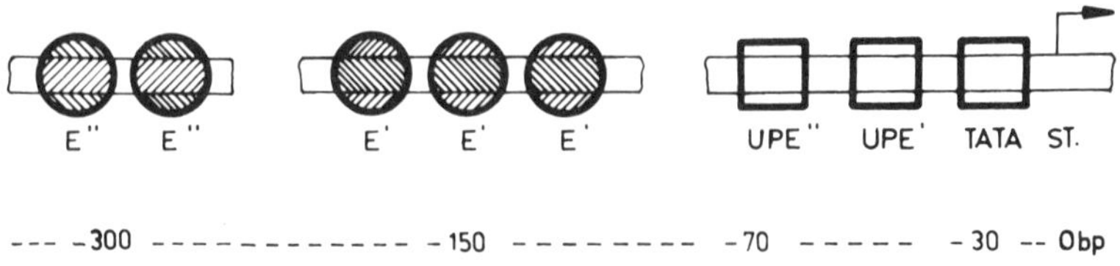

FIGURE 7. Spatial organization of control sequences (elements) of eukaryotic genes: St = RNA transcription start-point; TATA = the TATA box; UPE′ and UPE″ = upstream promoter elements. The numbers indicate distances in base pairs upstream from the transcription start point.

Promoters are specific sequences recognized by the RNA polymerases, and are located in the immediate upstream of the transcriptional startpoint (+1), which represent the minimal DNA elements capable of directing gene transcription. They are responsible for the precise and efficient initiation of the transcription. Promoters may be constitutive or regulated. The regulation of transcription can be positive or negative, and control is effectuated by the binding of proteins (activators or repressors) to the sequences within or close to the promoters. Contrarily, enhancers are regulatory elements involved in activation of the rate of gene transcription. They can increase the transcription rate given by promoters by up to 10^3 times. The enhancer acts cis-linked to the promoters, and can lie further upstream or downstream from the transcription startpoint. Both control sequences act in the presence of regulatory proteins by binding them to these sequences. The typical organization of this control sequence is depicted in Figure 7. The difference between promoters and enhancers is rather an operational one.

A typical promoter region generally has a first recognition sequence, called the TATA box (or Hogness box), between −19 and −27 bp upstream from the startpoint required for a correct initiation of the transcription. Somehow, identical elements to enhancers called upstream promoter elements (UPEs), each of 8 to 12 bp in length, are found in yeasts. The TATA box is similar to the Pribnow box found in bacterial promoters. The difference appears to be with respect to the position of the initiation startpoint: this is at −25 for the TATA box instead of −10 for the Pribnow box. In mammalian systems the TATA box appears to be strongly dependent on the sequences further upstream. Some accessory factors were identified to bind to them. The position of these sequences varies for different promoters, but they are located more than 40 bp upstream from the TATA box. When the TATA box is deleted, transcription initiation can occur, but the startpoint of initiation can be different from its usual location. In addition, the sequences around the TATA box appear to be important for an accurate initiation.

For an accurate initiation process, two other separate sequences are required: a middle region situated between −50 and −70, and a distal one which lies between −80 and −110. The frequency of the transcription se-

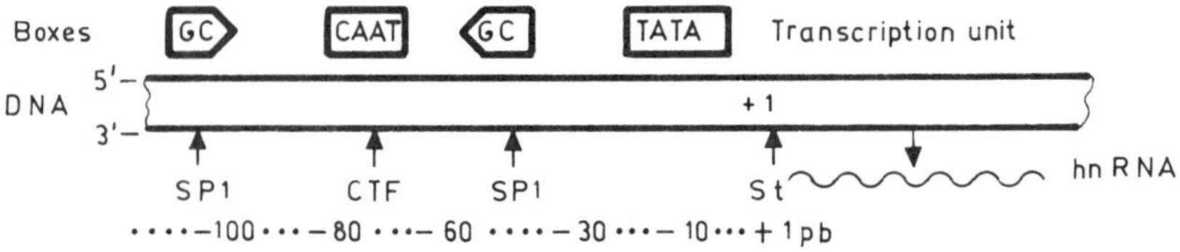

FIGURE 8. Different types of boxes in the control regions of genes: TATA, GC, and CAAT and different DNA binding elements for specific regulatory proteins (St = start point of transcription).

quences is strongly dependent on these upstream middle and distal sequence elements. RNA polymerase can bind directly to these control sequences which could assist the enzyme in initiating the transcription at the proper site, rich in AT sequences. The sequences neighboring the initiation point are also required to hold the RNA polymerase in a proper configuration, permitting it to recognize the precise startpoint. The sequences further upstream are more important for the binding of the enzyme.

There are also other boxes beside the TATA box. The CAAT box is found in the promoters of thymidine kinase (TK) and β-globin genes and seems to be responsible for the transcription rate. Another box, GC, appears in multiple copies and in either orientation in the promoter. In the TK promoter, all these elements could be found (Figure 8). The TATA box is located upstream from the startpoint; the GC box in the middle of the region of the promoter; the CAAT box and eventually a second GC box are present in the far region of the promoter. However, there are promoters lacking some of these elements.

The promoter is considered as an essential sequence for binding the transcription factors and RNA polymerase, but the ability of the promoter to initiate and transcribe the gene can be highly activated by the presence of the enhancer. The location of this sequence (usually of 72 bp repeats) can vary substantially and can function in either orientation. There are also some similar elements in yeasts, the so-called upstream activator sequences (UAS).[72] These UAS, which are located at a variable distance upstream of the promoter, may function in either orientation. The UAS have a regulatory role in the *gal* system where the regulatory protein is bound to UAS and activates the genes.

Multiple binding sites can exist within an enhancer region. The SV-40 early transcript gene enhancer has three DNA regions, each of them being required for its functionality: a region upstream from the transcription start, a 5′ binding site for the SP_1 protein (a transcription enhancing factor), and afterwards a repeated B — A — B — A sequence binding two different proteins, A and B. For the integrated mouse mammary tumor virus (MMTV) gene, the transcription is stimulated by steroid hormones and regulated by the glucocorticoid receptor protein (GRP). There is a group of three GRP binding sites at about 100 to 150 bp upstream from the RNA startpoint and another group of two GRP sites at about 300 bp. This multiple nature of control sequences seems to be common for the sites involved in the interaction of DNA with regulatory protein — interaction required for the active transcription of genes.[73,74]

Two mechanisms were put forward to explain the transcriptional stimulation of RNA polymerase II genes: upstream promoters and enhancer elements. In the looping model, the upstream enhancer interacts via the activating protein, with the proximal promoter. In the scanning model, a transcriptional factor binds to the upstream recognition elements and scans along the DNA until it reaches the promoter. The discovery of *in vivo* transcription stimulation by the SV-40 enhancer, noncovalently attached to a β-globin region via a streptavidin molecule, favors a looping model.[75]

Gene transcription is controlled by more than a single element. For instance, regulation of the immunoglobin genes requires at least three distinct types of control sequences: an enhancer specific to B cells (i.e., activated by a B cell protein), a lymphoid specific promoter or enhancer, and some intragenic sequences involved in posttranslational control. The B cell-specific enhancer consists of five distinct binding sites (O, E1, E2, E3, and E4), four of them binding the same B cell specific factor. Deletion of any of these sites decreases the enhancer activity but does not abolish it. For *Drosophila* genes, there are several different enhancers controlling the same gene; different enhancers are active in different tissues. For instance, the maximal expression of the alcohol dehydrogenase *Adh1* gene requires the activation of at least three DNA sites; a lower level of gene expression in other tissues requires the activation of only two.[71]

In *S. cerevisiae* yeasts the *ADR1* gene encodes a protein which plays the role of a positive transcription activator for the glucose-repressible *Adh2* gene.[76] The two zinc finger domains of the regulatory protein have shown that it binds upstream of both UAS1 and UAS2 sequences of the *Adh2* gene.

The promoter of the mouse albumin gene has at least six binding sites for no less than two different proteins.[77] For the rat albumin gene, at least four factors that interact with the promoter of this gene are known, one factor being similar or identical to the nuclear factor NF-1.[41]

A special regulatory transcription appears in the yeast mating type locus (MATa and MATα) which can include both positive and negative regulation.[9] Many strains of yeast can propagate either in haploid or diploid conditions. The passage between these conditions occurs by mating (fusion) and by sporulation (meiosis of the diploid to give haploid spores). Haploid types may be either MATa or MATα. Opposite type cells can mate, but the same type cannot. How does this rapid conversion or switching happen? In the proposed cassette model, two additional loci, *HML* (left) and *HMR* (right) were identified on the same chromosome containing the MAT locus. Thus, the MAT locus has a MAT active cassette of either α or *a* type. HML and HMR appear to have two cassettes (α and *a*, respectively) called silent cassettes. Silent and active cassettes have distinct regions (*Ya* or $Y\alpha$) which are flanked on either side by regions that are virtually identical (X or Z_1).

Both MAT*a* and MATα encode proteins which can activate or repress the expression of these genes. In α_1 cells, the specific genes for α are induced by the protein α_1. In *a* cells the expression of specific genes (MAT*a1*) which encode the protein *a1* occurs in a constitutive manner and the genes for α_1 are blocked. Prevention of the α_1 gene expression is accomplished by a repressor protein encoded by a gene within the MATα locus, the so-called α_2 protein. The α_2 protein binds to the operator and represses the specific genes responsible for producing a mating type. In diploid cells, genes appear to be silent. A third regulatory protein encoded by the MAT locus is required to start the meiosis or sporulation, and is called protein *a1*. But this protein needs the presence of the α_2 protein. Now we can see that the activity of *a1*-α_2 are involved in the regulation of many genes from *a*/α cells. The *a1*-α_2 complex represses the transcription of the RME1 gene which, in fact, transcribes other repressor proteins for meiosis and sporulation. These types of regulation activities (α_1, α_2 and *a1*-α_2) encoded by MAT genes are relevant for a new class of regulatory protein which is nonresponsive to the environment. Here, the locus for the yeast sexual type appears as the master locus in regulation of the gene expression.[71]

HML and HMR expression was identified to be dependent on four loci, the so-called silent information regulation (SIR). In (SIR), wild-type HML and HMR appear to be in a repressed status. If a mutation is induced in one of these four SIR loci (sir^-), both silent cassettes can be switched on and the transcription of both HML and HMR alleles starts. Thus, the same regulatory event can repress the silent cassette and can prevent it from being replaced by another cassette (Figure 9).

For the mating type switching process, a cell requires the HO endonuclease that makes a double-stranded cut at a specific DNA site in the MAT sequence locus. The DNA near the cut is then excised and replaced by a copy of the silent cassette of the opposite mating type (Figure 9).

One of the most puzzling problems is the location of the promoter for the genes encoding the 5S RNA and tRNAs. It was assumed that they were to be found upstream of the initiation startpoint. On the contrary, these promoters were discovered to lie within the gene about 55 bp downstream of the RNA initiation startpoint. For instance, using mutation procedure for 5S in *X. laevis,* the promoter was identified between positions +55 and +80 within the gene. Particularly for tRNA, the promoter consists of two separate regions of about 20 bp located between (+8, +30) and (+51, +72) in the reading frames of the gene.

However, there are some C genes which are controlled by regulatory sequences situated upstream from the transcription startpoint of the gene.

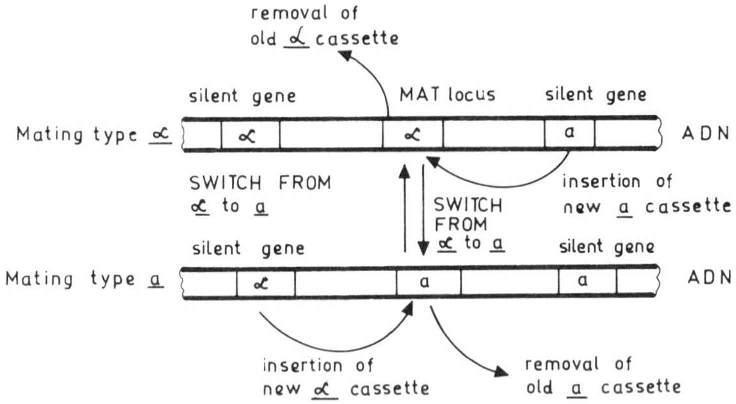

FIGURE 9. Cassette model of yeast mating-type switching. The mating-type switching process is triggered by H endonuclease, which induces a double-stranded incision at a specific DNA sequence in the MAT locus. The old α (or *a*) silent cassette is of the opposite mating type.

These regulatory elements seem to be similar to those which transcribe mRNA by polymerase II. Recently, it has been reported for both U6 and 7SK genes that they contain a sequence similar to the TATA box, situated 30 bp upstream from the transcription startpoint. The TATA box was demonstrated to be necessary for an efficient transcription in humans of U6 and 7SK RNA genes by polymerase III (or pol C). In fact, in all U6 genes of vertebrates and yeasts, the TATA box (at position −30) was found, and it was supposed that it can not be restricted only to these genes.[79] This special type raises the problem of recognition of these promoter sequences situated in unusual places. The binding of RNA polymerase III to each type of promoter requires different ancillary proteins, too.

A model for a silencer was suggested from the search undertaken for the gene encoding the antigene receptor of T cells. The genes of the T cell receptor (TCR) encode for $TCR\alpha\beta$, $TCR\gamma$, and $TCR\gamma\delta$ proteins. In transgenic mice (Tg) and γδ T cell hybridoma systems it was found that the γ and δ Tg gene are transcriptionally silent in αβT cells, and the repression of γ gene expression appears to be mediated by a γ silencer (cis-acting DNA element) located in the flanking sequence of the gene.[80]

To sum up, each gene of eukaryotic cells is controlled by several DNA elements (promoters, enhancers) responding to different activating factors. Enhancer elements have several binding sites, the maximal activating effect being obtained when all sites are occupied by specific proteins.

B. TRANSCRIPTIONAL FACTORS

The control of gene expression in the eukaryotic cell frequently takes place at the DNA transcription level. It was demonstrated that specific transcription initiation by RNA polymerase needs the coordinated action of the multiple regulatory proteins which activate or repress this process.[81,82]

In this category of ancillary proteins we can operationally distinguish two groups: general transcriptional factors and upstream element-binding proteins.[83] The proteins which compose the transcriptional apparatus are involved in several steps of genes activities, namely:

1. Commitment of the DNA duplex after binding the transcriptional factors;
2. Apparition of the active state of the gene under the action of other transcription factors and RNA polymerase II;
3. Accomplishment of the required energy (hydrolysis of the β-γ phosphate bond of either ATP or dATP);
4. Starting of the transcription (initiation start); and
5. Elongation of the transcript.

One may see that the existence of various steps within the complex process offers many opportunities for regulation of gene activities.

The transcription factors (TF) are proteins which bind to specific DNA sequences of the promoters. These consensus sequences can be found in all promoters, and in this case the factor would be a part of the general transcription apparatus, or it can be present only in certain groups of promoters and the factor would be specific. We use the term transcription factor for RNA polymerase and other regulatory proteins (ancillary proteins) involved in the initiation and regulation of *in vitro* DNA transcription. These proteins could be considered part of the transcriptional apparatus.

The cells of higher organisms have three types of RNA polymerases whose function requires several ancillary proteins for gene recognition, as well as for other processes. RNA polymerase I transcribes the 45S rRNA precursor; RNA polymerase II transcribes the various mRNAs; and RNA polymerase III transcribes the small RNAs (tRNAs, 5S RNA) and several viral RNAs.

Molecular mechanisms of transcription initiation by RNA polymerase I on the ribosomal RNA gene involve at least two steps: the binding of the factor termed TFID to the promoter sequence, which is followed by the binding of a second factor, TFIA. After these events, RNA polymerase I repeats the transcription several times. Using a human-mouse chimeric gene constructed by DNA recombinant technology, the TFID recognition site was located in the -12 to -40 sequence of the promoter (a region protected by partially purified TFID against DNase I digestion).

RNA polymerase II has the most complicated regulatory system. The transcription factors TFIIA, TFIIB, TFIID, TFIIE, and TFIIF are required for an accurate and efficient initiation of the adenovirus major late (*AdML*) gene, the *Drosophila* heat shock protein (*hsp*) gene, and the histone *H4* and *H2B* genes. Fingerprint analysis demonstrated that the B factor for *Drosophila*

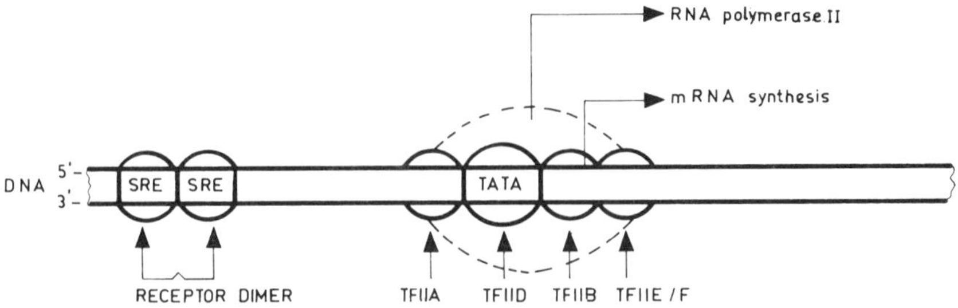

FIGURE 10. A general schema for basal expression of the gene. The transcriptional factors bind to the core promoter (upstream and downstream); *SRE* is an enhancer responsive to a serum factor.

and human factor TFIID are, in fact, TATA box binding proteins, but from one gene to another the fingerprint appears to be different. In the case of the *H4* gene, activation by TFIID requires a downstream sequence. The binding of these factors occurs at promoters and not at the RNA polymerase II level for basal expression of the gene (Figure 10).

Environmental studies[81] have indicated the existence of three hierarchical regulation types in the transcriptional control: basic (common), species specific, and tissue specific. These three kinds of regulation were revealed in a fibroin gene system. For instance, using a HeLa extract, the fibroin gene was transcribed, and in this system only the TATA box was necessary. This was proved by using a mutant gene having a deletion up to −31 bp upstream which still remained active in transcription. In order to have a more efficient transcription of the fibroin gene in other extracts (ovarian tissue, middle silk gland, embryonic cell line, and silkworm) another sequence centered at −72 bp is required. A more efficient transcription was considered to be supported by species specific and constitutive factors. For tissue-specific enhancement, a further upstream region (at −238 bp) appears to be necessary. The sequence −73 to −238 bp could be a tissue-specific enhancer-like element.

In yeast products transcribed by *GAL4* and *GAL80,* these genes are involved in the regulation of three other genes encoding the proteins: *GAL1, GAL7,* and *GAL10.* Protein encoding by *GAL4* appears to be a positive regulatory factor for the three mentioned genes. In contrast, the *GAL80* protein acts as a negative regulator.

The yeast transcription factor *GAL4* is a 99-kDa protein that binds with its terminal 74 residues to a specific sequence upstream from the *GAL1, GAL10,* and *GAL7* promoters.[84]

In the absence of galactose, the activating action of *GAL4* is inhibited by the *GAL80* factor, an inhibition which is reversed when galactose is added to the medium.[85] In this case, it is supposed that *GAL80* is not a true DNA binding protein but rather it prevents the association of *GAL4* to the DNA sequence.

Instead of the term transcriptional factors, several authors use the term nuclear factors (NFs), such as *NF-A1, NF-A2, NF-kB*. For instance, *NF-kB* exists in early pre-B cells, but it does not bind to the enhancer except after treatment with lipopolysaccharides (LPS) or phorbol esters (e.g., 13-*O*-tetradecanoylphorbol acetate, TPA), which induce the expression of the k gene through pretranscriptional modification. In B, T, or even HeLa cells treated with TPA, *NF-kB* appears to be activated, suggesting that the phosphorylation by the protein kinases C (PKC) could be involved in the process. The regulation of the *myc* gene is also relevant: a negative control element (or silencer) between -100 and -353 bp and also a positive control sequence having two *NF-i* binding sites were identified, both located near DNase hypersensitive sites.[81]

Recent data emphasize that a highly conserved upstream cis-element, the interferon consensus sequence (ICS), is involved in interferon (IFNs) transcription of the major histocompatibility complex (MHC) class I. A large number of nuclear factors that bind to the ICS are in fact either constitutionally transcribed or IFNs induced. The IFN regulated genes have a very similar motif to a part of the ICS. It was demonstrated that IFN genes and IFN-regulated genes are controlled by a shared cis motif. The IFN and ICS bind the same transcriptional factor (ICS binding protein: ICSBP). The ICSBP belongs to the gene family of interferon response factors (IRF) which plays a key role in transcriptional control of IFN regulated genes and also in gene regulation of cells involved in the immune response.[86]

In the mouse B cells, the (MHC) locus contains four co-expressed class II genes (Aα, Aβ, Eα, and Eβ). The expression of these genes is under the control of one dominant locus which encodes a cross-species trans-acting factor.[87] These are involved in eukaryote gene activity. Thus, a large number of these proteins are able to recognize the two boxes (X or Y) located upstream from class II genes of human and murine cells. It was demonstrated that the protein which binds the gene Aα-X box is different from the proteins that bind to Eα- and Eβ-X boxes. In several patients with immunodeficiency diseases, class II genes are not transcribed. In the cells of these patients, the protein which must bind to the X box is missing. We can mention that the control sequences for eukaryotic genes are subject to transregulation by multiple distinct regulating proteins. Liou et al.[87] consider that the multiplicity of the X box binding proteins, described above, is reminiscent of a large number of other nuclear factors such as the proteins that bind the *CAAT* motif, the Kruppel family of zinc finger proteins, and the *Drosophila* homeobox proteins.

Multiple regulation regions of a gene could be under the control of multiple trans-acting factors. In human hepatocellular carcinoma proteins cis-activated by an *HBV* promoter (or enhancer) insertion cannot be accepted as a general mechanism of cell transformation. The mechanism of trans-activation was suggested to be more plausible. This mechanism is supported by the discovery that the *HBV* encodes a transcription trans-activator factor which induces its own promoter and many others heterologous promoters in transfected cells. It was established that a *preS/S* sequence of *HBV* contains an enhancer-like element which is responsible for this stimulatory effect. This sequence appears to be essential for the biological functions of the *HBV* (e.g., regulation of virus assembly, its involvement in the attachment of the *HBV* to the hepatocyte, and regulation of the genetic restriction of the immune response).[88]

Under some conditions, these regions bind different factors, or a factor can bind to different regulatory regions. Evidence is available for the cooperative binding of transcription factors to such control sites,[54] for instance, the binding of the heat shock response element (HSE) in *Drosophila*. The binding of two dimeric HSTF molecules to both HSEs of the hsp 70 promoters is required to activate the transcription of the heat shock gene.[89] For the B cell enhancer elements discussed in the previous section, protein factors binding to three of the five sites also exist in other cells, but their presence is not enough for activation; the sites may be inaccessible in other cells or the protein factor must be activated before binding. The HeLa cell protein TEF-1 binds specifically and cooperatively to two enhancer motifs of SV-40.[90] The GRE/PRE DNA minisequence, responsible for gene induction by glucocorticoids in rats, and by progesterone in chickens, binds in a cooperative manner to two receptor protein molecules.[91] The AP-1 DNA sequence binds either a homodimer c-Jun protein or a heterocomplex of one c-Jun and one c-Fos protein molecule.[92] The human estrogen receptor, in the presence of estrogen, also binds as a dimer to its specific DNA binding sequence.[93]

The transcription of *Xenopus* 5S RNA genes in *in vitro* system requires a protein (37 kDa) which was called TFIIIA factor which binds at position +45 to +96 within the gene. For transcription *in vitro* of the U6 gene by polymerase III, at least two factors are required: TFIIIB and TFIIIC. In addition, it seems that *in vitro,* the transcription system also needs the TFIID. In this case it was assumed that TFIID may favor the association of TFIIIB, which by itself cannot bind to the DNA sequence. Contrarily, in the transcription of mRNA by polymerase II, the TFIID that bind to the TATA box ensures the formation of a preincubation complex which favors the binding of TFIIB.[94]

In human cell lines, the initiation may depend on three factors: A, B, and C. In yeast, only B and C factors are needed to transcribe the tRNA. In fact, B and C factors form a preinitiation complex which could be considered as a signal for the gene to pass in an active form. This type of initiation of gene transcription through a preinitiation complex could be considered as a general regulatory mechanism. The complex is stable and, by binding to the promoter, favors the binding of RNA polymerase, thus switching the gene on. By remaining bound to the promoter, the complex could stimulate many successive rounds of gene transcription. In this case, the zinc finger DNA binding motif is involved as the mechanism of protein-DNA recognition.

Several proteins with putative zinc finger domains have been identified in many nucleic acid binding proteins and they were considered to be essential for eukaryotic gene regulation.[95]

We can state that a gene may be regulated through the recognition of a promoter sequence by a specific transcriptional factor. The transcriptional factor is needed by RNA polymerase in order to initiate the gene transcription. However, a promoter may be regulated in various ways. These alternative pathways could be explained by the binding of each specific factor to a particular sequence of the promoter. To illustrate this principle, we can mention that one of the *Drosophila* shock genes is transcribed in two different conditions: at high temperature it is expressed in all tissues, and at normal temperature only in the ovarian tissues. Sequential analyses have shown that the region from −49 to −85 is required for gene expression during heat shock and the region upstream of −341 is needed for expression of the gene in ovarian tissue.

C. CELL TRANSFORMATION AS AN ALTERATION OF CONTROL MECHANISMS

The fundamental problem in oncogenesis lies in the elucidation of the malignant transformation and the mechanisms which underlie this process.

A number of experimental data support the assumption that cancer arises through alteration of the cell genome information. The cell transformation also could be considered as an alteration of the control mechanisms involved in cell division and differentiation.[96] The alteration appears as a consequence of various genetic and epigenetic mechanisms,[97] including inherited defects in certain genes.[98-100] This leads to a permanent and anarchic multiplication of transformed cells, according to a kinetic law, quantitatively different from those of the normal cells from which they originate.[101] The inherited transmittible modifications (from the neoplastic cell to its daughter cells) represent the final stage of a series of genetic, biochemical, morphological, and biological events, originating in a rare initial event at the molecular level, whose nature is not yet fully understood.

Epidemiological analysis and experimental data have confirmed that the natural history of human and animal cancers is usually a multistep process in its evolution and a multifactorial one in its etiology. At least three steps can be theoretically distinguished:[102] (a) initiation, (b) promotion, and (c) progression.

The only known exceptions for this multistep process are found in experimental oncovirology. Our desire is to briefly comment here on the significance of these steps in carcinogenesis (Figure 11) and the types of genes supposedly involved.[103-105]

1. Initiation Step

Concerning the initiation, DNA damage has been axiomatically claimed to be the key step in carcinogenesis. During this complex process, multiple

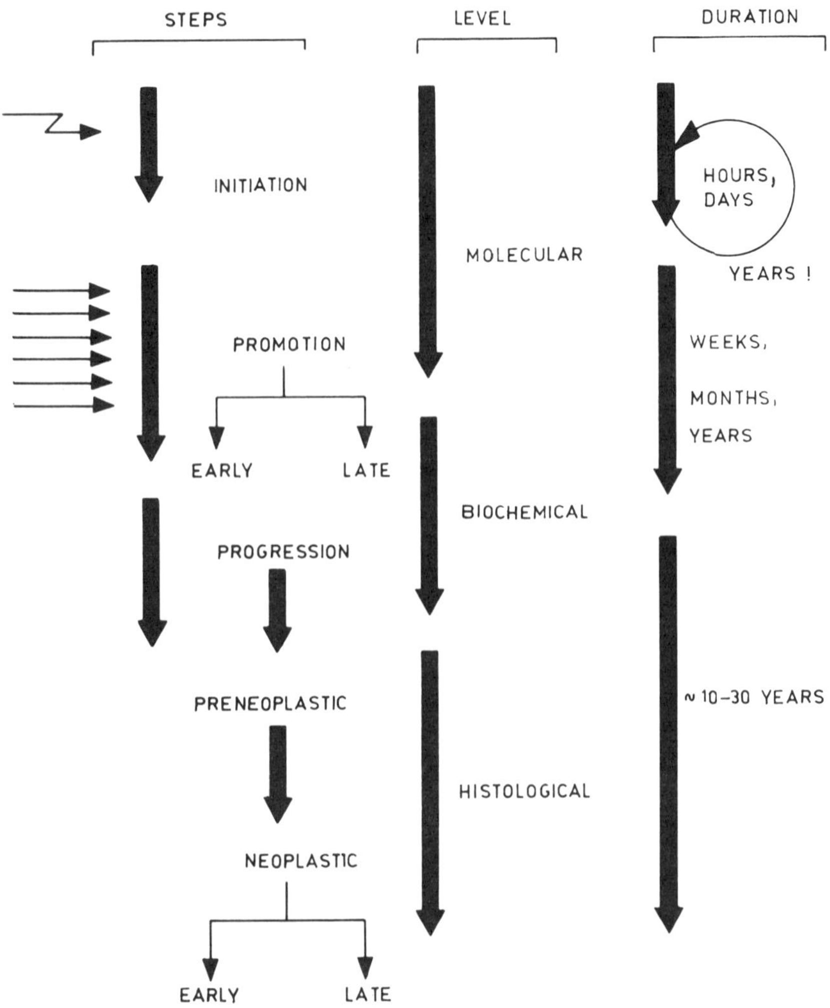

FIGURE 11. Schematic representation of the carcinogenic process.

modifications are produced in DNA structure and/or functions. They are reflected by important changes in the expression of a number of genes[105-108] — some of them still unknown — that contribute to cellular growth and differentiation.

It is generally accepted that the molecular substrate of malignancy lies in the genome of every normal human or animal cell.[109-111] During the last decade, the discovery of proto-oncogenes gave a powerful impetus to carcinogenic research, casting a new light on human cancer.[112-115]

When oncogenes isolated from human or animal DNA (spontaneously, chemically, or radiation induced) were transfected *in vitro,* in *HIH 3T3* cells

TABLE 2
Oncogenes Identified in Chemically Induced Tumors in Rats

Chemical agent	Induced tumor	Oncogene	Oncogene-containing tumors (%)
NMU	Mammary gland carcinoma	H-*ras-1*	86
DMBA	Mammary gland carcinoma	H-*ras-1*	23
DMN	Renal carcinoma	H-*ras-2*	40
MNU, ENU	Neuro/glioblastoma	*neu*	100
NMU	Schwanoma	*neu*	70
NMS	Nasal carcinoma	—	100

(a mouse fibroblast cell line), the latter acquired several features of malignant cells. These active transforming genes, once identified, were subsequently cloned using DNA recombinant technology and their sequences were established. It was clearly demonstrated that these transforming genes, called oncogenes, are modified versions of normal proto-oncogenes.

Cellular proto-oncogenes represent a direct target for chemical carcinogenic agents[103] and are also the sites for proviral or viral integration.[108] The modifications that can be induced may underlie the initiation step in the carcinogenic process (Table 2).

The first oncogene isolated from human tumor cells was a point-mutated cellular Harvey *ras* gene (H-*ras*) homologous to a retroviral oncogene.[111] The *ras* proto-oncogene suffers a single point mutation at either 12, 13, or 61 (and 59 or 63) position and it becomes an oncogene. It encodes a protein (with a molecular weight of 21 kDa, the so-called p-21) with a single amino acid substitution. Thus, because a point mutation occurs at a specific codon, it leads to the transcription of structurally and functionally altered proteins.

Among the proto-oncogenes activated by a single point mutation are cellular genes of the *ras* family (H-*ras,* K-*ras,* and N-*ras*). Identical point mutations were identified, not only in established human tumor cells lines, but also in human surgical tumor specimens, as well as in various types of animal tumors (chemically or radiation induced) (Table 3), and even in transduced viral genes.

Chemical carcinogens administered in rodents induce point mutations in *ras* proto-oncogenes, often in a specific manner. For instance, DMBA induces mutations through A to T transversion and ENU through G to A transition. For DMBA, the frequency of mutation is time dependent rather than dose dependent. Although such activated *ras* genes were identified in almost all types of human tumors, the frequency of these mutated genes is not high in human cancer (approximately 15 to 20%).

In addition, besides point mutation, the modulation of the functionality of proto-oncogenes may be induced by (1) an overexpression of these genes (Table 4); (2) an increased specific activity of some proteins transcribed by

TABLE 3
Oncogenes Identified in Chemically or Radiation-Induced Tumors in Mice or Detected in Human Tumors

Inducing agent	Induced tumor	Oncogene	Oncogene containing tumors (%)
DMBA	Skin carcinoma	H-*ras-1*	90
X-rays	Lymphoma	K-*ras-2*	57
NMU	Lymphoma	N-*ras-1*	85
3-MC	Fibrosarcoma	K-*ras-2*	50
Hepatic carcinogens	Hepatoma	N-*ras-1*	100
MNNG	Skin papilloma	H-*ras*	100
	Skin carcinoma	H-*ras*	100
Human tissues	Colon carcinoma	K-*ras*	20–40
	Pancreas carcinoma	K-*ras*	90–95

TABLE 4
A Summary of Reported c-onc Amplifications

Oncogene	Tumor cells	Amplifications
c-*myc*	HL-60 (acute promyelocytic leukemia)	20 ×
	Colo 320 (apudoma)	30 ×
	SK BR-3 (breast carcinoma)	10 ×
	Small cell lung cancer	Up to 80 ×
N-*myc*	Small cell lung carcinoma	Up to 80 ×
	Neuroblastoma	Up to 250 ×
L-*myc*	Small cell lung cancer	Up to 80 ×
c-*myb*	Colo 201/205 (colonic carcinoma)	10 ×
	ML1-3	5–10 ×
c-K-*ras*	Y1 (adenocortical tumor)	50 ×
c-*abl*	K562 (chronic myelogenous leukemia)	10 ×
c-*erbB*	A431 (epidermoid carcinoma)	15–20 ×
	Glioblastoma	Up to 50 ×

these proto-oncogenes; and (3) a perturbation of the transcription process and/or other quantitative alterations of gene products. Any of these alterations could be considered a trigger for initiation of the carcinogenic process. In this category could be mentioned the oncogenes which codify the proteins located in the nucleus which can act as regulatory factors of transcription, namely: c-*myc*, c-*fos*, c-*jun*, p-*53*, etc.[111-113]

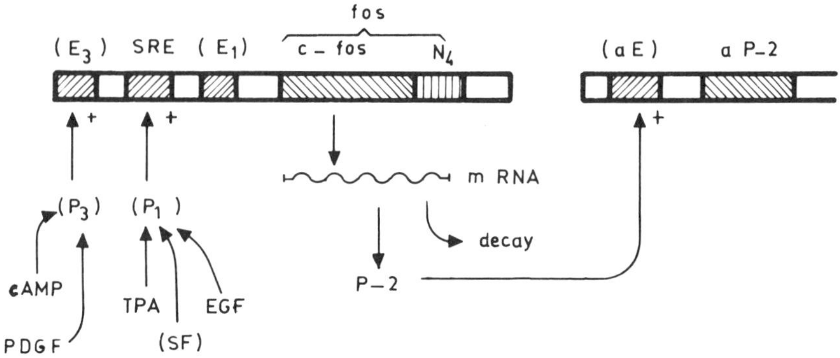

FIGURE 12. Control processes of the c-*fos* gene. *E1, SRE, E3* are the activating sequences of the c-*fos* gene, *aE* of the *aP2* gene (adipocytic gene), and *aE'* of the c-*jun* gene. The *N* sequence of the c-*fos* gene encodes a nucleotidic sequence enhancing the degradation of the c-*fos–mRNA*. The protein encoded by the c-*fos* and c-*jun* genes are c-*Fos* and *c-Jun*, respectively. *P1* represents the protein involved in c-*fos* activation by TPA (phorbol esters), SF (serum factor) and EGF (epidermal growth factor) activate the c-*fos* gene via the *E3* sequence. The c-*jun* gene is activated by a c-*Fos*–c-*Jun* heterodimer.

For instance, the c-*fos* proto-oncogene codifies a transcriptional factor that under constitutive conditions (at low level) can be enough to confer the transformant properties to the cells. The transcription of c-*fos* (or other genes from this category) in general, occurs rapidly (and transitory) being initiated by various stimuli in an extremely diverse cellular context.[116] Figure 12 summarizes the control processes studied for this proto-oncogene. The following regulatory sites of the *fos* gene were revealed: (1) a site (marked E_1 in Figure 12) at 100 bp upstream from the transcription start; (2) the serum responsive element (SRE); and (3) a negative control site (N_4) located after the transcription terminator site. SRE is activated by a specific protein (P_1) which probably mediates the effect of the epidermal growth factor (EGF), serum factor (SF), and phorbol esters (TPA). The SRE sequences of the c-*fos* gene are strongly conserved during evolution; similar elements are found in *Xenopus laevis*. The E_3 sequence is involved in the induction of the c-*fos* gene by the platelet-derived growth factor (PDGF) and by cAMP. The deletion of the N_4 sequence increases the stability of the transcribed *fos*-mRNA and is involved in the conversion of the normal proto-oncogene into the corresponding oncogene. The c-fos protein binds to a control sequence (aE) of the adipocyte gene *aP2*, activating it during the adipocyte differentiation. Serum growth factors and other inducers of the c-*fos* gene activate cellular kinases which produce protein phosphorylation. Factors which bind specifically to c-*fos* enhancers are found in HeLa cells. In A431 and BALB/3T3 mouse cells such factors appear only if the cell cultures were treated with EGF of PDGF, respectively.

The v-*jun* or c-*jun* oncogenes encode a protein with homology to a protein (GNC4) from yeast that regulates DNA transcription. Using DNA chimera of GCN4 in yeast, with substitute corresponding sequences, it was found that the yeast still makes a product that functions like GCN4.[117] But the v-*jun* oncogene and the human proto-oncogene, c-*jun*, encode DNA binding proteins structurally and functionally similar to the human transcription factor AP-1 (isolated from HeLa cells).[118] The AP-1 factor was shown to bind to DNA sequence in a manner identical to those of GCN4.[119] Therefore, the product of the c-*jun* gene and AP-1 have common antigenic and enhancer-binding properties suggesting that these proteins are closely related, if not even identical.[120] Actually, the human c-*jun* gene encodes one of the polypeptides present in affinity-purified preparations of AP-1.

The c-*fos* proto-oncogene is also associated with the AP-1 recognition sequence of the protein p39 that was demonstrated to be the product of c-*jun*.[121,122]

These proteins are produced by families of inducible genes and could play an important role in modulating gene expression at the end of a cascade of signals transduced directly from the cell surface. For example, c-Jun and c-Fos proteins bound together (in a 1:1 complex) are 25 times more efficient in interaction with the Ap-1 DNA binding site than the hemodimer of c-*jun*.[92] The gene product enhances c-*jun* activation function of TPA, probably also via a c-*Jun*-c-*Fos* protein complex. This complex is stabilized by a heptad repeat of leucine residues, the so-called leucine zipper (hydrophobic interaction).[123-125] This c-*Fos*-c-*Jun* complex is quoted to inhibit the transcription of the c-*fos* gene and to be responsible for the only transient activation of this gene to TPA induction. The c-*jun* proto-oncogene is also positively autoregulated by its product; c-*jun* and this positive loop are probably responsible for the prolongation of transient signals generated by the activation of protein kinase.[127] The c-*fos* and c-*jun* genes appear as a master switch in the signal transduction pathway which can convert short-term stimulation to long-term responses in such diverse processes as growth, differentiation, and proliferation.[123]

It is very difficult to assert, but we also assume that the intra- or interchromosomal molecular rearrangement of the nucleotidic sequences or genes could, under some conditions, be a starting point for the initiation of carcinogenic processes. The finding that a large number of chromosomal fragile sites can be induced by chemical carcinogens or that oncogenic viruses are preferentially integrated into chromosomes at the promoter regions of proto-oncogenes (c-*myc*, c-*erb*, c-*myb*, and *prim-l*) in mice and chickens support the previous assumption.[128]

Cancer can arise not only by activation of the oncogenes but also by inactivation of the growth suppressor genes. It was considered that the control system of normal cell growth must be altered in at least two or three points, before anarchic cell division may occur.[127] Therefore, intrinsic safety mechanisms of this system were assumed to exist — mechanisms able to constrain the activity of individual oncogenes. During recent years, some of these safety mechanisms were identified with the discovery of the growth suppression genes (or so-called anti-oncogenes, emerogenes, or orthogenes),[104,131] blocking the signal for cell division.[132] The existence of genes limiting cell growth has been presumed for a long time, since it had been noticed that cells somehow *feel* the situation when they should stop multiplying.[133] Typical in this respect is the contact inhibition of cells. This contact produces, probably by second messengers, a signal that reaches the nucleus and activates several genes which stop cell growth and division. Thus, the presence of such genes is required in the cell in order to counteract the activity of the proto-oncogenes or oncogenes.[114,115]

A first clue of this behavior resulted from the investigation undertaken on retinoblastoma. It was shown that either of the loss of both Rb alleles on chromosomes 13 or the alteration of their function (by mutation) leads to malignancy.[134,135]

In this field, a consequential finding that offers great hope for further insight to the genetic basis of human cancer is represented by the *p53* gene that can act as an anti-oncogene or as an oncogene.[136] The *p53* has been conserved during evolution and it has been isolated from human, rat, mouse, frog, chicken, and bony fish. It is present in low concentrations in normal cells, but high levels have appeared in a large number of tumors and tumor cell lines.[137]

Two particular characteristics have been noted in this gene: in the wild type it works as an anti-oncogene, and its half-life is short (6 to 20 min). In mutated cells, this protein (a 375-amino acid nuclear phosphoprotein) has transforming properties (so it appears as an oncogene) that have a much longer half-life (4 to 8 h), which in fact explains the high concentration of *p53* protein in tumor cells. Thus, the *p53* gene can be an oncogene in its mutant form or an anti-oncogene in its wild-type form.[138,139]

Like the Rb protein (110 kDa), the *p53* protein (wild type) is physically bound to the proteins of the different DNA tumor virus oncogenes. It was discovered that the Rb gene protein binds to the large T antigen of SV40, E1A, and E6 protein and the p53 protein binds to the large T antigen, E1B, and E7 gene proteins. In addition, the E7 protein was observed to abolish the ability of TGF-beta to inhibit the *myc* gene expression. Many experimental models strongly suggest that these interactions are important for the inhibition of the transforming activity of many DNA viral oncogene products.[136] The same proteins (of the wild-type genes) have the ability to bind to the DNA sequences and to act as transcriptional factors for expression of some genes.

The evidence suggests that the *p53* protein, in the wild state, is essential for normal cell growth, perhaps acting as a negative regulator, whereas in the mutated state it is probably involved in the development of malignancy.

These findings open new directions in molecular oncology for establishing the real distinction between oncogenes and anti-oncogenes and controlling these mechanisms in the aim of cancer prevention or reversion (treatment).[140]

2. Promotion Step

Interest in the promotion step was important because this stage of carcinogenesis accounts for most of the tumor latency period and for the reversibility of the process.

Promotion consists in events interfering with the genetic program of proliferation and especially differentiation.[141] The promoted cell no longer recognizes the usual differentiation signals removing it from the replicative population. The balance between the expression of programs for growth and differentiation is under stringent control in normal cells, whereas in promoted cells imperfections emerge, possibly resulting in (1) deficient control of cellular growth and differentiation; (2) autocrine production of growth factors; and (3) lowered sensitivity to growth inhibitors and increased sensitivity to growth factors secreted by other cells.

These phenomena, and possibly others, could lead to a selection advantage in the clonal expansion for preneoplastic and neoplastic cells.

The best-studied promoter is TPA. In a first stage, it inhibits proliferation and differentiation; afterwards, it induces a pleiotropic proliferative answer at an epidermal level, stimulating cell growth.[142] In contrast with initiators, promoters do not bind to DNA, their main target being the cell membrane. Their behavior is consistent with epigenetic mechanisms of action. For instance, the effect of TPA is mediated by PKC, considered as a potential receptor of this promoter. Following this interaction, PKC is activated and induces, by a cascade of second messengers,[143] a number of phenotypic effects that may be divided into three categories: (1) effects mimicking the transforming phenotypes; (2) inhibition or induction (modulation of differentiation); and (3) membrane effects regarding the control of transport, cell-cell communication, etc.[144]

In some experiments in which cells were treated with TPA, transient changes in the expression of virus oncogenes and cellular proto-oncogenes have been observed.

Weinstein,[145] using a cell line of rat fibroblasts which overexpresses the PKC β_1, observed alterations in morphology, growth, enhanced responses to promoters and certain growth factors, altered expression of ornithine decarboxylase and phorbin, and increased susceptibility to transformation of some proto-oncogenes.

The type of phosphorylation of several proteins involved in transduction signals to the nucleus is altered. This results in the modulation of specific gene expression, for instance, of the gene involved in the synthesis of plasminogene activator, prostaglandins, and possibly γ interferon, calcitonin, prolactin, and glycoforine. Several proto-oncogenes (c-*myc*, c-*fos*, c-*jun*, and c-*myb*) that transcribe nuclear proteins involved in cell proliferation, also represent potential targets for PKC. Like lymphocyte activation, PKC could induce the expression of other genes involved in growth factor activation, promoting a long-term cell proliferation.[146] Thus, PKC plays a critical role in growth factor control and in cell transformation, in spite of the fact that PKC can act as a tumor suppressor in some cell systems.

In certain mouse cell lines it was supposed that the genome contains specific gene sequences that make them highly sensitive to the action of TPA.[147] Thus, in neoplastic mouse epidermal cell lines (JB6) two genes were identified, separated, cloned, and sequenced. They were named *pro 1* and *pro 2*, due to their ability to confer tumor promotion responsiveness (such an event takes place by transfection of these genes in P^- cells lacking *pro 1* and *pro 2* genes); malignant transformation occurs in these cells subsequent to TPA treatment.[148] Thus, sensitive ($JB6P^+$) or resistant ($JB6P^-$) cells are known, according to the presence or absence of the *pro* genes, respectively.

These genes were also found in the human and murine genome, being preserved throughout evolution (like proto-oncogenes). Moreover, the presence of both genes was observed in DNA isolated from cell lines derived from a human nasopharyngeal carcinoma. Transfection of the segments of DNA extracted from this tumor to P^- mouse cells conferred their susceptibility to promotion, suggesting the involvement of *pro* genes in human carcinogenesis.[149]

The molecular mechanism of promoter action appears to be epigenetic. They induce the modification of cellular differentiation or the stimulation of proliferation as a progressive disorder in signal transduction. To sustain this assumption, we have to mention some properties of the promoter agents: they do not bind to the DNA; they induce cell surface alteration; they activate specific cellular receptors; they activate some genes; and they induce the modification of cellular differentiation and induce selective stimulation of cellular proliferation.

Further studies are needed to elucidate the mechanisms by which the promoters alter nuclear functions and gene expression and lead to tumor growth and progression.

3. Progression Step

It is difficult do define progression and usually it denotes entities about which little is known. The most important characteristics which have the greatest significance are the ability of tumoral cells to invade locally and to metastasize. The acquisition of these properties by neoplastic cells remains the fundamental definition of malignancy.

Thus, the other aspect of tumor progression consists of a tendency of malignant cells to increase in time their growth rate, because the cells escape from local growth regulation mechanisms. This process is a consequence of an increased proliferation fraction and not of a shortening of the cell cycle time. The tumor cell populations within a tumor society continue to proliferate actively instead of progressing in terminal differentiation or programmed cell death (apoptosis).[150]

Tumors often become more malignant in their characteristics and more aggressive in their behavior during their development, although the time course could be quite different. The cells show morphological, metabolic, genetic, and immunological alterations that are generally considered as a loss of differentiation.[151] At the present time, this phenomenon is considered to be a complex process by which specific genes (oncogenes) and their products involved in tumor genesis can function in certain lineages and stages of ontogeny.[152]

Evidence suggests that the events of tumor progression represent the result of sequential selection of various subpopulations from a common progenitor. But the clonal evolution of tumor cell populations might result from enhanced genetic instability which leads to new variants with a selective growth advantage over earlier forms. This process appears as a consequence of the high frequency of mitotic errors and variations of genetic markers.[153] The increased number of variants and continued presence of multiple subpopulations within malignant tumors remain at the basis of the so-called tumor heterogeneity.

Cytogenetic studies started to establish the molecular mechanisms of the phenomenon of clonal evolution, tumor progression, and tumor heterogeneity. Many experimental data supporting the clonal evolution concept are derived from chromosome studies of leukemia and lymphomas. The development of banding techniques allowed not only the rigorous identification of normal chromosomes (and established their bands),[154] but also the genesis of chromosomal translocation (or molecular rearrangement).

Cytogenetic analysis of animal or human tumors revealed many chromosomal disorders (translocation, deletion, inversion, addition, and numeric variation).[155-157] Particular attention has been given to leukemia, Burkitt lymphomas, and tumors associated with rare congenital chromosome changes. In human chronic myeloid leukemia (CML), the cytogenetic hallmark is Philadelphia (PH) chromosome 9 and 22, t(9,22) (q34, q11).[158] Molecular rearrangements result in the translocation of the proto-oncogene *abl* from the long arm of chromosome 9 to the long arm of chromosome 22 juxtaposed with a specific region termed bcr (break-point cluster region). This rearrangement leads to a mRNA which is structurally different from that observed in normal cells. This translocation generates a chimeric gene which produces a new mRNA and a new protein with a kinase activity specific for thyrosine residues. The fusion protein is 210 kDa, and contains a fragment (about 140 kDa) of the usual (145 kDa) c-Abl protein and a fragment (about 70 kDa) of the Bcr protein.[159] This phenomenon could be used by the cell to increase genetic information during the evolution.

Another example belongs to the B and T cell malignancies, particularly Burkitt lymphoma. In this case, the translocation involves chromosome 8 (q24) and chromosome 14 (q32) or 22 (q11). As a consequence of this translocation, a constitutive overexpression of oncogene *myc* takes place. The myc protein appears to be involved in DNA replication and — when it is overexpressed — it might lead and maintain an uncontrolled cell proliferation. Specific chromosomal translocations of the c-*myc* gene are also observed in acute or chronic leukemia with T cell lymphocyte. In this case, the c-*myc* gene remains on chromosome 8, but the fragment containing the alpha chain gene of the T cell receptor, which usually lies on chromosome 14, is translocated on chromosome 8, being juxtaposed to the c-*myc* gene.[111]

Other types of genes (*bcl-1, bcl-2,* and *tcl-1*) that are activated with high frequency by specific translocation were observed on other leukemia malignancies, such as chronic lymphocytic leukemia of B cells, acute B cell lymphoblastic leukemia, and T cell chronic lymphocytic leukemia.[11,152,159] In the case of *bcl-1* genes located on chromosome 11 and *bcl-2* on chromosome 18, respectively, they are translocated on chromosome 14, near the immunoglobulin heavy chain locus.

Bcl-2 is a particular gene which also could be involved in the initiation of cell transformation because it blocks apoptosis (programmed cell death and cell suicide). By blocking cell death, cell proliferation can be increased and under these conditions allow the cell to undergo mutations in other genes (proto-oncogenes or/and suppressor genes).[160,161]

Generally, the alteration of proto-oncogenes followed by their overexpression or somatic deletion in some gene (e.g., nm 23, a tumor suppressor gene) has been associated with more aggressive cancers. In this category may be mentioned breast cancer, where a number of proto-oncogenes have been observed to be altered or aberrantly expressed, such as: c-*erb B-2* and *hst/int-2*. The amplification of these genes has been statistically correlated with a bad prognosis of breast cancer.[109,162]

The genes c-*erb B-2* and c-*myc* appear to be preferentially amplified together in advanced stages of breast cancer (i.e., in the stages in which the tumors have lost receptors). Contrarily, the amplification of *hst/int-2* genes appears to occur independently.[162]

The phenomenon of coexpression or cooperation of two or more proto-oncogenes (oncogenes and possibly other genes, too) in the tumorigenesis process is well established.[105] This statement is based on the evidence that in many human tumoral tissues, or in tumor cell lines derived from human tumors, the expression or activation of multiple oncogenes were detected.[113] The activated N-*ras* gene coexists with an amplified c-*myc* gene in the HL-60 cell line. In human lung giant cell carcinoma or in pancreatic carcinoma, the point-mutated K-*ras* oncogene was accompanied by amplification of the c-*myc* gene.

In conclusion, we emphasize that:

1. In tumorigenesis processes (initiation-progression) at least three major classes of genes could be involved: oncogenes, tumor suppressor genes (anti-oncogenes), and the so-called modulator genes;
2. Tumor progression is a multistep process; each step may reflect the activation (by mutation, amplification, or chromosomal rearrangement) or loss of different genes;
3. At this moment it is difficult to establish the role of each type of gene or network of genes involved in every step of tumorigenesis; and
4. Investigation of the function of these groups (classes) of genes will be a crucial component to the understanding of molecular mechanisms of tumorigenesis.

To understand the molecular mechanisms of cancer, a better knowledge of the role of cellular oncogenes and of other modulator genes in normal development and differentiation processes is important. In future years, we expect to have such information that will permit us to elaborate the rational strategies for cancer prevention and develop more specific and efficient diagnosis and treatment methods.

D. NEGATIVE TRANSCRIPTIONAL CONTROLS

Several mechanisms are described for negative transcriptional control (downregulation of transcription). One mechanism consists in binding of the inhibitory protein to an activating transcription factor. A classical example is offered by the regulatory protein GAL80 which prevents GAL4, a positive activator, from binding directly to a specific sequence of *GAL* genes.[167]

Proteins encoded by anti-oncogenes (for example, those encoded by retinoblastoma or *p53* genes) act by (physically) complexing to the oncogene proteins encoded by different groups of DNA tumor viruses.[107,164]

The *Rb* gene spans a sequence of over 200 kbp containing 27 exons and encodes a mRNA of 4.6 kb, which is present in all normal tissues but absent or altered in retinoblastoma cells.[165] The product of this mRNA was identified to be a nuclear phosphoprotein of 110 kDa (pp 110^{Rb}). This protein has the ability to interact specifically with the large T antigen of SV40, E1A, and E7 oncoproteins and perhaps with other proteins encoded by several DNA tumor viruses,[137,166] blocking the transforming activity of these types. It was supposed that the RB protein may play a role in regulation of other genes and could mediate the oncogenic effects of some different viral-transforming proteins. For instance, the interaction with SV40 large T antigen takes place through two distinct regions and it has also been demonstrated that the RB protein has a DNA binding activity.[114,167] It is not clear whether this activity is mediated with other factors or whether the RB protein has its own DNA binding activity. A difference in DNA binding affinity between the wild-type and mutated RB protein is not yet established, but other experiments have shown that the hyperphosphorylated form of RB has a higher DNA binding activity than the hypophosphorylated form. These data suggested that phosphorylation of some proteins could play an essential role in regulating the RB function and many other transcriptional factors.[168] But the significance of the DNA binding activity of the protein encoded by the *Rb* suppressor gene in preventing cell transformation has to be elucidated.

The *p53* gene, like the *Rb* gene, appears to be a tumor suppressor gene. The product of the wild-type *p53* gene is essential for normal cell growth, perhaps as a negative regulator.[136] The mouse p53 protein can control SV40 DNA replication by blocking the binding of DNA polymerase and downregulating the T antigen DNA helicase activity.[168,169] Under these conditions the wild-type *p53* acts as an anti-oncogene, like the *Rb* gene.

However, mutant forms of *p53* (bearing point mutations) fail to bind SV40 large T antigen, and they also lose their ability to inhibit SV40 replication *in vivo* and *in vitro*.[170,171] Therefore, this mutant type of *p53* functions as an oncogene.

The *c-myb* proto-oncogene encodes a nuclear DNA binding protein that is expressed in a large range of concentrations in various stages of B cell development. The same variation was also observed in c-*myb* mRNA, in murine pre-B cell lymphomas and plasmocytomas, and was explained by a mechanism of downregulation of c-*myb* RNA: a blockage at the transcription elongation level.[172] The same phenomenon was observed in the case of c-*myc* mRNA transcription, where the blockage takes place in the first intron of the c-*myc* gene.

The c-*fos* proto-oncogene transcribes a nuclear phosphoprotein (62 kpb) that is associated with transcriptional factors, such as the members of the *Jun* family which, in turn, can regulate the transcription of genes containing activator protein AP-1 binding sites.[121] Regulation of c-*fos* at the transcription level implies, as in the case of c-*myb* and c-*myc* genes, a premature termination of RNA transcript that can also downregulate c-*fos* gene expression.[173]

Another physiological event that implies the downregulation mechanism appears to be the modulation of the number of cell surface receptors. For instance, in rat pituitary cells, the thyrotropin-releasing hormone (TRH) downregulates the number of TRH receptors by decreasing the mRNA activity.[174] A glucorticoid receptor, binding the hsp90 instead of the steroid hormone, is inhibited in its specific transcription activating function.[175] Generally, the steroid hormones exert a negative regulation, inhibiting the expression of many genes (e.g., collagen type I, IV, etc.).[176,177]

Another mechanism is the competition of the inhibitor and activator protein for the DNA control sequence. The thyroid receptor T_3 binds to the chicken vitellogenin II gene A-2 estrogen response element (ERE), in competition with the estrogen receptor protein, and inhibits the vitellogenin gene transcription in *in vitro* assay. In a homologous system, by using *in vitro* transcription competition assays, it was demonstrated that the chicken vitellogenin expression in the male is under negative transcriptional control and may be derepressed by estradiol administration.[179]

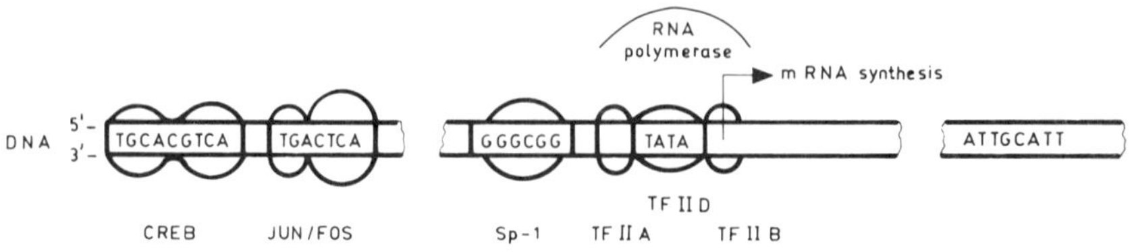

FIGURE 13. Schematic representation of the regulation of gene expression by various auxiliary proteins bound upstream and downstream from the initiation point of transcription.

The engrailed gene protein competes with the *ftr* gene protein for transcriptional enhancers specified for the ftr protein and counteracts *ftr* activation. The *Drosophila eve* gene protein is a repressor for *ubk, ftr,* and *wg* genes, but the repression does not seem to be due to competition with an activator protein. The negative regulator of T cells, responsible for downregulation of the TgA enhancer action, also seems to interact more directly with the IgA-coding genes.[181] Interaction with a silencer DNA sequence could be the mechanism in these cases.

Competitive downregulation of the transcription activated by a steroid hormone receptor, or by the presence of receptor proteins for other steroid hormones (and by these hormones) implies an intricate mechanism. Competition for a common, limitative, additional transcription factor could be the mechanism for this competitive inhibition.[182]

E. MOLECULAR ASPECTS OF GENE REGULATION

Molecular cloning of genes which encode DNA binding transcriptional factors in mammalian cells, related to the RNA polymerase II function, also allowed the analysis of the structure and function of these proteins.[183] DNA transcription is regulated by a very intricate mechanism, which includes several regulatory proteins. These proteins are bound mainly upstream, but also downstream of the transcription startpoint. Figure 13 illustrates this complexity.

The mechanisms by which these regulatory proteins activate transcription have not been completely elucidated. Specific protein-protein interactions between these transcription factors, bound to distal DNA elements, and the proteins of the RNA transcription-initiation complex, bring together such distal elements, looping the DNA double helix and producing conformational changes. All these events are required for transcription initiation. A scissors-like model was proposed for DNA recognition by C/EBP, a family of leucine zipper proteins.[184,185] These zipper proteins are Y-shaped, with the leucine zipper region as the stem and the basic one as arms. The DNA is situated between the arms, with the symmetry center of the binding site positioned at the bifurcation of the Y. The arms extend along the DNA, up to a point when each arm turns sharply, assuring the contacts (scissors-like grip). This model is supported by modeling with "ball and stick" structures, computer graphics simulation, biochemical protection, and labeling experiments.

We are now going to describe some features of the regulatory proteins and of their binding sites.[186] The DNA binding sites are oriented from upstream to downstream, in the 5′–3′ direction, giving the direction of transcription.

Many of the regulatory proteins in mammalian cells are represented by the nuclear receptors — a family of ligand-inducible transcriptional factors. The ligands activate the gene transcription through recognition of the DNA sequence represented by responsive elements (REs). The glucocorticoid receptor (GR) from rat liver, which is a protein of about 80 kDa, was shown to contain a ligand-inducible transcription activation function. It is a member of the receptor superfamily, which includes receptors for steroid hormones, retinoids, peroxisomal activators, vitamin D, thyroid hormones, and other proteins of unknown function.[187]

The GR free hormone is associated in the cytoplasm with the Hsp 90 protein. The DNA binding domain is characterized by a pattern of eight cysteine residues, coordinating two Zn^{2+} ions, each with tetrahedral geometry[187] (see Figure 14). The sequence of the corresponding DNA site is 5′-GGTACAN$_3$TGTTCT -3′ with N_3 a sequence of undetermined three-base pairs.

The c-Jun of HeLa cells is about 37 kDa and its DNA binding sequence is TGAC/GTCA. It is a representative of a protein family which binds at AP-1 sites, the latter conferring TPA inducibility. c-Jun can be associated as a heterodimer with c-Fos (and other members of the AP-1 family), via leucine-zipper motifs. Homologues of Jun and Fos are also found in *Drosophila* and yeast.

OCF-1 and OCF-2 are ubiquitous proteins in mammalian cells, with a molecular weight of 85 and about 60 kDa, respectively; the first was isolated from HeLa cells, and the second was identified especially in T and B lymphocytes. The DNA binding sequence for these proteins is ATTTGCAT. They bind specifically to octamer sequences of promoters located in the histone H2B and immunoglobulin heavy chain genes. The proteins contain the leucine zipper domain of serum factors that bind to the serum responsive element (SRE) on the DNA molecule. This interaction mediates responsiveness to serum growth factors (SGF), epidermal growth factors (EGF), tetraphorbolacetate (TPA), and insulin.[188]

Epidermal growth factor is a short, soluble peptide of 53 amino acids that has a distinct motif of 6 cysteine residues in positions 6, 11, 17, 26, 28, and 37, with disulfide bonds between the 6–17, 11–26, and 28–37 residues. There is a whole family of proteins containing EGF-type structures, probably with significant functional overlap. One of the functions of the EFG-repeat is to facilitate interaction between two proteins; a second one is to bind and respond to Ca^{2+} ions, with a dissociation constant of 10^{-5} to 10^{-4} *M*.

1. Dimerization as Control Mechanism for Transcriptional Regulation

Several promoter DNA sequences interact with an entire family (or superfamily) of structurally related regulatory proteins. The mechanism of interaction, transcriptional activation, or inhibition effect depends mainly on the homo- or heterodimerization of the regulatory molecules involved.[18]

The Jun and Fos families of regulatory proteins consist of at least six proteins.[109,190] They contain a leucine zipper motif which, upon dimerization, adopts a coiled-coil structure, stabilized by the interacting leucine side chains. The Jun-A and Jun-B proteins bind both to the AP-1 binding site as homodimers or heterodimers with the Fos protein. The consequences of binding to the AP-1 site differ. Jun A is able to activate transcription controlled by a promoter with a single AP-1 site;[191] Jun B binds, but does not activate transcription. Thus, by competition with Jun A, Jun B may act as a transcriptional repressor. In contrast with the c-*jun* family, the v-*jun* gene, which misses the negative-acting domain, has a high transcription activity and, in consequence, a neoplastic potential. Other proteins with basic leucine zipper motif for dimerization are Fos and CREB (CRE binding protein).[192]

Another motif implied in DNA binding and dimerization is represented by the basic helix-loop-helix motif (bHLH), found in a second large family of nuclear proteins involved in gene regulation.[186] The E12 and E47 proteins bind the immunoglobulin enhancers, MyoD, Myf 5, and myogenin which play a role in molecular differentiation. This motif is also found in proteins that regulate the development of the nervous system in *Drosophila*. The bHLH proteins also form homo- and heterodimers mediated by the helix-loop-helix structure. The short amphipathic helixes have hydrophobic residues separated by three or four positions in the peptidic backbone. The loop contains one or several helixes that break amino acid residues. DNA binding is mediated by the basic region which precedes the first amphipatic helix.

The role of dimerization in regulation of homo- and heterodimerization depends on the concentrations of the respective monomers. The regulation of the entire set of genes could be modified by an upset in the balance of subunits within the family, simply by changing the concentration of a component.

Competition between active and inactive proteins in heterodimerization is a general principle of negative regulation in mammalian cells.[186] As an example, we can mention the Jun_2 homodimer and heterodimer Jun-Fos — both bind to the AP-1 DNA site, but the latter does so with a 500 times larger association constant. Similarly, the MyoD · E12 and MyoD · E47 heterodimers bind better to the kEN enhancer site of the immunoglobulin chain than the three heterodimers. MyoDl increases its affinity for a sequence of the muscular creatine kinase enhancer when dimerized to E12 (a factor that binds particularly to the Ig heavy chain enhancer).

The DNA binding specificity is different for the same protein, when this forms different heterodimers. The Jun A · Fos heterodimer recognizes preferentially the AP-1 DNA site, while the Jun A · CREB complex prefers the CRE site.[193] Another consequence of dimerization appears in the mutant proteins with altered basic regions, which can act as negative regulators.

The regulation of the nervous system development in *Drosophila* includes four structurally related genes: *AS-C, da, h,* and *emc*. The first products of the two genes have a positive role in transcriptional regulation, while the last two have a negative one. The genetically defined interactions may be explained by different heterodimer formation.[186,190,193] The AS-C · da heterodimer protein binds and activates the KE2-DNA site more efficiently than the homodimer. The *emc · da* and *emc · AS-C* heterodimers repress transcription, although these heterodimers do not bind the DNA site; the emc-encoded protein has the HLH motif, but it does not contain the adjacent basic (b) domain. On the contrary, the *h* gene protein contains the b domain, but because it ends with a proline residue, the α helix character of the subsequent sequence in the HLH motif is broken. Under these conditions, therefore, the dimers of the *h* gene products could be inactive.

The constraints imposed upon transcriptional control by the existence of a single polymerase, and a limited selection of structural motifs involved in DNA binding, can be overcome by the existence of protein-protein interactions between multiple levels. Each intermediate component may be checked at its level of expression, modification, degradation, or competition, allowing a multiple control of genes. Several families of transcriptional factors seem to derive, at least in part, from differential splicing. Since the DNA binding domains rest unaltered, the nonconserved domains could be involved in the selection of different auxiliary proteins.[194]

2. Description of Some Regulator Protein-DNA Complexes

The eukaryotic transcription factor C/EBP contains a leucine zipper and a flanking basic domain.[195] Interferential experiments with DNA methylation and ethylation have detected major groove contacts over a full turn of the DNA helix on both the symmetric and asymmetric parts of the C/EBP binding site. The essential contacts for C/EBP bonding have a twofold symmetry, but differ significantly from the contacts of other dimeric proteins, including those with HLH motif and the restrictive endonuclease EcoRl.

For the glucocorticoid receptor, two crystal structures of its DNA binding domain complexed with DNA were reported.[187] The domain has a globular fold which contains two Zn-nucleated substructures of distinct conformation and function (see Figure 14). When it binds DNA, the domain dimerizes, placing the subunits in adjacent major grooves. In one complex, the DNA has a symmetrical consensus target sequence, in the second, the central spacing between the target's half-sites is one base pair larger. One subunit interacts specifically with the consensus target half side and the other acts nonspecifically with an unknown element. The DNA-induced dimerization fixes the separation of the subunit recognition surfaces, so that the spacing between the half sites becomes a critical feature of the target sequence identity. The crystal structure of a peptide from Zif268 (a mouse immediate early protein), containing the three zinc fingers and a consensus DNA binding site, was determined at a high precision level.[185] In this complex, the zinc finger binds in the major groove of DNA and wraps around the double helix. Each finger has a similar relation to the DNA and makes its primary contacts in a three-base-pair subsite. Residues from the amino terminal portion of an α helix contact the DNA bases, and most of the contacts are made by the guanine-rich strand. This structure provides a framework for understanding how zinc fingers recognize DNA and suggest that this motif may provide a useful basis for the design of novel DNA binding proteins.

Two 57-residue peptides containing one pair of zinc fingers from a human enhancer binding protein were prepared by solid-phase peptide synthesis and characterized by chemical and physical methods. Their DNA binding properties were studied by gel retardation.[196] One peptide, MBP-DF, contains the native sequence, the second, (Abu^{11})-MBP-DF has an α-aminobutyric acid residue substituted for the nonconservative cysteine residue at position 11. The addition of Zn^{2+} ions resulted in specific conformational changes in both peptides, indicating that the Cys^{11} residue is not involved in chelation. The stable folded structure formed during the gel retardation experiments has demonstrated that the peptides have the ability to interact with a 15-mer oligonucleotide, comprising a portion of the class I major histocompatibility complex enhancer sequence (R_1), and that the interaction is Zn^{2+} dependent. The dissociation constant for the (Abu^{11})MBP-DF · R_1 complex is 1.4×10^{-7} *M*, maximal binding occurring at a zinc-peptide ratio of 2:1. Thus, each finger domain binds one zinc ion. The binding specificity observed with respect to related enhancer sequences exhibits the same relative order as was noted previously for the whole protein. Point mutation experiments, using the major histocompatibility complex enhancer binding sequence, indicate that the last GCA trinucleotide in a four-guanine stretch plays a pivotal role in the interaction between the peptide and the DNA.

The Hsp protein of *Drosophila*[197] binds specifically to the HSP sequence, having a dissociation constant of 4×10^{-12} *M*, while when it binds to *E. coli* DNA the constant is 10^{-6} *M*. The specific factor that activates the adult chicken globin gene[198], bound to specific chromatin sites, has a dissociation constant of about 1.5×10^{-12} *M*; when bound to nonspecific sites the constant is 4×10^{-8} *M*. Point mutation in enhancer sequences usually decreases the transcription rate of the activated gene about 5 to 10 times, corresponding to a ΔΔG of 1 to 1.5 kcal/mol. Sometimes, this kind of mutation increase these rates about three times.

VI. HORMONE RECEPTORS. HORMONAL CONTROL OF GENE ACTIVITY

Steroid hormone receptors are proteins that regulate gene activity and also mediate the action of steroid hormones upon gene activity. Their mechanism of action is described in some recent handbooks.[199,200]

A. STRUCTURAL ASPECT AND MECHANISM OF ACTIVATION

Nuclear receptors for hormones comprise a superfamily of transcription factors, modulated by hormones as ligand, which regulate homeostasis, re-

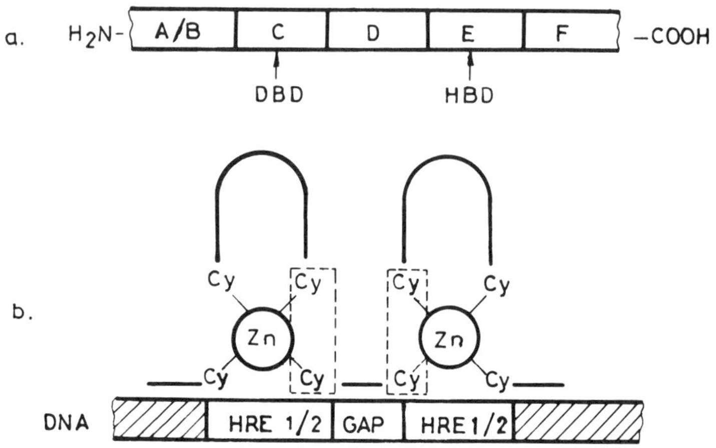

FIGURE 14. The interaction of a steroid hormone receptor with a DNA sequence: (a) the common domains of nuclear steroid hormone receptors — they are divided into six homologous regions (A to F); and (b) a fragment of the protein receptor, showing Cys-Zn connectivity and dimer interaction with the hormone responsive element (HRE) of the DNA. The HRE with the two half sites (HRE 1/2) and the GAP are represented.

production, and differentiation. These proteins recognize specific DNA sequences generally known as hormone responsive elements (HREs), which may belong to one of the following three classes:[201,202] (1) an estrogen-responsive element, ERE, with the nucleotide sequence for the transcribed strand (in a downstream direction before the transcription initiation point): 5′-GCTCAXXXTGACC-3′; (2) a glucocorticoid-responsive element, GRE: 5′-AGAACAXXXTGTTCT-3′; and (3) a thyroid hormone-responsive element, TRE: 5′-GGGTCATGACC-3′.

These consensus sequences are palindromic, suggesting that each site is recognized like a DNA binding domain, with binary symmetry, by a dimeric receptor. X represents different nucleotides that are able to confer significant hormone inducibility on target genes.

Steroid hormone receptors can be divided into six regions (A to F). Among these, two separate domains appear to be responsible for ligand and DNA binding (Figure 14). These regions were proved to be conserved between different steroids and related nonsteroid nuclear receptors. Being highly conserved, these regions play an important role in the interaction between DNA and receptor proteins. Thus, region C, the most highly conserved domain that contains two putative Zn fingers, corresponds to the DNA binding domain (DBD) and is important in the achievement of tight nuclear binding (Figure 14). This region is responsible for defining the target gene expression and it is the first finger that ensures the specificity. The hormone-binding (HBD) domain E is less conserved and is essential for an efficient activation of the target gene transcription.[204]

In addition, the N-terminal region A/B, the least well-conserved, may interact with the promoter and with cell-specific transcription factors. Region D, a hydrophilic sequence, is supposed to act as a ''hinge'' between DNA and hormone binding domains.[201,203]

The DNA binding domain of the glucocorticoid receptor contains two zinc finger motifs (as in the case of TFIIIA) and binds responsive elements composed of two half sites, separated by a gap of variable size[204,205] (Figure 14). The first finger of the receptor contains primary information for DNA site specificity and the second Zn finger is thought to stabilize the binding of the receptor to the DNA-hormone responsive element (HRE sequence) via interactions with the phosphate backbone of the HRE. The DNA binding specificity is accomplished by a combination of mechanisms. First, the discrimination between the half site sequences is mediated by the three amino acids of the first zinc finger. Second, responsive elements with different half site spacings can be discriminated by five amino acids which also act as a dimerization interface. A second dimerization signal is embedded in the ligand binding domain of several receptors. Thus, hormone (ligand) binding favors dimerization. Common to the steroid nuclear receptor superfamily is a short DNA binding domain that contains about 70 amino acid residues with many conserved cysteines:[202]

$$-\mathrm{Cys}-a_2-\mathrm{Cy}-a_{13}-\mathrm{Cys}-a_2-\mathrm{Cys}-a_{15-17}-\mathrm{Cys}-a_5-\mathrm{Cys}-a_9-\mathrm{Cys}-a_2-\mathrm{Cys}-a_4-$$

where a represents different amino acid residues.

The binding of receptor dimers to their DNA sites has a cooperational character. For example, the binding of the progesterone receptor dimer to one of the two tandem positioned GRE/PRE sites of the tyrosine aminotransferase gene, increases about 100 times the binding constant for the second receptor dimer. Such synergism was also observed for the receptor DNA sequence bindings.[203] For instance, the A/B region of ER acts synergistically with HBD to optimally activate certain promoters. In addition, the A/B region of GR contains a sequence that can act cooperatively with the activation function (TAF1) which is located in the framework of the C-terminal half of the receptor. As a paradox, we can mention that the steroid hormone receptor (for example, the human estrogen receptor) also has a transcriptional activator function in the absence of the hormonal ligand.[204,205]

It is now well established that receptor molecules are involved in modulation of the target gene transcription, following the specific and high affinity interaction between the receptor and its DNA response element. The mechanisms of gene regulation are thought to be (1) induction of the allosteric activation of the receptor protein by the hormone; (2) formation of a stable complex at the DNA enhancer site; and (3) attraction of the transcription factors and RNA polymerase II to the target gene startpoint.

B. INTERACTION OF RECEPTORS WITH HORMONAL LIGANDS. SPECIFICITY OF HORMONAL EFFECTS ON GENE ACTIVATION

For the human estrogen receptor, hER, the transcription-inducing response begins at 10^{-11} *M* estradiol concentration and reaches its maximum somewhere in the range of 10^{-9} to 10^{-8} *M*. Serum estradiol levels in men and postmenopausal women are in the range of 3 to 15 $\times$ 10^{-11} *M*, which suggests that in *in vivo* conditions, hER is a partially constitutive transcriptional activator.[205]

The various receptors for steroid hormones do not present differences too large in their structure-affinity requirements, even for the naturally coexisting hormones. For instance, in a study on relative affinities (binding constants) of various progesterone and testosterone substitute derivatives for the progestin receptor (PR), the androgen receptor (AR), and the glucocorticoid receptor (GR) extracted with cytosol from homogenized target organs, and with the use of the competition method for labeled steroids, the following results were obtained:[206] the progesterone (PR hormone) presents, for AR, a dissociation constant only 5 times larger than testosterone and a dissociation constant 1.15 times even lower for GR than corticosterone.

Target tissue specificity is sometimes enzymatically mediated and not receptor mediated. For instance, the mineralcorticoid receptor has, *in vitro*, the same affinity for aldosterone (the physiologic hormone) as for glucocorticoids, cortisol, and corticosterone with higher plasmatic concentration levels.[207] In the mineralcorticoid target tissues, the enzyme 11-β-HO steroid dehydrogenase converts glucocorticoids to inactive 11-ceto-derivatives, but not also aldosterone.

Hormonal antagonists form an interesting class of drugs. We will mention here only the estrogen antagonists that are used specifically in the therapy of female cancers.

Tamoxifen derivatives (Figure 15) are used, for example, in breast cancers.[208] They act as competitive inhibitors with estradiol. While induction of hER effects by estradiol begin at a 10^{-11} *M* hormone concentration, in the presence of 10^{-9} *M* tamoxifen, 10^{-8} *M* estradiol is required for the initial induction effect. The ER binding affinity of tamoxifen and its derivatives is quite large; for example, the half maximal inhibiting effect of the fixed ring trans tamoxifen derivative I (Figure 15) is in the range of 5 $\times$ 10^{-11} *M*, while the fixed ring tamoxifen metabolite II is an estrogen agonist with half maximal effect at about 5 $\times$ 10^{-12} *M*. Structure function studies demonstrated that minor alterations in the estrogen or tamoxifen derivatives can transform an agonist into a partial agonist or even antagonist. For antiestrogenic tamoxifen derivatives, the presence of a side chain in the trans position (i.e., position 4 of the α' ring; see Figure 15) with a single pair of electrons (such as oxygen or nitrogen) is necessary. Competition studies on estradiol and its antagonists binding to the ER receptor, and the blockage of various sites by

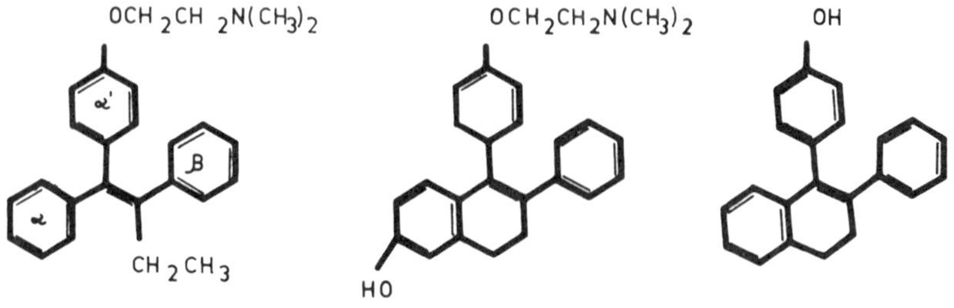

TAMOXIFEN I II

FIGURE 15. The tamoxifen formula and its two derivatives.

specific monoclonal antibodies, indicate that both agonists and antagonists bind the same site or nearby overlapping sites. Subtle changes in the ER protein conformation may explain the resulting agonistic or antagonistic effect of ligand binding.[208] Antihormones act by binding to the monomer receptor, but they have the possibility of becoming active if they follow an enzymatic activation.

Inhibiting effect may also appear by competition between steroid hormone receptors for factors that mediate their enhancer function.[209] For example, transcription stimulation of some reporter genes by the progesteronic receptor was inhibited in transfected HeLa cells by coexpression of the estrogenic receptor on a receptor and estrogen dose-dependent manner.[209]

Several mathematical models regarding gene activation by hormones have been elaborated to promote an easier understanding of these complex mechanisms.[210-212]

VII. CONCLUDING REMARKS

Regulation of gene expression is one of the most intensely studied topics in molecular biology. The regulatory processes are organized as a network and are multistep in development.

In bacteria, chromosomes consist of a highly organized circular DNA which is supercoiled, exhibiting a large number of loops. The active genes lie on the external part of loops and this location is favorable to the DNA binding proteins involved in regulation. In prokaryotes, the genes are colinear, but in eukaryotes they are interrupted. In bacteria and lower types of eukaryotes, the most common type of gene regulation occurs by induction and repression. The control of gene expression in eukaryotic cells takes place at the DNA transcription level. The initiation of transcription is a complex, multistep process, governed by the interaction between RNA polymerases, promoter sequences (other control elements), and multiple accessory transcription factors.

Cell transformation is considered to be an alteration of control mechanisms of gene expression. The alteration process appears as a consequence of various genetic and epigenetic mechanisms.

The recognition problem is one of the most important mechanisms. The binding of the regulatory proteins to specific sequences of DNA may occur by three pathways: sliding of the protein on the DNA chain; intersegment transfer, in which the protein passes from a certain segment of DNA to another; and by a hybrid process in which the regulatory proteins undertake both the sliding and the intersegment transfer.

The helix-turn-helix structure, the zinc finger, the leucine zipper, and other motifs have been well described. The DNA binding proteins, which control the expression of eukaryotic genes are associated with great selectivity to some particular site of the DNA sequence and the binding occurs with high affinity for the correct site and low affinity for other DNA sequences.

Many of the regulatory proteins in mammalian cells are represented by nuclear receptors — a family of ligand-inducible transcription factors. The ligands activate the gene transcription through recognition of the DNA sequences represented by responsive elements. From this class, those members of the steroid receptor superfamily whose prototypes are the glucocorticoid receptor and the estrogen receptor were discussed here.

ACKNOWLEDGMENT

I wish to thank Ion Niculescu-Duvăz, Simona Vasilescu, and Liliana Puiu for useful comments on the manuscript, and Monica Nagy for typing the manuscript.

REFERENCES

1. **Marinescu, I. and Voiculetz, N.,** Modelling cell transformation, *Arch. Roum. Pathol. Exp. Microbiol.*, 50(3), 79, 1991.
2. **Deuschle, U., Hipskind, R. A., and Bujard, H.,** RNA polymerase II transcription blocked by *Escherichia Coli Lac* repressor, *Science*, 248, 480, 1990.
3. **Alberts, B., Bray, D., Leavis, L., Raff, M., Roberts, K., and Watson, J. D.,** *Molecular Biology of the Cell*, 2nd ed., Garland Publishing, New York, 1989.
4. **Revzin, A.,** Nonspecific binding and the lactose operon, in *The Biology of Non-Specific DNA-Protein Interactions*, CRC Press, Boca Raton, FL, 1990, chap. 4.
5. **Ostrander, E. A., Benedetti, P., and Wang, J. C.,** Template supercoiling by a chimera of yeast GAL4 protein and phage T7 RNA polymerase, *Science*, 249, 1261, 1990.
6. **Record, M. T. and Spolar, R. S.,** Some thermodynamic principles of nonspecific and site-specific protein-DNA interactions, in *The Biology of Non-Specific DNA-Protein Interactions*, Revzin, A., Ed., CRC Press, Boca Raton, FL, 1990, chap. 2.

7. **Berg, O. G.,** The role of non-specific interactions of gene-regulatory proteins in their search for specific target sites, in *The Biology of Non-Specific DNA-Protein Interactions,* Revzin, A., Ed., CRC Press, Boca Raton, FL, 1990, chap. 3.
8. **Lewin, B.,** *Genes,* 3rd ed., John Wiley & Sons, New York, 1987.
9. **Watson, J. D., Hopkins, N. H., Roberts, J. W., Steitz, J. A., and Weiner, A. M.,** La regulation de la synthese et de la function de proteines chez les bacteries, in *Biologie Moleculaire du Gene,* Interedition, Paris, 1989, chap. 16.
10. **Lesser, D. R., Kurpiewsky, M. R., and Linda Jean-Jacobson,** The energetic basis of specificity in the EcoRl endonuclease-DNA interaction, *Science,* 250, 776, 1990.
11. **Record, M. T., Jr. and Mossing, M. C.,** Physical chemical origin of stability, specificity and control of protein-DNA interactions, in *RNA Polymerase and Regulation of Transcription,* Reznikoff, W. et al., Eds., Elsevier, New York, 1987, 61.
12. **Schleif, R.,** DNA binding by proteins, *Science,* 241, 1182, 1988.
13. **Mascotti, D. P. and Lohman, T. M.,** Thermodynamic extent counterion release upon binding oligolysines to single-stranded nucleic acids, *Proc. Natl. Acad. Sci. U.S.A.,* 87, 3833, 1981.
14. **Barkley, M. D.,** Salt dependence of the *lac* repressor-operator interaction: role of non-operator DNA in the association reaction, *Biochemistry,* 20, 3833, 1981.
15. **Singer, P. and Wu, C. W.,** Promoter search by *E. coli* RNA polymerase on a circular DNA template, *J. Biol. Chem.,* 262, 14178, 1987.
16. **Scheek, R. M., Zuderweg, E. R. P., Klappe, K. J. M., Van Boom, J. H., Kaptein, R., Ruterjans, H., and Beyreuther, K.,** Lac repressor headpiece binds specifically to half of *lac* operator: a proton nuclear magnetic resonance study, *Biochemistry,* 22, 228, 1983.
17. **Ptashne, M.,** Gene regulation by proteins acting nearby and at a distance, *Nature,* 322, 697, 1986.
18. **Nielsen, P. E.,** Sequence selective DNA recognition by synthetic ligands, *Bioconjugate Chem.,* 2, 1, 1991.
19. **Simon, Z.,** *Quantum Biochemistry and Specific Interactions,* Abacus Press, Tunbridge Wells, Kent, U.K., 1976.
20. **Von Hippel, P. H. and Mcghee, J. D.,** DNA-protein interactions, *Annu. Rev. Biochem.,* 41, 232, 1972.
21. **Simon, Z.,** Intermolecular forces in specific protein-nucleic acid interactions, *Stud. Biophys. (Berlin),* 28, 179, 1971.
22. **Lang, M. and Felsenfedd, G.,** Preferential interactions of polyLys and polyArg with specific base sequences in DNA, *Proc. Natl. Acad. Sci. U.S.A.,* 56, 1325, 1966.
23. **Paoletti, D. and Le Pecq, J. B.,** Resonance energy transfer between ethidium bromide molecules bound to nucleic acid. Does intercalation wind or unwind the DNA helix?, *J. Mol. Biol.,* 59, 43, 1971.
24. **Bianchi, M. E., Beltrame, M., and Pionessa, G.,** Specific recognition of cruciform DNA by nuclear protein HMGl, *Science,* 243, 1056, 1989.
25. **Marx, J. L.,** A crystalline view of protein DNA binding, *Science,* 229, 846, 1985.
26. **Neidle, S. and Abraham, Z.,** Structural and sequence dependent aspects of drug intercalation into nucleic acids, *CRC Crit. Rev. Biochem.,* 17, 73, 1984.
27. **Wakelin, L. P. G. and Waring, M. J.,** The binding of echinomicin to deoxyribonucleic acid, *Biochem. J.,* 157, 721, 1976.
28. **Riggs, A. D., Suzuki, H., and Bourgeois, S.,** Lac repressor-operator interaction. I. Equilibrium studies, *J. Mol. Biol.,* 48, 67, 1970.
29. **Zakrzeneska, K., Lavery, R., and Pullman, B.,** Theoretical studies on selective binding to DNA of two nonintercalating ligands: netrospin and SN 18071, *Nucl. Acids Res.,* 11, 8825, 1984.
30. **Gilbert, W. and Maxam, A.,** The nucleoside sequence of the lac operator, *Proc. Natl. Acad. Sci. U.S.A.,* 70, 3881, 1973.

31. **Maniatis, T. and Ptashne, M.,** Multiple repressor binding at the operators in bacterophague lambda, *Proc. Natl. Acad. Sci. U.S.A.*, 70, 1531, 1973.
32. **Simon, Z. and Ghetie, V.,** Molecular size and specificity in immunology, *Rev. Roum. Biochem.*, 8, 251, 1971.
33. **Sadler, J. R. and Smith, T. E.,** Mapping of the lactose operon, *J. Mol. Biol.*, 62, 139, 1971.
34. **Maniatis, T., Goodburn, S., and Fischer, J. A.,** Regulation of inducible tissue specific gene expression, *Science*, 236, 4837, 1987.
35. **Gurskii, G. B. and Zasedatelev, A. S.,** Exact relations for calculating binding of regulatory proteins and other network ligands to double stranded polynucleotides, *Biophisika (Moscow)*, 23, 932, 1978.
36. **Bruskov, B. J., Polsov, R. V., and Polteo, V. I.,** On the role of nonspecific complexation in protein-nucleic acid recognition, *Biophisika (Moscow)*, 23, 610, 1978.
37. **Wolfe, A. P.,** New approaches to chromatin function, *New Biol.*, 51, 89, 1982.
38. **Igo-Kelemes, T., Hörz, W., and Zacchau, A. G.,** Chromatin structure in eukaryotes, *Annu. Rev. Biochem.*, 51, 89, 1982.
39. **Moyne, G., Freeman, R., Savayosti, S., and Yaniv, M.,** A high resolution EPR study of nucleosomes from SV-40 chromatin, *J. Mol. Biol.*, 149, 735, 1981.
40. **Thoma, F., Koller, T. H., and Klug, A.,** Involvement of H_1 in the organisation of the nucleosome and in the salt dependent superstructure of chromatin, *J. Cell Biol.*, 83, 403, 1980.
41. **Adolph, K. W., Cheng, S. M., Pauson, S. R., and Laemmli, I. I.,** Isolation of a protein scaffold from mitotic HeLa cell chromosome, *Proc. Natl. Acad. Sci. U.S.A.*, 74, 4937, 1977.
42. **Ericson, C., Mehlin, H., Björkroth, B., Lamb, M. M., and Dancholt, B.,** The ultrastructure of upstream and downstream regions of an active Balbreni ring, *Cell*, 56, 631, 1989.
43. **Croston, G. E., Kerrigan, L. M., Marshak, D. R., and Kadonaga, J. T.,** Sequence specific antirepression of histone H_1 mediated inhibition of basal RNA polymerase II-transcription, *Science*, 251, 634, 1991.
44. **Comnungs, D. E. and Riggs, A. D.,** Molecular mechanism of chromosome pairing, folding and function, *Nature*, 233, 48, 1991.
45. **Record, M. T. and Spolar, R. S.,** Some thermodynamic principles of nonspecific and specific protein DNA interaction, in *The Biology of Nonspecific DNA-Protein Interactions*, Revzin, A., Ed., CRC Press, Boca Raton, FL, 1990, 33.
46. **Jacob, F. and Monod, J.,** Genetic regulatory mechanisms in the synthesis of proteins, *J. Mol. Biol.*, 3, 318, 1983.
47. **Gilbert, W. and Miller-Hill, B.,** Isolation of the *Lac* repressor, *Proc. Natl. Acad. Sci. U.S.A.*, 56, 1891, 1966.
48. **Ptashne, M.,** Specific binding of the λ phage repressor to DNA, *Nature*, 214, 232, 1967.
49. **Ebright, R. H., Pendergrast, P. S., and Gunasekera, A.,** Conversion of a helix-turn-helix motif sequence-specific DNA binding protein into a site-specific DNA cleavage agent, *Proc. Natl. Acad. Sci. U.S.A.*, 87, 2882, 1990.
50. **Menon, K. P. and Lee, N. L.,** Activation of *ara* operons by truncated Ara C protein does not require inducer, *Proc. Natl. Acad. Sci. U.S.A.*, 87, 3708, 1990.
51. **Lowson, R. C., Jr. and York, S. S.,** Stoichiometry of *lac* repressor binding to nonspecific DNA: three different complexes form, *Biochemistry*, 26, 4867, 1987.
52. **Berg, O. G. and Von Hippel, P. H.,** Selection of DNA binding sites by regulatory proteins, *Trends Biochem. Sci.*, 13, 207, 1988.
53. **Raiband, O. and Schwartz, M.,** Positive control of transcription initiation in bacteria, *Annu. Rev. Genet.*, 18, 173, 1984.

54. **McGuire, W. R.,** Mechanism and control of transcription in prokaryotes, *Annu. Rev. Biochem.,* 54, 171, 1985.
55. **Hochschild, A. N., Irwin, N., and Ptashne, M.,** Repressor structure and the mechanism of positive control, *Cell,* 32, 319, 1983.
56. **Shanblatt, S. H. and Revzin, A.,** Two catabolite activator protein molecules bind to the galactose promoter region of *Escherichia coli* in presence of RNA polymerase, *Proc. Natl. Acad. Sci. U.S.A.,* 80, 1594, 1983.
57. **Decrombrugghe, B., Bushy, S., and Buc, H.,** Cyclic AMP receptor protein: role in transcription activation, *Science,* 224, 831, 1984.
58. **Gilbert, W. and Miller-Hill, B.,** *Lac* operator is DNA, *Proc. Natl. Acad. Sci. U.S.A.,* 58, 1891, 2415, 1967.
59. **Heinemann, U.,** Molekulare Basis der Genkontrolle, *Nachr. Chem. Technol. Lab.,* 32, 712, 1984.
60. **Sadler, J. R. and Smith, T. F.,** Mapping of the lactose operon, *J. Mol. Biol.,* 62, 139, 1971.
61. **Jordan, S. R. and Pabo, C. O.,** Structure of the lambda complex at 25 Å resolution. Details of the repressor-operator interaction, *Science,* 242, 893, 1988.
62. **Aggarval, A. K., Rodgers, D. W., Drottar, M., Ptashne, M., and Harrison, S. C.,** Recognition of a DNA operator by the repressor of phage 434: a view at high resolution, *Science,* 242, 899, 1988.
63. **Bass, S., Sorrells, V., and Youderian, P.,** Mutant *trp* repressor with new DNA binding specificities, *Science,* 242, 240, 1988.
64. **Amowyal, M., Mortensen, L., Buc, H., and Hamma, K.,** Single and double loop formation when deoR-repressor binds to its natural operator site, *Cell,* 58, 545, 1989.
65. **Riggs, A. D., Suzuki, H., and Bourgeois, S.,** The *Lac* repressor-operator interaction. I. Equilibrium studies, *J. Mol. Biol.,* 48, 67, 1970.
66. **Riggs, A. D., Bourgeois, S., and Cohn, M.,** The *lac* repressor-operator interaction. III. Kinetic studies, *J. Mol. Biol.,* 53, 401, 1970.
67. **Gierer, A.,** Model for DNA-protein interaction and operator function, *Nature,* 212, 1480, 1966.
68. **Y. Lin, S. and Riggs, A. D.,** *Lac* repressor binding to DNA not containing the *lac* operator and the synthetic poly daT, *Nature,* 228, 1184, 1970.
69. **Wallace, M. R., Marchuk, D. A., Anderson, L. B., et al.,** Type 1 neurofibromatosis gene: identification of a large transcript disrupted in three NF-1 patients, *Science,* 249, 181, 1990.
70. **Muskavitz, M. A. T. and Hogness, D. S.,** Hypersensitivity to DN-ase of transcribed DNA sequences, *Cell,* 29, 1941, 1982.
71. **Maniatis, T., Godburn, S., and Fisher, J. A.,** Regulation of inducible and tissue specific gene expression, *Science,* 236, 4837, 1987.
72. **Guarante, L.,** The UASs and enhancers: common mechanism of transcriptional activation in yeast and mammals, *Cell,* 52, 303, 1988.
73. **Echols, H.,** Multiple DNA-protein interactions governing high precision DNA transcriptions, *Science,* 233, 1050, 1986.
74. **Graydon, R. D., Costa, R. H., Xanthopoulos, K. G., and Darnell, J. E.,** One factor recognizes the liver specific enhancers in α_1-antitryptisyn and transthyretin genes, *Science,* 239, 786, 1988.
75. **Müller, H. P., Sogo, J. M., and Schefner, W.,** An enhancer stimulates transcription in *trans* when attached to the promoter via protein bridge, *Cell,* 58, 767, 1989.
76. **Parraga, G., Horvath, S. J., Eisen, A., Taylor, W. E., Hood, L., Young, E. T., and Klevit, R. E.,** Zinc-dependent structure of a single-finger domain of yeast ADR_1, *Science,* 241, 1489, 1988.
77. **Lichsteiner, S., Wuarin, J., and Schibler, U.,** The interplay of DNA binding proteins on the promoter of mouse albumin gene, *Cell,* 51, 983, 1987.

78. **Cheregrin, S., Raymondjean, M., Carranca, A. G., Herbonnel, Ph., and Yaniv, M.,** Factors involved in control of tissue specific expression of albumin gene, *Cell,* 50, 627, 1987.
79. **Margottin, F.,** Participation of the TATA factor in transcription of the yeast U6 gene by RNA pol C, *Science,* 251, 424, 1991.
80. **Ishida, I., Verbeek, S., Bonnevill, M., Itohara, S., and Berns, A.,** T-cell receptor $\gamma\delta$ and transgenic mice suggest a role of a γ gene silencer in the generation of $\alpha\beta$ T cells, *Proc. Natl. Acad. Sci. U.S.A.,* 87, 3067, 1990.
81. **Muramatsu, M.,** U.S.-Japan seminar on oncogenes in relation to developmental and transcriptional control, *Jpn. J. Cancer Res. (Gann),* 78, 1287, 1987.
82. **Wigender, E. and Sefart, K. H.,** Transkription in Eukarioten die Rolle von Transcriptionkomplexen und ihren Komponenten, *Angew. Chem.,* 99, 206, 1987.
83. **Van Dyke, W. M., Roeder, R. G., and Sawadago, M.,** Physical analysis of transcription preinitiation complex assembly on a Class II gene promoter, *Science,* 241, 1335, 1988.
84. **Kagan, L., Gill, G., and Ptashne, M.,** Separation of DNA binding from the transcription activating function of a eukaryote regulatory protein, *Science,* 231, 699, 1986.
85. **Ma, J. and Ptashne, M.,** The carboxyl-terminal 30 aminoacids of GAL 4 are recognized by GAL 80, *Cell,* 50, 137, 1987.
86. **Driggers, P. H., Ennist, D. L., and Gleason, S. L.,** An interferon γ-regulated protein that binds the interferon inducible enhancer element of MCH class I genes, *Proc. Natl. Acad. Sci. U.S.A.,* 87, 3743, 1990.
87. **Liou, H. C., Boothby, M. R., and Glinicher, L. H.,** Distinct cloned class MHC DNA binding proteins recognize the X box transcription element, *Science,* 242, 69, 1988.
88. **Casselmann, W. H., Mayer, M., Kekulé, A. S., Lauer, U., Hotschneider, P., and Koshy, R.,** A transactivator function is generated by integration of HBV pre S/S sequences in human hepatocellular carcinoma DNA, *Proc. Natl. Acad. Sci. U.S.A.,* 87, 2970, 1990.
89. **Wu, C., Wilson, S., Walker, B., David, I., Paisley, T., Zimarino, V., and Ueda, H.,** Purification and properties of the *Drosophila* activator protein, *Science,* 238, 1237, 1987.
90. **Davidson, J., Xiao, J. H., Rosaes, R., Staub, A., and Chambon, P.,** The HeLa cell protein binds specifically and cooperatively to two SV 40 enhancer motifs of unrelated sequence, *Cell,* 54, 931, 1988.
91. **Tsai, S. Y., Duke, J. C., Weigel, N. L., Dahlman, K., Gustaffson, J. A., Tsai, M., and O'Malley, B. W.,** Molecular interaction of a steroid hormone receptor with enhancer element: evidence for receptor dimer formation, *Cell,* 55, 361, 1988.
92. **Hlazonetis, T. D., Georgeopoulos, K., Greenberg, M. E., and Leder, Ph.,** C-Jun dimerises with itself and with c-Fos, forming complexes of different DNA binding affinity, *Cell,* 55, 917, 1988.
93. **Kumar, V. and Chambon, P.,** The estrogen receptor binds tightly to its responsive element as a ligand induced homodimer, *Cell,* 55, 145, 1988.
94. **Greenblatt, J.,** Roles of TFIID in transcriptional initiation by RNA polymerase II, *Cell,* 66, 1067, 1991.
95. **Gradwohl, G.,** Menissier de Murcia, J., Molinete, M., and Simon, F., The second zinc-finger domain of poly (ADP-ribose) polymerase determines specificity for single stranded in DNA, *Proc. Natl. Acad. Sci. U.S.A.,* 87, 299, 1990.
96. **Mason, M. D.,** Differentiation and cancer, *Eur. J. Cancer,* 293, 255, 1992.
97. **Barrett, J. C.,** Genetic and epigenetic mechanism in carcinogenesis, in *Mechanism of Environmental Carcinogenesis,* Vol. 1, Barrett, J. C., Ed., CRC Press, Boca Raton, FL, 1987, chap. 1.
98. **Bishop, J. M.,** The molecular genetics of cancer, *Science,* 253, 305, 1987.
99. **Temin, H. M.,** Evolution of cancer genes as a mutation driven process, *Cancer Res.,* 48, 1697, 1988.

100. **Mars, W. M. and Saunders, G. F.,** Chromosomal abnormalities in human breast cancer, *Cancer Metastasis Rev.*, 9, 35, 1990.
101. **Voiculetz, N., Balaban, Al., Niculescu-Duvăz, I., and Simon, Z.,** *Modeling in Cancer Genesis and Prevention,* CRC Press, Boca Raton, FL, 1991.
102. **Beremblum, I.,** Sequential aspects of chemical carcinogenesis of skin, in *Cancer: A Comprehensive Treatise,* Vol. 1, 2nd ed., Becker, F. F., Ed., Plenum Press, New York, 1982.
103. **Voiculetz, N.,** Some new aspects in the molecular mechanism of tumorigenesis, *Folia Oncologica,* 15, 103, 1992.
104. **Patterson, H.,** Approaches to proto-oncogene and tumor supressor gene identification, *Eur. J. Cancer,* 293, 258, 1992.
105. **Hunter, T.,** Cooperation between oncogenes, *Cell,* 64, 249, 1991.
106. **Weinberg, R. A.,** The action of oncogenes in the cytoplasm and nucleus, *Science,* 230, 770, 1985.
107. **Weinberg, R. A.,** Binding the anti-oncogene, *Sci. Am.*, 258(9), 34, 1988.
108. **Peters, G.,** Oncogenes at viral integration sites, *Cell Growth Differ.*, 1, 503, 1990.
109. **Theillet, C., Adnane, J., Szepetowski, P., Simon, M. P., and Jeanteur, P.,** BCL-1 participate in 11q13 amplification found in breast cancer, *Oncogene,* 5, 147, 1990.
110. **Varmus, H. E.,** The molecular genetics of cellular oncogenes, *Am. Rev. Genet.*, 18, 553, 1984.
111. **Nishimura, S. and Sekiya, T.,** Human cancer and cellular oncogenes, *Biochem. J.*, 243, 313, 1987.
112. **Klein, G.,** *Cellular Oncogene Activation,* Marcel Dekker, New York, 1988, 302.
113. **Bishop, J. M.,** Molecular themes in oncogenesis, *Cell,* 64, 235, 1991.
114. **Marshall, C. J.,** Tumor suppressor genes, *Cell,* 64, 313, 1991.
115. **Weinberg, R. A.,** Tumor suppressor genes, *Science,* 254, 1138, 1991.
116. **Piechaczyk, M.,** Le control de la translocation nucleaire des proteines: un niveau supplementaire de régulation de l'activité des genes, *Mini-Synth. Med.*, 6, 803, 1990.
117. **Struhl, K.,** The Jun onco-protein a vertebrate transcription factor activates transcription in yeast, *Nature,* 332, 649, 1988.
118. **Bohman, D., Bos, T. J., Adma, A., Nishimura, T., Vogt, P. K., and Tjian, R.,** Human proto-oncogene *c-jun* encodes a DNA binding protein with structural and functional properties of transcription factor AP-1, *Science,* 238, 1386, 1987.
119. **Turner, R. and Tjian, R.,** Leucine repreats and an adjacent DNA binding domain mediate the formation of functional c-Fos-c-Jun heterodimers, *Science,* 243, 1689, 1989.
120. **Short, N. J.,** Regulation of transcription. Fraudulent promotion, *Nature,* 331, 393, 1988.
121. **Sassone-Corsi, P., Ransone, L. J., Lamph, W. W., and Verna, I. M.,** Direct interaction between Fos and Jun nuclear oncoprotein: role of the leucine zipper domain, *Nature,* 336, 629, 1988.
122. **Curran, T. and Franza, B. R., Jr., Fos and Jun,** the AP-1 connection, *Cell,* 55, 395, 1988.
123. **Roux, P., Blanchard, J. M., Fernandez, A., Lamb, N., Jeanteur, P., and Piechaczyk, M.,** Nuclear localization of c-Fos, but not v-Fos proteins, is controlled by extracellular signals, *Cell,* 63, 341, 1990.
124. **Kouzarides, T. and Ziff, E.,** The role of the leucine zipper in the Fos-Jun interaction, *Nature,* 336, 646, 1988.
125. **Gentz, R., Rauscher, F. J., Abate, C., and Curran, T.,** Parallel association of Fos and Jun leucine zipper juxtaposes DNA binding domains, *Science,* 243, 1695, 1989.
126. **Maelicke, A.,** Jahresuberblick Molekularbiologie 1988, Oncogene, *Nachr. Chem. Technol. Lab.*, 37, 166, 1989.
127. **Angel, P., Hattori, K., Smeal, T., and Karin, M.,** The *jun* proto-oncogene is positively autoregulated by the product jun-AP-1, *Cell,* 55, 875, 1988.

128. **Yunis, I. I. and Hoffman, W. R.,** Nuclear enzymes, fragile sites, and cancer, *J. Gerontol. Biol. Sci.,* 44, 37, 1989.
129. **Land, H., Parada, L. F., and Weinberg, R. A.,** Tumorigenic conversion of primary embryo fibroblasts require at least two cooperating oncogenes, *Nature,* 304, 596, 1983.
130. **Todaro, G. I.,** Tumor Growth Factor in *Accomplishments in Cancer Research,* 1986, Gen. Motor Cancer Res. Found., Fortuer J. G. and Rhoads J. E., Eds., Lippincott, Philadelphia, 1987, 149.
131. **Lupu, F. and Voiculetz, N.,** Antioncogene sau gene supresoare ale creşterii tumorale, *Oncologia,* 4, 241, 1989.
132. **Klein, G.,** The approaching era of the tumor suppressor genes, *Science,* 238, 1539, 1987.
133. **Pelech, S.,** How anticancer genes turn off tumors, *Science,* 29(1), 39, 1989.
134. **Knudson, A. G. J.,** Hereditary cancer, oncogenes and antioncogenes, *Cancer Res.,* 45, 1437, 1985.
135. **Harbour, J. W., Lai, S. L., Wang-Peng, L. F., Gazdar, A. F., Minna, J., and Kaye, F. J.,** Abnormalities in structure and expression of the human Rb gene in SCLC, *Science,* 241, 353, 1988.
136. **Lane, D. P. and Benchimol, S.,** Oncogene or antioncogene?, *Genes Dev.,* 4, 1, 1990.
137. **Stratton, M. R.,** The p 53 gene in human cancer, *Eur. J. Cancer,* 28, 293, 1992.
138. **Wileock, D. and Lane, D. P.,** Localization of p 53, retinoblastoma and host replication proteins at sites of viral replication at herpes-infected cells, *Nature,* 349, 429, 1991.
139. **Gannon, J. V. and Lane, D. P.,** Protein synthesis required to anchor a mutant p53 protein which is temperature sensitive for nuclear transport, *Nature,* 349, 802, 1991.
140. **Hollingworth, R. E. and Wen-Hwa, Lee,** Tumor suppressor genes: new prospects for cancer research, *J. Natl. Cancer Inst.,* 83, 91, 1991.
141. **Carter, T. H.,** Regulation of gene expression by tumor promoters, in *Mechanisms of Environmental Carcinogenesis,* Barrett, J. C., Ed., CRC Press, Boca Raton, FL, 1987, 47.
142. **Fusing, N. E. and Samsel, W.,** Growth-promoting activity of phorbol ester TPA on cultured mouse keratinocytes fibroblasts and carcinoma cells, in *Carcinogenesis,* Vol. II, Slaga, T. J., Sivak, A., and Bontwell, R. K., Eds., Raven Press, New York, 1978, 203.
143. **Yarnold, J. R.,** What are cancer genes and how do they upset cell behavior?, *Eur. J. Cancer,* 293, 251, 1992.
144. **Dimond, L.,** Tumor promoters and cell transformation, in *Mechanisms of Cellular Transformation by Carcinogenic Agents,* Prunberger, G. and Goff, S., Eds., Pergamon Press, Elmsford, New York, 1987, 73.
145. **Weinstein, I. B.,** Multistage carcinogenesis: the role of environmental factors, *Eur. Cancer News,* 4(4), 2, 1991.
146. **Nishizuka, I.,** Perspectives on the role of protein kinase C in stimulus-response coupling, *J. Natl. Cancer Inst.,* 76, 159, 1987.
147. **Yuspa, S. H.,** Tumor promotion, in *Accomplishments in Cancer Research 1986,* Gen. Motor Cancer Res. Found., Fortner, J. G. E., Ed., Lippincott, Philadelphia, 1987, 169.
148. **Colburn, N. H.,** The genetics of tumor promotion, in *Mechanism of Environmental Carcinogenesis,* Barrett, J. C., Ed., CRC Press, Boca Raton, FL, 1987, 81.
149. **Lerman, M. I., Hegamayer, G. A., and Colburn, N. H.,** Cloning and characterization of putative genes that specify sensitivity to induction of neoplastic transformation by tumor promoters, *Int. J. Cancer,* 37, 293, 1986.
150. **Ford, R. and Maizel, A., Eds.,** *Mediators in Cell Growth and Differentiation,* Raven Press, New York, 1986.
151. **Nowell, P. C.,** Mechanisms of tumor progression, *Cancer Res.,* 46, 2203, 1986.

152. **Croce, C. M. and Nowell, P. C.,** Molecular basis of human cell neoplasis, *Blood,* 65, 1, 1985.
153. **Klein, G. and Klein, Eva,** Evolution of tumors and the impact of molecular oncology, *Nature,* 315, 190, 1985.
154. **Mitelman, F.,** Catalog of chromosome aberatives in cancer, *Cytogenet. Cell Genet.,* 36, 1, 1983.
155. **Sandberg, A. A.,** Role of chromosome changes in carcinogenesis, in *Mechanisms of Environmental Carcinogenesis,* Barrett, J. C., Ed., CRC Press, Boca Raton, FL, 1987, 96.
156. **Solomon, E., Borrow, J., and Goddard, A. D.,** Chromosome aberrations and cancer, *Science,* 254, 1953, 1991.
157. **Rhim, S. J., Fujita, I., and Park, J. B.,** Activation of H-ras oncogene in 3-Me transformated human cell lines, *Carcinogenesis,* 8, 1165, 1987.
158. **Graffen, J., Stephenson, I. R., and Heisterkamp, N.,** Philadelphia chromosomal breakpoints are clustered within a limited region, *bcr,* in chromosome 22, *Cell,* 36, 93, 1984.
159. **Wiedemann, L. M. and Morgan, G. J.,** How are cancer associated genes activated or inactivated?, *Eur. J. Cancer,* 293, 248, 1992.
160. **Brada, M.,** BCL 2 gene: current relevance to clinical oncology, *Eur. J. Cancer,* 293, 270, 1992.
161. **Pegoraro, L., Palumbo, A., Erikson, J.,** CMA t (14) and a t(8;14) chromosome translocation in a cell line derived from an acute B-cell leukemia, *Proc. Natl. Acad. Sci. U.S.A.,* 81, 7166, 1984.
162. **Adnane, J., Daudray, P., and Simon, M. P.,** Proto-oncogene amplification and human breast tumor phenotype, *Oncogene,* 4, 1389, 1989.
163. **Ma, J. and Ptashne, M.,** Converting an eukaryotic transcriptional inhibitor into an activator, *Cell,* 55, 443, 1988.
164. **Green, M. R.,** When products of genes and antioncogenes meet, *Cell,* 56, 1, 1989.
165. **McGee, T. L., Yandell, D. W., and Dryia, T. A.,** Structure and partial genomic sequence of the human retinoblastoma susceptibility gene, *Gene (Amst.),* 80, 119, 1989.
166. **Dyson, N., Howley, P. M., Munger, K., and Harlow, E.,** The human papilloma virus-16 E7 oncoprotein is able to bind to the retinoblastoma gene product, *Science,* 243, 934, 1989.
167. **Lee, W. H., Shew, I. Y., Sery, T. W., Donaso, I. A., Young, L. J., Bookstein, R., and Lee, E. Y. H.,** The retinoblastoma susceptibility gene encodes a nuclear phosphoprotein associated with DNA binding activity, *Nature,* 329, 642, 1987.
168. **Mitchell, P. J. and Tjian, R.,** Transcriptional regulation in mammalian cells by sequence-specific DNA binding protein, *Science,* 245, 371, 1989.
169. **Gannon, J. V. and Lane, D. P.,** *p53* and DNA polymerase *alpha* compete for binding to SV40 T antigene, *Nature,* 329, 456, 1987.
170. **Sturzbecher, H. W., Brain, R., Mainets, T., Addison, C., Rudge, K., and Jenkins, J. R.,** Mouse p53 blocks SV40 DNA replication *in vitro* and downregulators T antigen DNA metilase activity, *Oncogene,* 3, 405, 1988.
171. **Braithwaitem, A. W., Sturzbecher, H. W., Addison, C., Palmer, C., Rudge, K., and Jenkins, J. R.,** Mouse p53 inhibits SV40 origin-dependent ADN replication, *Nature,* 329, 458, 1987.
172. **Bender, T. P., Thompson, C. B., and Kuohl, W. M.,** Differential expression of *c-myb* mRNA in murine B lymphomas by a block to transcription elongation, *Science,* 237, 1473, 1987.
173. **Lamb, N. C. J., Fernandez, A., Tourkine, N., Jeanteur, P., and Blanchard, J. M.,** Demonstration in living cells of an intragenic negative regulatory element within the rodent *c-fos* gene, *Cell,* 61, 485, 1990.

174. **Oron, Y., Straub, R. E., Traktman, P., and Gershengorn, M. C.,** Decreased TRH receptor mRNA activity precedes homologous downregulation: assay in oocytes, *Science,* 238, 1406, 1987.
175. **Picard, D., Salser, S. J., and Yamanato, K. R.,** A movable and regulable function within the steroid binding domain of the glucocorticoid receptor, *Cell,* 54, 1073, 1988.
176. **Akerblom, J. E., Slater, E. P., Beato, M., Baxter, I. D., and Nellon, P. L.,** Negative regulation by glucocorticoids through interference with cAMP responsive enhancer, *Science,* 241, 350, 1988.
177. **Glass, C. K., Helloway, J. M., Devary, O. V., and Rosenfeld, M. H.,** The thyroid hormones receptor binds with opposite transcriptional effect to a common sequence motif in thyroid hormone and estrogen response elements, *Cell,* 54, 313, 1988.
178. **Vaccaso, M., Pawlak, A., Jost, J. R.,** Positive and negative regulatory elements of chicken vitellogenin II gene characterized by *in vitro* transcription competition assays in a homologous system, *Proc. Natl. Acad. Sci. U.S.A.,* 87, 3047, 1990.
179. **Jaynes, J. B. and O'Farrell, P. H.,** Activation and repression of transcription containing proteins that bind at a common site, *Nature,* 336, 744, 1988.
180. **Biggin, M. D. and Tjian, R.,** A purified *Drosophila* homeodomain protein represses transcription *in vitro, Cell,* 58, 433, 1989.
181. **Yu, H., Porton, B., Shan, L., and Eckhardt, L. A.,** Role of the octamer motif in hybrid cell extinction of immunoglobuline gene expression: extinction is dominant in a two enhancer system, *Cell,* 58, 441, 1989.
182. **Meyer, N. M., Gronemeyer, H., Turcotte, B., Bocquel, M. T., Lasset, D., and Chambon, P.,** Steroid hormone receptors compete for factors that mediate their enhancer function, *Cell,* 57, 432, 1989.
183. **Mitchell, P. J. and Tjian, R.,** Transcriptional regulation in mammalian cells by sequence specific DNA binding proteins, *Science,* 245, 471, 1989.
184. **Vinson, C. R., Sigler, P. B., and McKnight, S. L.,** Scissor-grip model for DNA recognition by a family of leucine zipper proteins, *Science,* 246, 911, 1989.
185. **Parletich, N. P. and Pabo, C. O.,** Zinc finger-DNA recognition: crystal structure of Zif 268-DNA complex at 2.1 Å, *Science,* 252, 809, 1991.
186. **Jones, N.,** Transcriptional regulation by dimerisation: two sides of an incestuous relationship, *Cell,* 61, 9, 1990.
187. **Luisi, B. F., Xu, W. X., Otwinowski, Z., Frechman, L. P., Yamamoto, K. R., and Sigler, P. B.,** Crystallographic analysis of the interaction of the glucocorticoid receptor with DNA, *Nature,* 352, 497, 1991.
188. **O'Malley, B.,** The steroid receptor superfamily: more excitement predicted for the future, *Mol. Endocrinol.,* 4(3), 363, 1990.
189. **Dawis, C. G.,** The many faces of epidermal growth factor repeats, *New Biol.,* 3, 410, 1990.
190. **Chin, R., Boyle, W. L., Meek, J., Smeal, T., Hunter, T., and Karin, M.,** The c-Fos protein interacts with c-Jun/AP-1 to stimulate transcription of AP-1 responsive genes, *Cell,* 54, 541, 1988.
191. **Auwerx, J. and Sassone-Corsi, P.,** IP-1: a dominant inhibitor of Fos/Jun whose activity is modulated by phosphorylation, *Cell,* 64, 983, 1991.
192. **Benbrook, D. M. and Jones, N. C.,** Heterodimer formation between CREB and Jun protein, *Oncogene,* 5, 295, 1990.
193. **Foulkes, N. S., Borrelli, E., and Sassone-Corsi, P.,** CREB gene: use of alternative DNA-binding domains generates multiple antagonists of cAMP-induced transcription, *Cell,* 64, 739, 1991.
194. **Shaw, P. E.,** Multicomponent transcription factors: the exception or the rule?, *New Biol.,* 2, 111, 1990.

195. **Nye, J. A. and Graves, B. J.,** Alkylation interference identifies essential DNA contacts for sequence specific binding of the eukaryotic transcription factor C/EBP, *Proc. Natl. Acad. Sci. U.S.A.,* 87, 3992, 1990.
196. **Sukaguchi, K., Appella, E., Omichinski, J. G., Clarc, C. M., and Gronenborn, A. M.,** Specific DNA binding to a major histocompatibility complex on enhancer sequence by a synthetic 57-residue double zinc finger peptide from human enhancer binding protein, *J. Biol. Chem.,* 266, 7306, 1991.
197. **Wu, C., Wilson, S., Walker, B., David, J., Paisley, T., Zimarino, V., and Ueda, H.,** Purification and properties of the *Drosophila* heatshock activator protein, *Science,* 238, 1247, 1989.
198. **Emerson, B. M., Lewis, C. D., and Felsenfeld, S.,** Interaction of specific nuclear factors with the nuclease hypersensitive region of the chicken adult beta-globine gene. Nature of the binding domain, *Cell,* 41, 21, 1985.
199. **Roy, A. K. and Clark, J. H., Eds.,** *Gene Regulation By Steroid Hormones III,* Springer-Verlag, New York, 1987.
200. **O'Malley, B. W.,** A review of the cellular biology of steroid hormone action in the chick oviduct, in *The Role of Receptors in Biology and Medicine,* Gotto, A. M. and O'Malley, B. W., Eds., Raven Press, New York, 1986.
201. **Forman, B. M. and Samuels, A. H.,** Dimerisation among nuclear hormone receptors, *New Biol.,* 2, 587, 1990.
202. **Berg, J. M.,** DNA binding specificity of steroid receptors, *Cell,* 57, 1065, 1989.
203. **Tsai, S. Y., Tsai, M. J., and O'Malley, B. W.,** Cooperative binding of steroid hormone receptors contributes to transcriptional synergism at target enhancer element, *Cell,* 57, 443, 1989.
204. **Wang, L. H., Tsai, S. Y., and Cook, R. G.,** COUP transcription factor is a member of the steroid receptor superfamily, *Nature,* 340, 183, 1989.
205. **Tora, L., White, J., and Bron, C.,** The human estrogen receptor has two independent non acidic transcriptional activation functions, *Cell,* 477, 1989.
206. **Mornon, J. P.,** Approaches to drug development using receptors, in *The Role of Receptors in Biology and Medicine,* Gotto, A. M., Jr. and O'Malley, B. W., Eds., Raven Press, New York, 1986.
207. **Funder, J. W., Pearce, P. T., Smith, R., and Smith, A. J.,** Mineralcorticoid action: target tissue specificity is enzyme, not receptor mediated, *Science,* 242, 583, 1988.
208. **Murphy, G. S. and Jordan, V. C.,** The biological significance of the interaction of estrogen agonists and antagonists with the estrogen receptor, in *Receptors,* Litwack, G., Ed., Humana Press, New York, 1990.
209. **Meyer, M. E., Gronemeyer, H., Turcotte, D., Bocquel, M. T., Tasset, D., and Cheunbin, P.,** Steroid hormone receptors compete for factors that mediate their enhancer function, *Cell,* 57, 433, 1989.
210. **Simon, Z. and Voiculetz, N.,** Trigger models for gene activity regulation in carcinogenesis, in *Modeling of Cancer Genesis and Prevention,* Voiculetz, N., Balaban, A. T., Niculescu-Duvăz, I., and Simon, Z., Eds., CRC Press, Boca Raton, FL, 1991.
211. **Simon, Z., Vlad, A., and Mihalaş, G. I.,** Positive gene control and steroid hormone receptor protein interactions. A mathematical model, *Stud. Biophys. (Berlin),* 125, 137, 1988.
212. **Simon, Z. and Mihalaş, G. I.,** Steroid hormone regulation and gene autoactivation, analytical and computer simulation studies, *Rev. Roum. Biochim.,* 28, 173, 1991.

Chapter 5

BIOPOLYMER SPATIAL CONFIGURATION AND THE FIDELITY OF DNA REPLICATION, TRANSCRIPTION, AND TRANSLATION

Nicolae Voiculetz, Simona Vasilescu, Mariana Marin, and Zeno Simon

TABLE OF CONTENTS

0-8493-5398-X/93/$0.00 + $.50

I. INTRODUCTION

DNA replication and transcription and mRNA translation are the most important processes occurring in prokaryotes and eukaryotes. Each of them is a complex, multistep process, governed by interactions between various enzymes and several accessory proteins with important regulatory activities. Being so complex and with rapid kinetics, the general material and its products require not only an extremely accurate mechanism for DNA replication (e.g., during this process, about 500 nucleotides per second are incorporated in bacteria and about 50/s in mammals), but also a mechanism for repairing the multiple accidental lesions that continually occur during this process.

The fidelity of this process is important in understanding how DNA is perpetuated through innumerable generations and how its products (mRNA and final proteins) are accurately maintained. This problem is obvious and has already been discussed in many textbooks.[1-5]

Due to the statistically inherent error probability, in the case of DNA template duplication or transcription there is a p error probability for a false monomer to be incorporated in the resulting products. If the length of the last product is not small with respect to a threshold length ($N = 1/p$ monomer units), most duplicate or transcript products will contain one or more false monomer units. If we consider the error probability in the interactions of base pairing, we would expect errors to take place in the bacterial genome at a frequency of 10^{-4} to 10^{-5} per replication and base pair, and/or in other terms, with an approximate rate of 10 per gene, per generation. The accuracy of DNA replication for primitive eukaryotes corresponds to 10^{-11} mutations per replication and base pair[6] (as a reminder, diploid cells have 6×10^9 bp).

The error in the entire process of protein synthesis was calculated to be less than 10^{-4} per amino acid residues[1,4] and the length of polypeptide chains usually is no more than 10^4 amino acid residues.

The accomplishment of these required fidelities is not a trivial problem. For example, in the Watson-Crick complementary base pair sequence, guanine (G) has a proton acceptor-donor-donor arrangement of hydrogen bonding groups involved in complementary base bonding. For enolic guanine there is a proton donor-acceptor-donor arrangement, and G (enol) can build three hydrogen bonds with the false thymine (T) nucleotide. HMO calculation indicates that the G (enol) tautomer is only about +5.5 kcal/mol less stable than the standard G (keto) tautomers. More accurate calculations[1,2] indicate a 8.4 kcal/mol loss of stability for the G (enol) T pair as compared to the correct G (keto) C pair. According to the equilibrium described in Chapter 1, the error probability for incorporating a false nucleotide (when concentrations of false and complementary nucleotides are similar) is $p = \exp(-\Delta\Delta G/RT)$, where $\Delta\Delta G$ is the affinity loss for the false pair. For normal temperatures ($2.3\ RT = 1.4$ kcal/mol), the probability of the false G (enol) T pair formation will be 10^{-4} to 10^{-6}, while the high fidelity corresponding to a $p = 10^{-11}$ for point mutations in eukaryotes requires a $\Delta\Delta G \simeq 15$ kcal/mol.

The Watson-Crick model of complementary base pairing, by itself, cannot explain the high fidelity required by DNA replication. Thus, the high fidelity of DNA replication was established to be assured enzymatically, either by a presynthesis control of errors that improves the specificity of complementary base selection and/or by a control that appears during synthesis, that in case of a misfit there is immediate removal of the added nucleotide (by the so-called proofreading activity).

The mechanisms of replication, transcription, and translation require more complicated interplays of intermolecular forces, steric complementarity, and spatial configuration of the macromolecules involved.

In recent years, there have been some significant advances, especially in the study of DNA replication, in the interactions involving tRNA molecules, and in establishing the significance of the so-called second genetic code in the translation process.[7,8]

In this chapter, the emphasis will be directed to these aspects, without neglecting the more classical ones. Biopolymer spatial configuration, enzymes involved in DNA replication, and mechanisms ascertaining the required high fidelity will be discussed, as well as the mechanism of transcription and RNA maturation, tRNA structure, and interactions with complexes involved in the translation process of mRNA in the final products: proteins.

II. PROTEIN AND NUCLEIC ACID STRUCTURE

At present, there is a large number of proteins whose primary structures have been established and a significant, but much smaller, number of proteins whose spatial configurations are known. The 1968 edition of the *Atlas of*

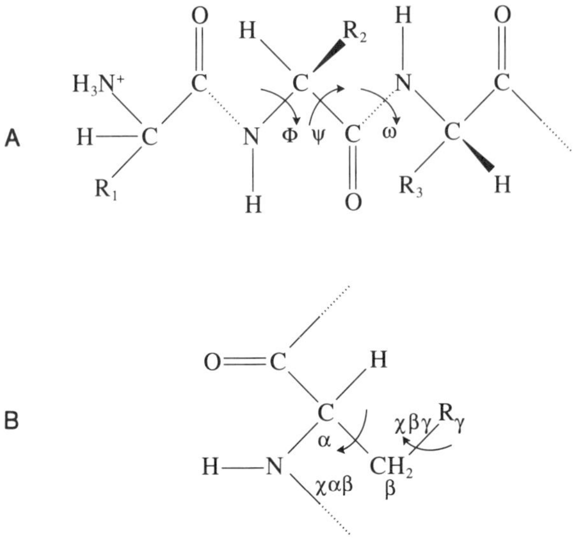

FIGURE 1. A: Spatial configuration of amino acid residues in the polypeptidic chain; B: Orientation of the side chain. The orientation in space of amino acid side chains could be characterized by the two angles $\chi_{\alpha\beta}$ and $\chi_{\beta\alpha}$ which exprime the torsion around C_α-C_β bonds and C_β-C_γ respectively.

Protein Sequence and Structures[9] lists the amino acid sequences for over 200 proteins. A more recent publication[10] cites more than 300 proteins with known spatial configuration and about 10,000 with known primary structure. Determination of the three-dimensional structure of proteins and nucleic acids is now possible by using nuclear magnetic resonance spectroscopy.[7] In addition, knowing the primary structure may measurably increase the possibilities of characterizing the spatial structure of other proteins and other macromolecules.

A. GEOMETRIC CHARACTERIZATION OF BIOPOLYMER SPATIAL CONFORMATION

The spatial conformation of biopolymers is determined by orientation toward the bonds in the molecular backbone. Each monomeric unit contains a number of backbone bonds, and the orientation may be characterized by the values of the corresponding angles.

The spatial orientation of amino acid residues in the polypeptide chain (except the terminal amino acids) is characterized by free angles: ψ, ϕ, and ω corresponding to the twisting angles around N-C_α, C_α-C, and C-N bonds, respectively[8] (Figure 1a). The CN bond being partially double owing to conjugation, the ω angle is fixed either at 0° (trans configuration) or at 180° (cis configuration). In polypeptides, trans is the most common configuration. In the case of proline, the N-C bond belongs to the pyrolidone ring, and the

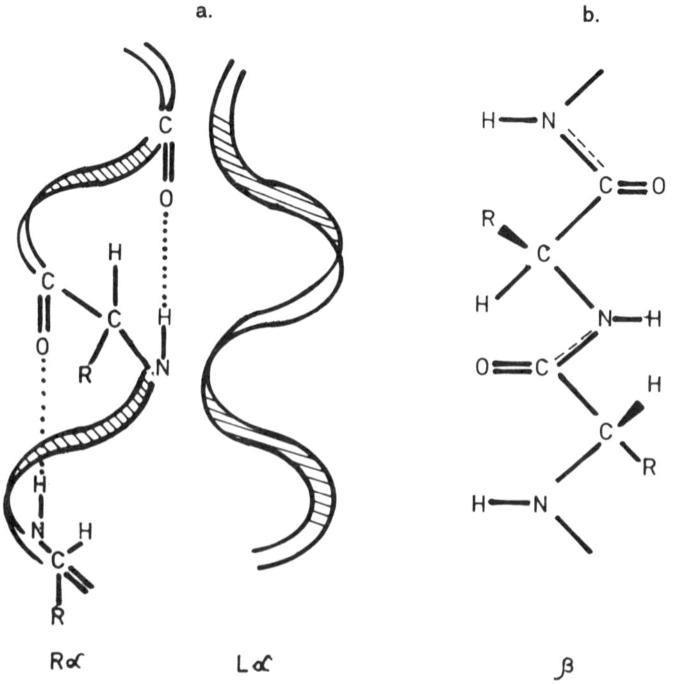

FIGURE 2. Secondary structure of polypeptides — a: R_α, α right-handed helix and L_α, α left-handed helix; b: β extended structure.

ϕ angle is fixed at 30°. When the ϕ and ψ angles are equal to zero, the orientation is the corresponding trans conformer.

The side chains usually are bound to the C_α atoms of the main polypeptide backbone by one CH_2 group, and its orientation can be characterized by two angles, $\chi_{\alpha\beta}$ and $\chi_{\beta\gamma}$ (Figure 1b).

The usual secondary structures for polypeptides (Figure 2a) are (1) the α helical structure (right-handed, R_α and left-handed, L_α), characterized by $\phi = \psi = +120°$ and $-120°$, respectively; (2) the β structure with $\phi = 25°$, $\psi = -25°$; and (3) the β turn (coil) (Figure 2b).

The dimensional characteristics for the α helix are one turn of 3.7 amino acid residues 5.44 Å along the helix axis, and a van der Waals thickness of about 21 Å. For the extended β structure, a length of 3.63 Å corresponds per amino acid residue. For the β turn or coil, there is a reverse turn consisting of 4 amino acids with 4.8 to 5.7 Å between the first and the last aliphatic C_α atom.

In the case of the nucleic acids, there are several simple bonds in the biopolymer backbone. According to a nomenclature system advanced by Sundaralingam,[11] there is an angle χ which characterizes the orientation of the base plane and six angles (ψ, ϕ, T_3, θ_2, and θ_1) for the phosphate sugar backbone. Their spatial orientations are depicted in Figure 3.

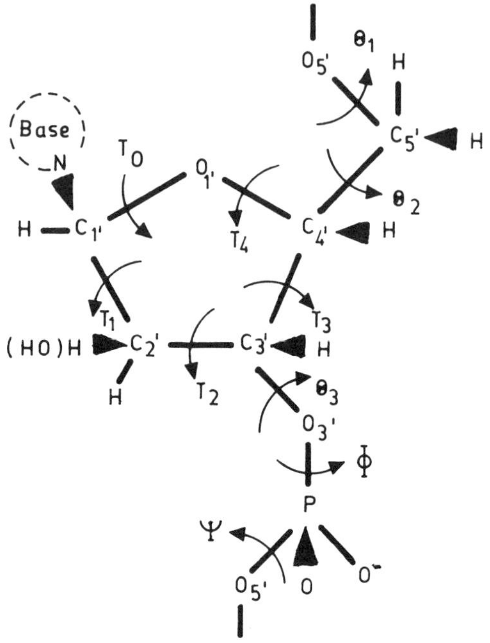

FIGURE 3. Spatial orientation of the nucleoside residue. There is a χ angle which characterizes the orientation of the base plane and six angles.

Examination of the structures by X-ray crystallography and steric models for polynucleotides, nucleotides, and related phosphoesteric derivatives has led to the conclusion that, for the helical nucleic acid backbone, the possible conformations are surprisingly limited, and the preferred orientation of the nucleotide unit in polynucleotides and nucleic acids is the same as was observed in the crystalline structure of the nucleotide. The DNA helix turn is 3.36 Å or 36°, and the van der Waals thickness is about 20 Å. The pentosophosphoric chain is at the outward extremity of the helix and the pentose rings are parallel to the longitudinal axis of the double helix. The N(2)(pyrimidine)–N(6)(purine) distances are approximately 11 Å, the distances between the C'_1 atoms in the deoxyribose rings of the opposite, antiparallel chains are 11 Å. The lengths of the hydrogen bonds range from 2.8 to 2.95 Å. The wide major groove is approximately 22 Å across, measured between phosphates along the helix axis, while the minor (shallow) groove is approximately 12 Å across. Both grooves are approximately 7 Å deep, measured from the surface of the enveloping cylinder. The base pairs of the DNA double helix are depicted in Figure 4. Some special structures of RNA chain segments, hairpin loops, internal loops, and bulges are depicted in Figure 5.

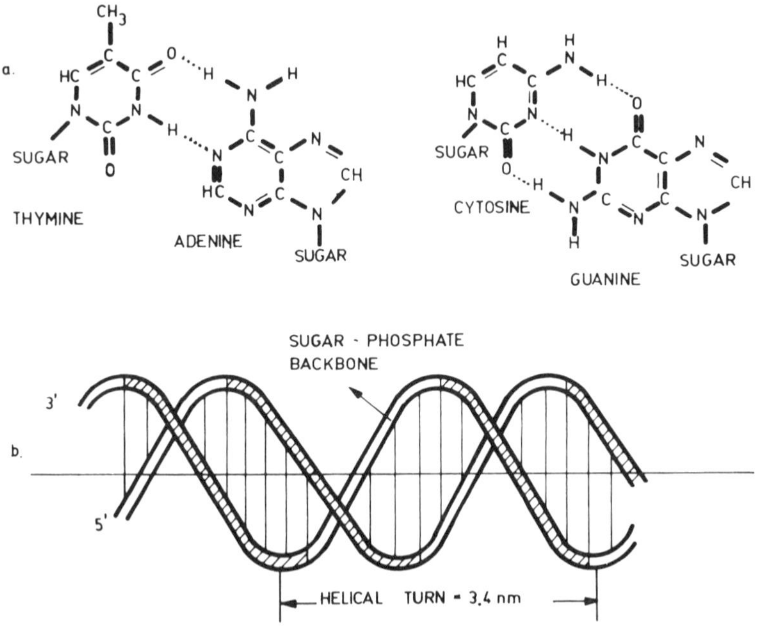

FIGURE 4. a: Complementary base pairing involves the formation of two hydrogen bonds between T and A, and of three hydrogen bonds between C and G; b: The two strands of DNA form a double helix.

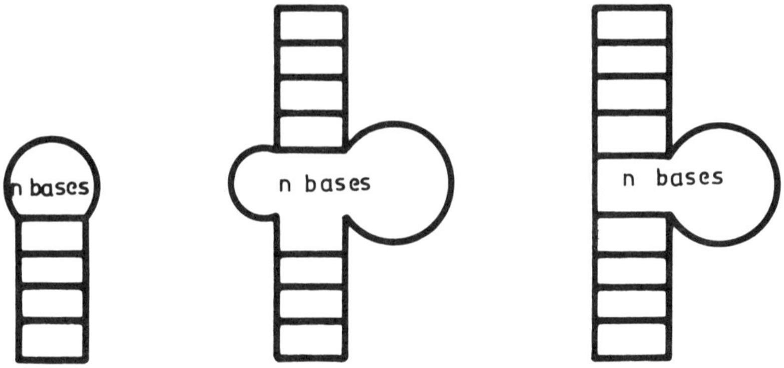

FIGURE 5. Single-stranded RNA folds back on itself to form a duplex hairpin and loops by pairing between complementary sequences.

Triple helix formations also occur in DNA and RNA strands, due to the formation of Hoogsteen's hydrogen bonds.[12,13] It is promoted by high concentrations of cations (e.g., 1 *M* Na^+), necessary for shielding the negative phosphate backbones, but it also can occur at physiological salt concentration and pH in the presence of spermidine (1 m*M*). It is a versatile approach to sequence selective DNA recognition.[13]

B. ENERGY CALCULATIONS REGARDING BIOPOLYMER CONFORMATION

Some results of energy calculations for biopolymer conformation, using molecular mechanic interaction potentials, were discussed in Chapter 2, Sections II.C and II.D. Such calculations require extensive use of computers and are successful in predicting spatial conformation of organic molecules, oligopeptides, oligonucleotides, and even in refining the spatial conformation of a polypeptide if one starts from a conformation near enough to the real one. They are not successful in predicting spatial conformation of polypeptides or global spatial structures of proteins, starting from the primary structure, because each amino acid residue has two simple bonds with three energy minima for twisting, which require energy computations for 9^N possible conformations. Also, one can not be sure if the spatial conformations of native proteins correspond to an absolute free enthalpy (G) minimum, or to some local one directed by kinetics of the folding of nascent polypeptidic chains on ribosomes.

''Energetic map'' calculations for oligopeptides require potential energy dependence on ϕ and ψ angles for each amino acid residue. These calculations also take into account interactions between nonbound atoms of the polypeptide and those between these atoms and water molecules.[8,14,15]

On the basis of some computations for homopolypeptides, dipeptides, and isolated amino acid units, Liquori[8] introduced a so-called ''stereochemical code''. Accordingly, amino acid residues in polypeptides may exist in one of the five spatial conformations corresponding to local energy minima. The spatial configuration of an N peptide may thus be described by a ''word'' of N-2 letters of a five-letter alphabet (the amino acids from the chain ends having no fixed configurations). In order to calculate the energy of the interaction, only the (E_{dis} + E_{tors}) terms (see Chapter 2, Section II.C, Equation 14) were considered. The results of such calculations for alanine and polyalanine are listed in Table 1. The five local minima in the energetic maps appear at about the same positions. The energy of regions outside the minima are somewhat higher than +4 kcal/mol for alanine and +2 kcal/mol for polyalanine. For the dipeptide, the position of the minima for the considered residue is practically independent of the orientation of the second unit. Calculations for other amino acids indicate very similar minima, independent of the side chain $-CH_2R$, with the exception of proline and glycine.

For proline $\phi = 30°$, is fixed by the pyrrolidinic cycle, while ψ has two minimum energy values, at $\psi = -12°$ and $+140°$. For glycine, the energetic map is centrosymmetric and very smooth. In all cases, the most stable configuration for the peptide group is the trans one ($\omega = 0$). If intrapeptidic carbonyl imine hydrogen bonds are included in the potential energy function,

TABLE 1
Energetically Favorable Conformations for Ala and Poly-Ala

	Ala				Poly-Ala		
	ϕ	ψ	V (kcal/mol)		ϕ	ψ	V (kcal/mol)
R_a	117°	131°	−0.4	R_α	119°	133°	−3.2
R_b	25°	−25°	0.7	R_β	25°	−25°	0.4
R_c	30°	−110°	0.0	R_γ	30°	−110°	−0.3
R_d	105°	−25°	−0.1	R_δ	105°	−25°	−0.4
L_a	−128°	−120°	0.1	L_α	−128°	−124°	−2.7

Note: R_α and L_α refer to α helixes and to the β extended structure in Poly-Ala.

the energetic map contains several minima, all of which are in the high energy region of the (E_{dis} + E_{tors}) map, except for two minima at ϕ, ψ = +135°, +120°, and −135°, −120°, respectively, which fall near the positions of R_α and L_α helical conformations. This explains the preference of polypeptides for α helical conformations.

Calculations of van der Waals and torsional energies (E_V + E_T) were performed for several chains, with a torsional barrier of 2.8 kcal/mol for nonterminal C–C bonds next to aromatic cycles. The conformation of each bond within the side chain results close to χ = 60°, 180°, and −60°.

According to Liquori,[8] each amino acid residue in a polypeptide should be found in one of the five low energy conformations (R_a, b, c, d, and L_a). Which of these orientations is preferred depends on the interaction between amino acid residues and water molecules. A test of this hypothesis on myoglobin demonstrated that most of the 153 amino acid residues are grouped, with orientation fairly close to the five predicted minima, and the statistical analysis gave satisfactory results.[8] The idea of this code has not been generally adopted.[14]

Several papers deal with a more complete description of protein configurations — such as the α helix or β sheet or coil conformation. One must also take into account that the global shape of the polypeptide can be strongly affected by small but systematic deviations from the five predicted conformations.[8]

Calculations for nucleotides and nucleic acids have been performed since the early 1970s.[11,16] The pentosophosphate backbone contains five bonds per monomer, around which several conformational isomers are possible.

More recently, molecular dynamic simulations for polymers were also performed, for example, for nylon fiber,[17] with interesting results regarding the free movement of methylene groups near the melting temperature. Relatively recently, conformational energy calculations for the folding of a polypeptide chain into a parallel structure were performed,[18] and a computer program, PACK, based on the finite differences method, was established for energy minimization in interactions of secondary structures in proteins.[19] An analysis of current methodologies for the conformational searching of complex molecules is presented by Howard and Kollman.[20] Molecular dynamics and computer graphics were applied to construct a three-dimensional model for prorenin starting from the known pepsinogen structure.[21]

C. SECONDARY STRUCTURE PREDICTIONS

Secondary structure prediction from the primary structure is a first step for predicting the steric structure. The first attempt at predicting the secondary structure was made by Prothero,[22] followed by Kotelchuck and Scheraga,[23] Ptitsyn and Finkelstein,[24] and several other groups. More recently, this work has been continued by Pachamabham et al.[25] and by Cohen et al.[26] The logic of these efforts is to analyze proteins with known secondary structure, to search for amino acids more frequently found within a given type of secondary structure, and to recognize types of short amino acid sequences that determine a given type of secondary structure. Table 2 lists such tendencies of the individual amino acids to form or break the α helix, β sheet, and α turn structures according to Meltzer as quoted by Chou and Fasman.[27]

It is clear from this type of research that several amino acids determine the structure of a sequence. The relative frequencies of a given amino acid within a secondary structure agree fairly well with the results obtained by different authors. The rules for secondary structure determination by a sequence of amino acids are generally different, depending upon the "training set" of the protein used. The results of some other authors, testing such rules on different sets of proteins, are usually less satisfactory.[28]

Various programs for "learning" secondary structure rules were devised for computers; for example, there are methods simulating the learning neuronal networks, starting from sequences of up to 9 amino acids with up to 220,000 analyzed sequences.[26] The precision of prediction is only 64% and probably cannot be improved because of interactions between amino acid residues which are distant in the primary structure, but which come into direct contact due to the spatial structure of the entire protein.[28]

Concerning nucleic acids, like the molecules of RNA and synthetic polyribonucleotides, the double helix-forming tendency is less evident than for polydeoxyribonucleotides.[29] The double helix conformation with the usual B form can also adopt the A form (11 base pairs per turn, rise per residue = 2.82 Å), and the C form (9.3 base pairs per turn, 3.32 Å rise per residue).[30] For oligoribonucleotides and deoxyribonucleotides, there are also some rules concerning the length and base composition of the chain and the tendency to form double helix structures.

D. SPATIAL PROTEIN STRUCTURE

Prediction of spatial protein structure from a primary structure puts the question if the spatial structure of protein in solution corresponds to the absolute G minimum or if the structure is determined by foldings of the nascent polypeptide chain. According to studies of mild denaturation of some

TABLE 2
Classification of Amino Acid Residues According to Their Tendency to Form α Helix, β Structure, and β Turns Secondary Structure

Amino acid	α Helix P_α	tendency	β Structure P_β	tendency	β Turn P_t tendency
Glu^-	1.53	+ +	0.26	br +	0.44
Ala	1.45	+ +	0.97	w	0.57
Leu	1.34	+ +	1.22	+	0.533
His^+	1.24	+	0.71	br	0.69
Met	1.20	+	1.67	+ +	0.67
Gln	1.17		1.23	+	0.56
Trp	1.14	+	1.19	+	1.11
Val	1.14	+	1.65	+ +	0.30
Phe	1.12	+	1.28	+	0.71
Lys^+	1.07	w	0.74	br	1.01
Ile	1.00	w	1.60	+ +	0.58
Asp^-	0.98	i	0.80	+ +	1.26
Thr	0.82	i	1.20	+	1.00
Ser	0.79	i	0.72	br	1.56
Arg^+	0.79	i	0.90	i	1.00
Cys	0.77	i	1.30	+	1.17
Asn	0.73	br	0.65	br	1.68
Tyr	0.61	br	1.29	+	1.25
Pro	0.59	br +	0.62	br	1.54
Gly	0.53	br +	0.81	i	1.68

Note: P_α, P_β, P_t, frequencies of the amino acidic residue in α, β, or turn structure, compared to medium frequency of residues in the corresponding regions; + +, +, the forming tendency (very strong; strong) in sequences with other amino acids; br, br +, breaking tendency in sequences; w, i, weak forming or inhibiting tendency in sequences.

enzymes, their activity is recovered after renaturation, and the same is true for reduction-reoxidation of disulfide bonds,[9] but not for any protein denaturation-renaturation process.

A theory of the way a polypeptide chain reaches its natural spatial structure was formulated in the early 1970s. De Cohen,[31] reporting results based on computer simulations and conformational energy calculation, claimed that during the synthesis of the polypeptide chain the ribosome proceeds to the formation of a peptide having a length of about six amino acid residues. This is the most thermodynamically stable conformation, the rest of the polypeptide chains begin coiling on this starting core.

A similar mechanism was proposed by Lewis et al.[32] who considered that some amino acid segments, based upon nearest-neighbor interactions, have a strong tendency to adopt a certain spatial conformation and serve as nucleation for the rest of the chain. Evaluation of the time frame for random evolution toward the most stable conformation indicates a period much longer than the duration of the denaturation and renaturation processes.[33] Protein folding also may depend on so-called "molecular chaperones", auxiliary factors (proteins) which favor this phenomenon.[34] Such auxiliary factors seem to be the mitochondrial hsp-60 protein and ubiquitin. Conclusively, protein folding is largely kinetically determined along with the biosynthesis of the polypeptide chain on the polyribosome.[35] These protein-folding rules seem rather complicated. Mutations at certain amino acid sites within a certain P22 bacteriophage protein produce a global suppression of protein folding.[36]

Various sophisticated methods have been tried to predict three-dimensional structures from amino acid sequences, such as the networks developed to simulate logical processes.[37] The result of the predictions for the amino acid residue based upon geometrical constraints and hydrophobicity, were well correlated with experimentally derived interdistances. The correlation coefficients appeared to be between $r = 0.5$ and $r = 0.9$ for different types of residues.[38] Nevertheless, using predictions based on sequence homologies, the data base of about 300 to 400 proteins with three-dimensional structures can be extended to about 1,800 proteins, as compared to more than 12,000 proteins with known sequences.[39]

Discussions of rules for spatial protein structure on a more semi-empiric basis, were given by Welscher[40] and Schultz.[41] The energy balance for polypeptide folding is of the type:

$$\Delta G = \Delta H_{bond} + \Delta G_{hph} - T\Delta S_{conf} \tag{1}$$

ΔH_{bond} corresponds to intrachain hydrogen bonding and disulfide bonds; ΔG_{hph}, to hydrophobic effects for the formation of inner cores; and ΔS_{conf}, for the loss of conformational entropy due to the fixing of a given conformation. The first two terms act in favor of folding in the structure formation, but the last one acts against it. With two simple CC_α and $C_\alpha N$ bonds, each with three conformational minima per amino acid residue at 300 K, the conformational entropy will be

$$-T\Delta S_{conf} \cong -TR\ln 3^2 = 1.2 \text{ kcal/mol of amino acid residue} \tag{2}$$

Two important indices for the tendency of association are the mean hydrophobicity, $\overline{H\phi}$, the sum of Tanford hydrophobicity increments (see Chapter 2, Table 3) per number of amino acids; and $\overline{FC}$ (the mean fractional charge) — the sum of side chains with ionized groups at $pH = 7.0$ per number of amino acids. Proteins with a high hydrophobicity ($\overline{H\phi} > 0.91$ kcal/mol) tend to adopt a globular form, those with a high $\overline{FC}$ are stretched, having a low tendency for association. The proteins which are monomeric in solution have high values of $\overline{FC}$ (e.g., myoglobin, $\overline{FC} = 0.36$; cytochromes, $\overline{FC} = 0.33$), whereas those which are oligomers have rather low $\overline{FC}$ values (α and β globins, $\overline{FC}$ = 0.26; heavy and light immunoglobulin chains, $\overline{FC}$ = 0.19).

In globular proteins, most ionic side chains are completely exposed to the aqueous solvent, but so are about 50% of the nonpolar residues, at least partially, at the exterior. The interior also contains about 50% of the polar groups, especially the –CO– and –NH– groups of the peptidic backbone, which are involved in internal hydrogen bonds to a fraction of about 90%.

When the same protein is analyzed in different species on the phylogenetic tree, one observes various substitutions, but also a lot of deletions and insertions. Modifications in the interior of the protein are usually conservative, with polar-to-polar or apolar-to-apolar amino acid substitutions, with the loss of methyl groups usually compensated for by the appearance of a supplementary methyl group in the vicinity.

Separate problems represent the surface and the inner volumes of proteins. For globular monomeric proteins, the accessible surface, A (in A^2) and the number n_B of completely masked amino acid residues can be calculated according to the equations:[42]

$$A = 11.1\ M^{2/3} \tag{3}$$

$$n_B^{1/3} = n^{1/3} - k \tag{4}$$

in which M is the molecular weight, n is the number of amino acid residues, and the constant B = 1.62 for an accessible surface of A = 20 Å. The specific volume of a protein in solution slightly exceeds (*circa* 3%, conformational volume) the sum of specific volumes of the amino acid residues.[43] This additional volume is related to the solubility of lipophilic molecules in protein solution and to the fact that the structure of globular proteins is not completely rigid (''crystallized'') in solution. A review of the flexibility of receptor sites on proteins and on conformational and allosteric changes produced by binding of effector molecules is given by Neuberg and Thomson.[44]

E. SOME SPECIAL SEQUENCE MOTIFS

Some special sequence motifs are important for specific and nonspecific dimerization of proteins and for their specific binding to regulatory DNA sequences.[45] The following are the more important sequences.[46]

The leucine zipper motif is found as a specific contact surface of eukaryotic proteins that allows specific dimerization. It consists of α helical regions with a leucine residue, from 7 to 7 positions (heptads); all of the hydrophobic Leu side chains protrude only on one side of the helix (seven amino acid positions correspond almost exactly to two α helical turns) and allow a specific zippering with a similar heptad[47,48] (Figure 6). This motif was first described for the mammalian enhancer binding liver-specific protein C/EBP. E/EBP shares a bipartite region consisting of about 30 amino acids

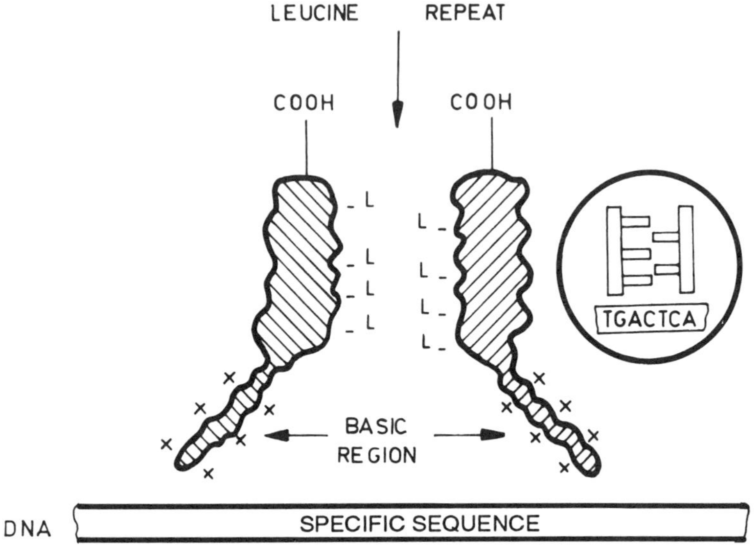

FIGURE 6. Schematic representation of the leucine zipper motif. Region containing 4 leucine residues (each residue separated by 7 other amino acids) allows a specific interaction between dimers and favors the attachment to a specific sequence of DNA. In the circle: the Fos and Jun proteins are able to dimerize by interdigitation of their zippers.

with a substantial net basic charge, immediately followed by a region containing 4 leucine residues positioned at intervals of 7 amino acids. The group of regulatory proteins having a similar DNA binding domain includes the Jun family (Jun B and Jun D), the Fos family (Fos B, Fra 1, and Fra 2), and CREB proteins.[49,50]

The zinc finger motif has been identified in the case of the mammalian transcription factor Spl and in a variety of other regulatory proteins. This motif is formed from about 30 amino acids with 2 histidine residues that stabilize the domain by tetrahedral coordination of a Zn^{+2} ion[51,52] (see Figure 14, Chapter 4, Section V.E). Another example of this type is the DNA binding domain of the steroid hormone receptor superfamily, which includes receptors for vitamin D and retinoids.[53] These proteins are homologous in six regions. The central portion contains a highly conserved DNA binding domain that consists of two zinc fingers. This motif uses two pairs of cysteines rather than the cysteine-histidine arrangement of Spl. Although the highly conserved amino acids in the zinc finger motif confer a structural framework for the binding domain, the determinants of binding specificity must lie elsewhere. For example, the slightly different DNA binding specificities of the progesterone and estrogen receptors have been found to be dependent on differences in several nonconserved amino acids at the base of the finger region. The zinc finger motifs also were established for an important number of oncoproteins, namely, Myc/RaR_{α}, ErbA, Evi-l, Vav, and Gli.[49]

The homeodomain motif (HDMs) represented by a region of about 60 amino acids first identified as a conserved protein segment in several regulators of *Drosophila* embryogenesis. It was also found in genes of vertebrate organisms, such as the mouse Hox-5 complex. The homeodomain, as a primary sequence, is related to the helix-turn-helix DNA binding structures of prokaryotic repressors. Homeodomains of different proteins present a conserved region which encompasses basic and hydrophobic residues. Some homeodomain-containing proteins bind to certain AT-rich elements of DNA control regions. HDM are found in the octamer ubiquitous transcription factor Oct-1, and in the predominantly B cell transcription factor Oct-2, and the pituitary specific transcription factor Pit-1/GHF-1. Oct-1 and Oct-2 bind to the octamer sequence lying adjacent to a variety of genes, and the Pit-1/GHF-1 proteins bind to sites upstream from growth hormone and prolactin genes.

Methionine bristles appear as methionine residues on the apolar side of an amphipathic α helical portion of a protein. Based upon their flexibility and hydrophobicity, the Met side chains recognize a variety of hydrophobic signals on the same or similar proteins. These bristles probably produce a weaker and less specific binding than the leucine zipper.[48]

The basic helix-loop-helix motif (HLH) often contains amphipathic helixes followed by basic amino acids, and can produce dimerization and DNA binding (see Chapter 4, Section III.D). In this group may be mentioned the formation of the functional heterodimer between the E12/E47 HLH proteins (which finally bind the E2 immunoglobulin light chain enhancer DNA sequence) and the MyoD and myogenin.[49] The Myc family of oncoproteins (N-Myc and L-Myc) can also be mentioned.[54]

Recent evidence regarding several other cloned mammalian transcription factors indicates that the number of DNA binding domains is not limited to the above five types. For example, the factor AP-2 contains no obvious similarities to the binding domains described here. Another DNA binding domain of the transcription factors has been localized to the NH_2-terminal of CTF protein. This region forms an α helical structure and possesses a high density of basic amino acids consistent with a DNA binding structure; it has no features which might include it among the types described above. In any case, these data suggest that there are a number of structures and/or sequences that have a role in DNA-protein interaction and recognition but remain undiscovered at present.

1. Activating Domains of Proteins

The activating domain is a distinct surface of 30 to 100 amino acids that can interact with another protein (target protein) to promote initiation of transcription. Two[55] types of activating domains have been identified: the acidic domains and nonacidic domains. The acidic domain is rich in glutamate and aspartate residues; it is present in many yeast activators and was first defined at the yeast factors GAL4 and GCN4. The acidic regions of these factors consist of relatively short stretches of amino acids that have two

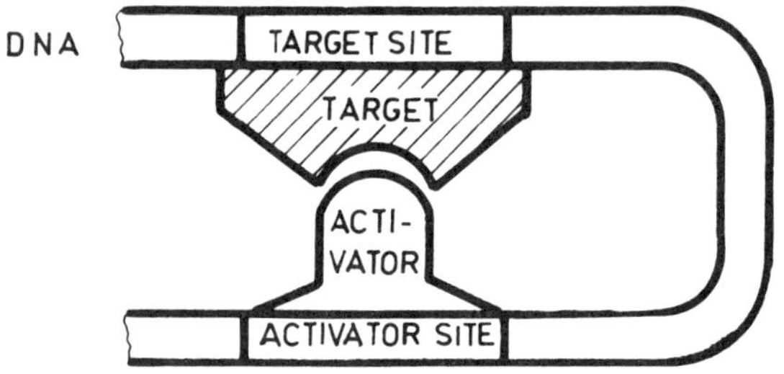

FIGURE 7. Schematic representation for the universal activator — a: The activator binds to its cognate DNA sequence changing the conformation of a universally conserved target protein, looping out the DNA; b: The activating region and DNA binding domain are present on separate molecules that are both required for activation.

common features: they are regions with a significant negative charge and can form amphipathic α helix structures. Both the acidic character and the conformation of this region are important. It has been proposed that acidic activating domains might facilitate initiation by interacting in a nonspecific manner with a general component of the initiation complex, such as the TATA binding TFIID or possibly RNA polymerase II itself.[56,57] The nonacidic domain includes the glutamine-rich (e.g., Spl)[58] and the proline-rich activation domains (e.g., CTF/NF-1).[59]

An activator can present other activating domains that are separable from the DNA binding domain. For example, the transcription factor, Spl, possesses four separate regions that contribute to transcriptional activation; all of them lie outside the zinc finger DNA binding domain.[60]

The experiments suggest that a "universal" activator must have two surfaces involved in transcriptional activation: the DNA binding domain and the acidic activating region (e.g., GAL 4). Activators bearing only one of these domains can work only in cells containing molecules that provide the missing region.[61]

2. Mechanism of Action

During the last several years, many studies have been dedicated to the mechanism by which the activators act.[62,63] We can now distinguish two major types of mechanisms which are in concordance with the structural features of the activators. Figure 7a is a model for the universal activators which share both activation regions (DNA binding domain and acidic activating surface). The activator, bound at its cognate through its DNA binding surface, brings to the DNA and/or changes the conformation of a "universal" conserved target protein that binds near the 5′ end of the gene, looping out the intervening DNA sequences. GAL is an example of this mechanism. The acidic activating

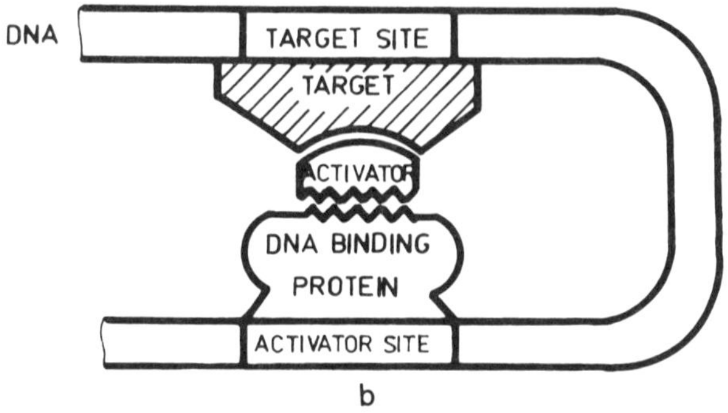

FIGURE 7. (continued)

region and DNA binding domain are present on separate molecules both of which are required for activation. An example is the complex Oct-1-VP17 (Figure 7b).

An alternative proposal for the mechanism by which these transcriptional activators function is by antagonizing histones, which otherwise would wrap up the promoter DNA into a repressed chromatin state.[62] In this model the activator would not directly activate a target factor in the transcription machinery, but would prevent repression rather than actually stimulating transcription. A special case of the nonacidic activators that work only in certain cell types are Spl and Ela. For these, there might be three possible pathways: (1) the nonacidic activator might bind another activator which presents an acidic region that interacts with TFIID, as described by Stringer et al.;[63] (2) it might contact directly or via an intermediate (co-activator) a part of TFIID different from that recognized by the acidic activators;[60] and (3) it might contact (directly of indirectly) some components of the transcription machinery other than TFIID. Actually, there is no basis for a choice between these models.

The data presented suggest the following generalizations:

1. Activators are modular proteins which present at least two domains involved in transcriptional activation: The DNA binding domain and an activating domain.
2. Whereas DNA binding regions are specific structures, each having a defined specificity for its cognate DNA binding site, activating regions are less precisely defined structures, generally characterized by an excess of acidic residues.
3. Activators with stronger activating regions more readily work at greater distances on the DNA.

F. SPATIAL STRUCTURE OF NUCLEIC ACIDS

Due to its monostranded structure, the RNA molecule exhibits a great variety of secondary structures, from coils to stacking base helixes. Tinoco et al.[64] have issued rules in order to determine the secondary structure of ribonucleic acids as a consequence of their primary structure. In order to estimate it, both thermodynamic and configurational entropy considerations and measurements of the melting equilibrium of helical sequences were used. Various forms of partially paired strands which may arise at the closing of RNA chain segments are shown in Figure 5. The ΔG (kcal/mol) for the closure of a m-nucleotide segment with λ loops (of lengths $m_1, m_2, \ldots, m_\lambda$ bases) and β bulges (of lengths $m_1, m_2, \ldots, m_\beta$) at 25°C and pH = 7, and at an ionic strength $\mu = 1\ M$ is

$$\Delta G = -1.2N_{AU} - 2.4N_{GC} + 5.9\lambda + 4.1\beta + 2\left[\sum_{i=1}^{\lambda} \lg(m_i + 1) + \sum_{j=1}^{\beta} \lg(m_j + 1)\right] \quad (5)$$

N_{AU} and N_{GC} are the number of AU and GC pairs in double helix structure. The assumption was made that all m_i, $m_j > 4$. For some tRNA molecules, the conformation indicated by X-ray studies really corresponds to the lowest ΔG, as calculated by Equation 5. Among other ΔG calculations for various nucleic acid conformations are those of Delsi and Rothers.[65]

At renaturation of nucleic acids, the nucleation consists of the formation of three base pairs in two stages, the addition of the third part being rate determining. The rate of poly-(dAT) renaturation in aqueous solution by pH modification at 10°C is 0.26 s^{-1}; for the nucleation of yeast alanine tRNA it is about 10^4 s^{-1} (at 15 to 85°C); while the growth of the double helix by stacking is much faster, approximately 10^7 s^{-1} per base pair.

Energy calculations for base pairing and folding in ribonucleic acids were performed, based upon a special algorithm by Zucker.[66] The spatial structure predictions are rendered difficult by the fact that, sometimes, completely different base pairings within a given RNA molecule have very similar energies. A discussion of tRNA structures will be given in Section V.

The base pairs which form double helix regions of RNA are not perfectly complementary; contrarily, the DNA duplex must be maintained as a perfect double-stranded structure and the complementary principle must be respected. A deviation from this principle might occur during the DNA replication. The double-stranded configuration must be site-to-site temporally separated and stabilized in a single-stranded state. Synthesis in both strands must be initiated and the incoming bases for the correct complementarity with the template bases must be respected. The fidelity is much higher than was calculated by simple base pairing considerations. In the next section we shall discuss the mechanisms by which the integrity of DNA is maintained during DNA replication.

III. FIDELITY OF DNA REPLICATION

Being the chemical memory of life, the DNA base has two main biological functions. First, it contains the information necessary for the synthesis of all proteic molecules which carry out the architectural and physiological functions of the cell. Second, it represents the genetic heritage of living bodies, and thus, it has to be copied with great fidelity. In both cases, the DNA molecules serves as a template and any mistake in its sequence can threaten the existence of the organism.

Thus, DNA replication has to be done with great fidelity. The semiconservative character of the replication has a solid support in the Watson-Crick double-stranded structure. By any means, it cannot justify the great accuracy of DNA replication, which in mammalian cells appears to be in the order of 10^{-10} to 10^{-11} false incorporations per replicated base. There are enzymes which control each replication step and also ensure the mismatch repairing activity. The central role of enzymes in maintenance of the integrity of the informational message is demonstrated by the great number of errors which occur when DNA synthesis is nonenzymatic — one error for every 100 bases — whereas in the presence of the enzymes, the fidelity rises to 10^{10} bases.

It is known that at least two enzymatic mechanisms are providing great fidelity to the replicative process, namely: an error-avoidance mechanism which occurs during replication and is represented by the enzymes involved both in selection and proofreading of the last added nucleotide, and an error-correcting mechanism which involves many enzymes that act after finishing the replication by removing the mistakes.

The maintenance of DNA integrity depends on special properties of many enzymes and especially on the DNA polymerases that form a multienzymatic complex that guides the replication process in both an accurate and fast fashion. For this reason, we will analyze first the structure and function of the enzymes that are involved.

A. DIVERSITY OF DNA POLYMERASE ACTIVITIES

1. The Polymerases

The discovery of an enzyme able to synthesize new DNA molecules on a single-stranded DNA template was reported by Kornberg et al. in 1955 (quoted by Delbrück and Stent[67]). *E. coli* polymerase I is encoded by the *polA* locus and is formed by a single polypeptide of 109 kDa, with several activities:

Polymerase activity — This function makes the addition of the deoxyribonucleotide units to the 3′-OH end of the forming DNA strand. The nucleotides addition is not random, those which are forming the Watson-Crick base pairs with the template base being favored.

3′-5′ Exonuclease activity — Before continuing the polymerization, a proofreading mechanism allows it to remove the last added nucleotide if this is an unpaired or mismatched one. This activity is specific for the single-stranded DNA,[68] requiring a 3′-OH terminus from which the attack begins, whereas the final 3′-P groups are resistant. It is one of the major functions of the DNA pol I, demonstrated also by the fact that after tryptic digestion of the enzyme this activity remains coupled with polymerase activity in the largest fragment (Klenow fragment).

5′-3′ Exonuclease — This is a postreplicative activity which occurs in double-stranded regions. The attack is not specific for the terminus, both 5′-OH and mono-, di-, and triphosphates being hydrolized. The main biological significance is the removal of the damaged sequences, up to ten nucleotides long, such as RNA primers or pyrimidine dimers induced by UV light. It should be mentioned that whereas the DNA priming oligonucleotides are removed, the single-stranded DNA and other small oligonucleotides are resistant to the enzymatic attack. It was also established that after protein digestion with trypsin, only the small fragment presents a 5′-3′ exonuclease activity.[68]

The studies made on *E. coli* mutants have demonstrated that DNA pol I is not the true replicative enzyme.[69] The mutant strain called *pA1* had the same multiplying rate as the normal strain, but the polymerizing activity was very reduced. More than that, the resistance of *pA1* to different DNA damaging agents (such as UV light) is much reduced compared with that of the normal strains. These data strongly suggested that DNA pol I is especially involved in damaged DNA and polymerization processes for filling the short single-stranded regions from the double-stranded DNA, such as the gaps between the Okazaki fragments.

After the isolation and purification from mutants of two enzymes called DNA pol II and III, it was revealed that both participate in the polymerization process, having the same activities as DNA pol. Whereas the role of DNA pol II in *in vivo* replication is not yet fully understood (it may have a key role in DNA repair after chemical or physical damage), the DNA pol III is considered to be the principal replicative enzyme. These polymerases are encoded by the *pol C (dna E), dnaN, X, Z* genes, having a molecular weight of 120 and 140 kDa, respectively.

Polymerase III appears to be a heteromultimeric, probably asymmetric, system with a molecular weight of about 800 kDa. Probably, its asymmetry allows it to synthesize simultaneously both leading and lagging strands at the replication fork. The speed of the polymerization is due to its great processivity.[3-6]

Whereas DNA pol I is a single polypeptidic chain in *E. coli,* the DNA pol III has eight subunits. This might explain its highly catalytic potential concomitant with a great fidelity. The main subunits are α, ϵ, and θ which form the core of the enzyme, developing the enzymatic activity, and the β subunit which holds the complex on the template and does not leave until the whole sequence is copied.[70]

2. Eukaryotic Polymerases

In eukaryotes, the family of polymerases includes about 30 enzymes. The main enzymes are

1. Pol α, with four subunits, having the same structure and activity among the eukaryotes. Its presence is strongly correlated with the replicative state of the cell, being the most important replicative enzyme. Viral replication systems have demonstrated that pol α also presents a primase activity,[71] which allows it to synthesize primers for both leading and lagging strands. In mammalian cells it has a rate of polymerization tenfold less than that of the bacterial DNA pol III; the lower rate is due to its interference with the nucleosomes.
2. Pol β, a small repairing enzyme, its complex with DNase V having a similar function with *E. coli* pol I. The activity of the complex appears to be induced by some DNA mutagens. Both enzymes manifest RNase activity removing the RNA primers and restoring the continuity of the DNA strands.
3. Pol γ, responsible for the replication of the mitochondrial genome. Interestingly, after replication the small pieces of RNA primers remain integrated in the closed circular DNA.
4. Pol δ, with an endogenous exonuclease activity and manifests a strong dependence on a peptide which regulates the cell cycle — the proliferative cell nuclear antigen (PCNA). This protein binds the polymerase allowing an equilibrated synthesis for both leading and lagging strands.[72] In its absence, only short Okazaki fragments are synthesized.
5. Pol ϵ, which also presents a 3′-5′ exonuclease activity; its separation from polymerase α was done with difficulty.

It was supposed that the eukaryotic pols, especially pol β, might also possess a 5′-3′ exonuclease activity in order to remove primers, but the experiments in which strictly purified enzymes were used have not demonstrated it.[73]

In proliferative cells all the polymerases have an enhanced activity; any modifications in their catalytic center determine an increase in the number of mistakes in DNA replication.[74]

The replicative proteins from yeast and mammalian cells were isolated and tested for their activities. The results have shown that the DNA primases which synthesize the primers are closely associated with α, but not δ polymerases. It is very probable that in eukaryote α and δ pols large multisubunit proteins form an asymmetric complex equivalent with DNA pol III from *E. coli*. Whereas pol α and the associate primase might synthesize the lagging strand, pol δ, together with PCNA, is responsible for leading strand formation. A concomitant synthesis of both strands at a single locus might occur.[75]

Despite the universal semiconservative character of DNA replication, different eukaryotic cell types may use different combinations of such polymerases. In addition, DNA polymerases do not have identical activities within the phylogenetic tree. The explanation resides in the particularities of eukaryotic replisomes, which require more complicated mechanisms of DNA synthesis.

B. ERROR AVOIDANCE SYSTEM

1. Nucleotide Triphosphates Selection

The fidelity of DNA replication depends on the efficiency of the DNA pols in complementary Watson-Crick base pairs formation (AT or GC). In order to give rise to a new, complementary strand, the DNA pols have first to discriminate among the four deoxyribonucleoside triphosphates which are present in quite similar concentrations.

According to Chapter 1, Equation 8, a $\Delta\Delta G$ difference of about 15 kcal/mol exists between the affinities of the DNA base site/enzyme complex to the correct and false deoxyribonucleoside triphosphates, respectively.

Another energetic control comes down on monophosphate incorporation. If the nucleotide is not complementary, perhaps because it presents an unfavorable tautomer which does not fit well on the template base, the reaction is reversed, the nucleotide triphosphate being restored. After nucleotide selection, the error frequency is 1 in 10^5 bases.

This level of accuracy cannot be provided by the classical Watson-Crick-based pairing mechanism. The problem is manifested especially in the low energy increase for the lactam-lactim tautomerism in guanine and cytosine; the lactim tautomer of guanine readily forms three hydrogen bonds with thymine, as does the lactam tautomer with the complementary cytosine (Figure 8).

According to experimental tautomerism studies,[76] and also to some theoretical calculations,[77] replacing the correct AT pairs with the G(enol)-T pair brings about an energy loss of 3 kcal/mol, and the probability of false base pairing may reach fairly high values, up to 10^{-4} or even higher. Thus, the error avoidance mechanism may operate through two pathways.

A model for DNA replication which copes with these difficulties has been proposed by Kubitschek and Henderson.[78] This model assumes that the precursors of DNA pair the deoxyribonucleoside triphosphates by means of hydrogen bonding of the bases. In addition to the structural code of DNA there also exists a replication code for the triphosphate pairs selection by DNA polymerase, based upon the 5 and 6 positions of pyrimidine and 6 or 7 of purine, namely:

CH_3 ; C......H_2N ; N for the TA pair
N ; NH_2......O ; CH_3 for the AT pair
H ; NH_2......O ; N for the CG pair
N ; O ......H_2N ; H for the GC pair

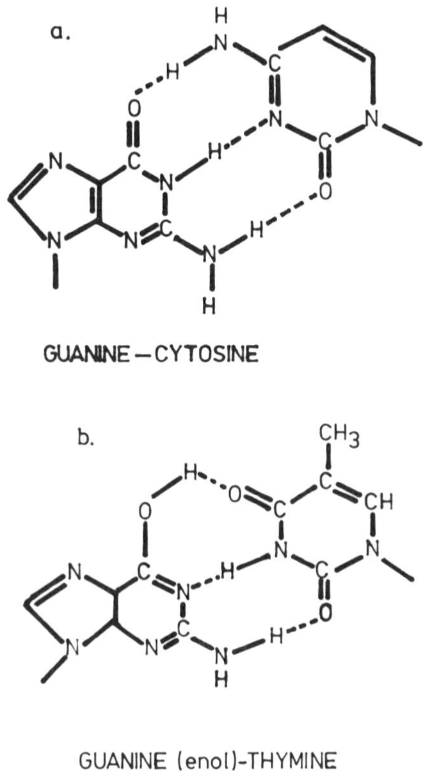

FIGURE 8. a: Correct pairing by hydrogen bonds between guanine (in normal form) and cytosine; b: False pairing between guanine (enol) and thymine which forms three hydrogen bonds.

The replicate would be an allosteric oligomer with the configuration depending both on the base pair from the "old" strands which are to be replicated and on the replication code selecting the monomer pair, which has to be incorporated in the new growing strands.[79] In order to have the same replication code like a correct base pair, a false pair (A:C, C:A, G:T, T:C) has to present both bases in odd tautomeric and/or protonated forms. For example, in order to respect the replication code which selects already paired triphosphates, the TG false pair has to reach a configuration similar to the normal CH_3; O . . . O; N one. Guanine 1-6 enolization produces a CH_3; O . . . HO; N configuration, with one proton less than the TA pair. To achieve a normal configuration, supplementary protonation, CH_3; O . . . H_2O^+; N is necessary. While according to experimental measurements and to theoretical calculations ΔG for guanine and cytosine tautomerization is probably smaller than 5 kcal/mol, in the tautomerization of adenine and thymine ΔG should be about 11 kcal/mol. We have to add to the ΔG for tautomerization the ΔG

value for protonation and deprotonation, which will be about 5.5 kcal/mol at pH = 7. The appearance of a positive charge can also decrease the affinity of the false base pair toward the enzyme.

2. Proofreading Activity

The second mechanism which also occurs during replication involves proofreading of the last added nucleotide and its elimination if it is not the right one. The mechanism was described after the discovery of the 3′-5′ exonuclease activity as a proofreading mechanism. The fidelity of DNA replication depends on the balance between the polymerasic and exonucleasic activities. If a wrong nucleotide was added to the nascent strand, the polymerasic activity of the DNA pol III would be inhibited. In contrast, the proofreading 3′-5′ exonucleasic activity will start to remove the noncomplementary nucleotide. In any case, the excision of a correct nucleotide (whose incorporation takes place much faster) is a rare event. The error is hereby reduced to 10^{-7} bases per replication cycle.[80]

Proofreading is possible because the free, incoming triphosphates are those which deliver the energy necessary for chain elongation, whereas the nascent chain has to possess an inactivated 3′-OH end. Any hydrolytic process of proofreading which eliminates the most recently attached, wrong deoxyribonucleoside leaves the nascent chain with another 3′-OH end. This is free to carry out another nucleophilic attack on another triphosphate which could become attached due to the polymerasic activity of the enzyme. If a terminal group was activated — for example, a phosphate group — any hydrolysis at this level would stop the polymerization until the terminal end of the nascent chain could be activated again.

The process of proofreading also has a structural support in the following experiment: when the Klenow fragment was isolated and analyzed, the great fragment resulted after tryptic digestion of DNA pol I presented both polymerasic and 3′-5′ exonucleasic activity. Structural analysis of the fragment revealed that it contained three distinct domains:[5] (1) the great one, which presents a deep depression with a diameter of 20 Å, positively charged, in which the DNA is fixed; (2) the smaller one, which probably represents the site of nucleotide attachment; and (3) a flexible subdomain which closes the depression immediately after the DNA becomes attached. Being fixed in the enzymatic structure, the DNA will now be able to make only a forward-backward movement through the depression.

When a wrong base is incorporated, the helical structure is distorted by the occurrence of a "bulky" lesion. Having a diameter greater than 20 Å, the molecule isn't capable of slipping through the enzymatic "tunnel". Thus, the nascent DNA has to come back, contacting the exo 3′-5′ enzymatic domain which removes the wrong base.

Some "bulky" regions also occur naturally in DNA molecules, having great importance in the regulation of gene expression, initiation of replication, and DNA packaging. Any mistake at this level could have disastrous consequences and has to be avoided. The same error-avoiding systems are implied in the preservation of these regions.

C. ERROR CORRECTING SYSTEM

1. Mismatch Repair Pathway

The errors which escape from the control of error-avoidance enzymatic systems are recorded in the helical structure as mutations. In order to avoid alteration of the genetic information they have to be eliminated. For this, another control system, a postreplicative one, is available: the mismatch repair.

A great help in understanding the repairing processes is a special class of *E. coli* strains known as strong mutators.[81] An abnormally increased rate of spontaneous mutations demonstrate that they are defective in some enzymes involved in DNA repair. The classical example of a mutator gene is *mutD* which codifies a modified subunit (exo 3′-5′) of DNA pol III, altering the correction system. *In vitro* systems containing DNA pol III with a modified subunit have a low efficiency in removing the unpaired bases. Other mutator genes encoding some proteins involved in repairing mechanisms are *mutU, mutH, mutS,* and *mutL. In vitro* experiments revealed the function of each Mut protein.[82] MutU is a topoisomerase which unwinds the two strands, mutS and mutL are both involved in mismatch recognition, and mutH is a nuclease which revolves the wrong sequence.

In spite of the great variety of DNA damage, a general pathway for DNA mismatch-repair can be adopted: (1) the recognition of the lesion; (2) removing the damaged sequence by nucleases; (3) filling the ''gap'' with the right sequence by polymerases; and (4) restoring the right sequence by DNA ligase, which seals the nicks (Figure 9).[83]

2. Error Recognition Process

A most exciting question is ''How could an enzyme distinguish between the wrong base and the right one''? or ''Which is the parental, undamaged DNA strand and which is the new, mutant strand''?

The first hypothesis was suggested by Meselsohm in 1975 (quoted by Radman and Wagner[84]). He considered that, in bacteria, the mismatch-repair enzymes could easily make the discrimination between new and parental strands.[85]

It is now well established that the signal that directs the mismatch-repair apparatus to the new strand is a GATC sequence containing an N^6-methyladenine on the old strand and an unmethylated adenine on the new one. The unmethylated GATC sequence guides the cleavage of the MutH protein in *E. coli*. The distance between this type of sequence and the mutant region could even be several thousands of bases, because a long-distance communication is established. There are two proposed models concerning the length of the

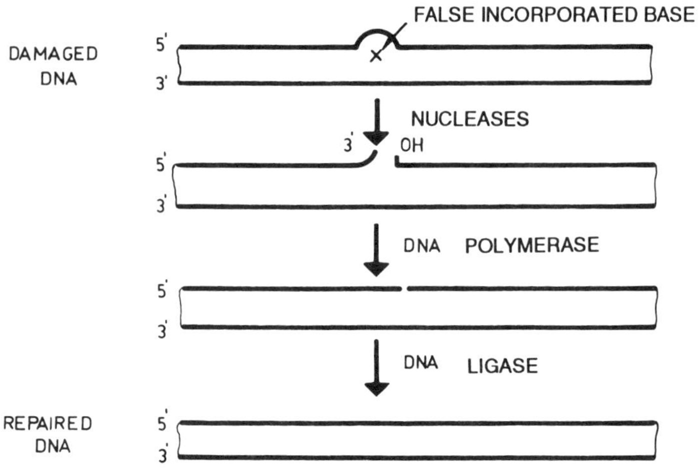

FIGURE 9. The general pathway for DNA mismatch repair which comprises at least four steps: (1) recognition of the DNA lesion, (2) removal of the damaged sequence, (3) filling the resulting "gap", (4) restoring the integrity of the DNA strand.

excised segment: one model considers that it lies between two adjacent GATC sequences, whereas the other proposes that the cleavage takes place at the mismatch itself and the nearby unmethylated GATC sequence.

The methylation phenomenon is widely described in bacteria, but in higher organisms no methylation at GATC level has been detected. In mammalian cells methylation is unequal, cytidine being more methylated than adenine. Methylation appears to mark especially inactive genes and those that have a low transcription rate. Thus, methylation could not be the criterion in error recognition.[4]

Two possibilities for lesion recognition in higher organisms were proposed: one theory considers that in the genome having a large number of replicons, the replication is discontinuous. The great number of nicks from the new strand distinguishes the new chains from the old ones. Another theory states that the possibility of error repairing in higher organisms may be the recombination between the sister chromatid DNA strands with elimination of the error.[86]

Being the result of semiconservative replication of DNA, the sister chromatids of each chromosome of an old DNA strand and a new one are genetically identical. Any mismatch which occurs in one of the two helixes representing the sister chromatid leads to a distortion on the outside of the helix, quickly recognized and excised by nucleases. An exchange between the intact and "nicked" DNA molecules may restore the integrity of the latter one.

It is very probable that the process takes place before chromatinization. In this case, the probability of improper recombination is almost eliminated, the phenomenon being reduced to short sequences just behind the replication fork.

Sister chromatid exchanges occur spontaneously in eukaryotes, usually with no genetic consequences. Giemsa staining of human chromosomes replicated in the presence of bromdeoxyuridine has demonstrated this.[87] Besides their probable role in cell differentiation, sister chromatid exchanges may also lead to the expelling of mismatched segments which have escaped from previous controls.

3. Other Enzymes Involved in DNA Correction

Several types of enzymes identified both in prokaryotes and eukaryotes are involved in recognition and excision of damaged sequences.[81] The first category is represented by apurinic endonucleases which scan the sugar phosphate residues of molecules that lack purine bases. This is by far the most common lesion which occurs in the DNA, due to the thermal fluctuations which break N-glycosyl linkages between purines and deoxyribose. For example, a human cell loses 5000 to 10,000 purine bases daily.

Another category is represented by the glycosylases which recognize altered or unconventional bases and remove them, initiating the DNA base excision repair pathway.[88] We have to make special mention of the uracil DNA glycosylase that excises the uracil residues from DNA. Uracil could be incorporated in DNA during replication as a failure of DNA pol III, or it can arise as a result of cytosine deamination. It is not a common base for DNA because, during evolution, the uracil was replaced by thymine. The motivation could be that the latter has greater photochemical resistance: UV light induces thymine dimers, which are efficiently reversed at a greater irradiation wavelength (photoreactivation). Some studies on synthetic substrates such as oligodeoxyribonucleotides of varying lengths and composition revealed that the substrate recognition and thus, the efficiency of uracil excision, depends directly on the length of the sugar-phosphatase backbone.[89]

In conclusion, after a strictly controlled replicative process, the error-correction systems provides a supplementary control in order to maintain the genetic stability of DNA. An equilibrium between the error-avoiding and error-correcting enzymes is established, each of them being more efficient when the other is more inept. A DNA molecule replicated with great accuracy ensures the template for a correct DNA transcription.

IV. TRANSCRIPTION PROCESS

Living cells have the ability to switch on or off the expression of their genes. This ability is known as the transcription process. Transcription of DNA involves the production of each type of RNA: messenger RNA (mRNA), ribosomal RNA (rRNA), and transfer RNA (tRNA). In the case of mRNA, the product is an intermediate whose function requires translation. In the case of rRNA and tRNA, the transcripts are the final products of this process.

During the last few years, the key role of the naturally occurring antisense RNA and the importance of RNase H and E in regulation, replication, and transcription have been shown both in prokaryote and eukaryote cells.[90]

A. STEPS AND FACTORS INVOLVED IN TRANSCRIPTION

Transcription presents three important steps: (1) The initiation step, which includes the initial recognition of duplex DNA, unwinding to generate a short single-stranded region and incorporation of the first nucleotide in the RNA chain; (2) the elongation step, which involves the forming of the RNA-DNA hybrid and starts when a few bases are incorporated into the RNA chain; and (3) the termination step, which is the capacity to recognize the point at which no further bases should be added to the RNA chain — at the moment that the last base is added to the chain the RNA-DNA hybrid is disrupted, the DNA reforms to the duplex state, and the enzyme and RNA are both released from it. The sequence of DNA required for this step is called the terminator.

The transcription process is catalyzed by RNA polymerases. In prokaryotes there is a single type of RNA polymerase which is responsible for the synthesis of all RNA species: mRNA, rRNA, and tRNA. In an *E. coli* cell the total number of RNA polymerase molecules is about 7000.

The transcription machinery of eukaryotic cells is more complex and less well defined than that of the prokaryotic cell. In an eukaryotic cell there are three nuclear RNA polymerases, occupying different locations, each with a complex structure. Each of these three enzymes is responsible for transcribing a different class of RNA:

- **RNA polymerase I (Pol A)** — A nucleolus enzyme that catalyzes the transcription of rRNA; it represents the most preeminent RNA synthesizing activity (relative activity = 50 to 70%).
- **RNA polymerase II (Pol B)** — A nucleoplasm polymerase responsible for transcribing mRNAs (relative activity = 20 to 40%).
- **RNA polymerase III (Pol C)** — A nucleoplasm enzyme that transcribes tRNA and 5S RNA (relative activity being about 10%).

Eukaryotic cells contain between 2×10^4 to 2.3×10^7 RNA polymerase molecules. Mammalian cells contain approximately 4×10^4 molecules of RNA pol I and about the same number of RNA pol II molecules, and approximately 2×10^4 molecules of RNA pol III. However, the number of molecules is regulated individually according to the state of growth and differentiation.[91]

In eukaryotes, the initiation of mRNA synthesis is a complex, multistep process which involves interactions between RNA polymerase II, promoter sequences, and multiple accessory transcriptional factors. Studies of transcription systems derived from rat liver, HeLa cells, and yeasts have revealed that the transcription factors can be classified into at least two classes:[92] "general factors" that are essential for initiation and assist RNA polymerase

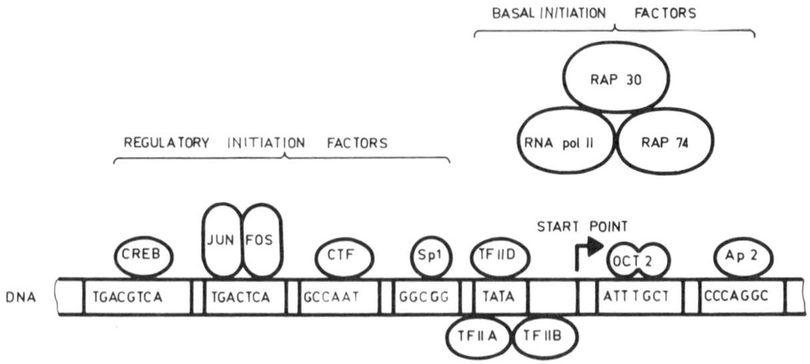

FIGURE 10. A schematic model representing a combination of cis elements that form the promoter and enhancer regions of a gene transcribed by RNA pol II.

II to recognize and bind to the core regions of promoters; and "regulatory factors" that interact with upstream and downstream promoter elements controlling the rate of initiation; they are thought to act in conjunction with RNA polymerase II and general transcription factors (see Chapter 4, Section V.B).

The general factors identified in human cells include TFIIA, TFIIB, TFIID, TFIIE, and TFIIF, which are necessary and sufficient for initiation of core promoters. Recently a new factor was described: TFIIG, which can substitute for the TFIIA in the core promoter function.[93]

Some promoter elements such as TATA, GC, and GCAAT boxes are common to many genes transcribed by RNA polymerase II. In contrast to the paradigm of prokaryotic cis-elements at -10 and -35 nucleotides relative to the transcription initiation point, the regulatory elements for RNA polymerase II transcription in eukaryotes can be scattered both upstream and downstream of the RNA start site for a gene.

In eukaryotic cells, some genes present multiple transcription initiation sites. These sites are each regulated by binding sites for different sets of factors. The combinations of ciselements arranged in unique configuration probably confer on each gene a specific spatial and temporal transcription program. A model for a combination of cis-elements that form the promoter and enhancer regions of a gene transcribed by RNA pol II is presented in Figure 10.

The promoter elements can be divided into two groups: initiators, that direct the initiation of transcription from a specific start site, and modulators, that influence the rate of transcription. The most common initiator is the TATA box. Promoter elements that have a modulating role include the GC box, the GCAAT box, and the octamer motif.[94]

TFIID, the eukaryotic nuclear protein, is an initiation factor and it has similarity with a domain unique to sigma 70 factors, binding the TATA box promoter element and regulating the expression of most eukaryotic genes transcribed by RNA pol II.[95] The TFIID-promoter complex directs the ordered assembly onto the promoter of RNA pol II and to other general initiation factors (TFIIA or TFIIG, TFIIB, TFIIE, and TFIIF) in order to create a multiprotein complex capable of transcriptional initiation.[96]

The binding of TFIID to the TATA box promoter element is the first step in the assembly of the transcription complex to the promoter.[97] Recent studies show that the human TFIID polypeptide has 339 amino acids and a molecular weight of 37,745 Da.[98] Under some conditions, this transcriptional complex is stable only in the presence of a second general initiation factor, either TFIIA or TFIIB.

The association of TFIIB in the presence or absence of TFIIA creates a preinitiation complex that can recruit RNA pol II, either alone or in association with the small subunit RAP 30 or/and RAP 74, known as TFIIF (Figure 10). After the addition of ATP, an activated complex is generated that, in the presence of nucleotide triphosphates, undergoes chain elongation.

These interactions represent the basic transcription machinery necessary for the initiation of transcription. Besides these factors, the transcription process involves other interactions between this "basic machinery" and modulating elements (such as cell type specific factors) which activate the transcription. These interactions have lately gained the attention of many molecular biologists.[99,100]

B. TRANSCRIPTION ERRORS

The error in the entire process of protein synthesis was found to be smaller than 10^{-4} per amino acid residue,[4] and the error per nucleotide residue in DNA transcription cannot be higher than this figure. The classical complementary base pairing mechanism (see Section III) explains, at least qualitatively, the synthesis of a RNA strand complementary to the transcribed DNA strands. The problem is by what signals the RNA pol II recognizes the promoter region, the starting point of the transcribed synthesis gene (SG), and its terminator.

Transcriptional errors present some difficulties in the use of the polymerase chain reaction (PCR) in recombinant DNA technology.[101] This reaction presents a higher error rate than the usual molecular cloning procedure, and one must take into account this higher error rate when synthetic oligonucleotides are used as templates.[102]

Enzymatic incorporation of new base pairs into DNA or RNA may extend the genetic alphabet. New base pairs, respecting the Watson-Crick complementarity rules but with different hydrogen bonding patterns than the AT and GC pairs, may be included in DNA and RNA double helixes by the corresponding polymerases (Figure 11).[103]

Nevertheless, one may doubt if other structural particularities probably recognized by the polymerases, would not annihilate this interesting idea. The accuracy of the viral reverse transcriptases is about 3×10^{-5} to 6×10^{-5} per nucleotide for polymerases isolated from myeloblastosis or the murine leukemia virus,[104] but about 5×10^{-4} or even 1.4×10^{-2} (for certain

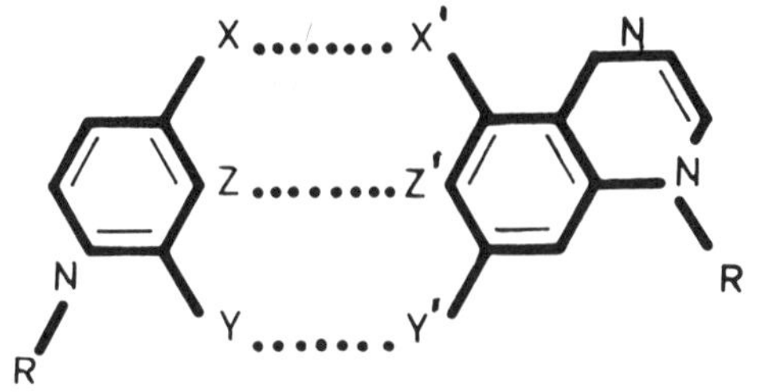

FIGURE 11. New types of base pairs: X, X′; Z, Z′; and Y, Y′ are H donor and acceptor groups, respectively.

mutational hot-spot template positions) for the HIV-1 reverse transcriptase. This last enzyme does not correct errors by exonucleolytic proofreading activity. The correction is done by mismatching, in which the specificity decreases in the order A:C >> A:G >> A:A.[105,106]

From the inhibition of polyA synthesis (or some oligothymidilic sequences on *E. coli* DNA), the affinity of a site on the transcribed DNA for the complementary NTP is about of $\Delta G = 4$ to 5.5 kcal/mol, greater than for the false nucleoside triphosphates.

C. POSTTRANSCRIPTIONAL CONTROL

At the molecular level, there are four major steps for controlling gene expression: (1) activation of the gene structure; (2) initiation of transcription; (3) processing transcripts; and (4) translation of mRNA. An important step is represented by processing. In transcription processing, the precursors (or the primary transcripts) of tRNA, rRNA, and mRNA (the last one is called pre-mRNA) are synthesized. The synthesis of a pre-mRNA is characteristic only for eukaryotic cells. The precursors are submitted to a series of reactions — cutting, trimming, and splicing — so-called processing, which releases the mature RNA molecule.[3] The most labor of all RNA processing is in the production of mature mRNA, which involves the removal of introns and joining of exons by RNA splicing.[107]

1. RNA Splicing

The splicing process must occur with pinpoint accuracy, because an error of even one nucleotide may result in a RNA transcript that cannot be translated into a functional protein.

Experiments have emphasized two aspects of RNA splicing: (1) the self-splicing of RNA, which has been discovered in organelle pre-mRNA, and (2) the splicing of nuclear pre-mRNA. These two types have raised a fundamental question: what catalyzes nuclear mRNA splicing — RNA, proteins, or a collaboration of both? The best answer seems to come from a comparison of the self-splicing introns with the remarkable architecture of nuclear pre-mRNA. A number of self-splicing introns have now been identified. These

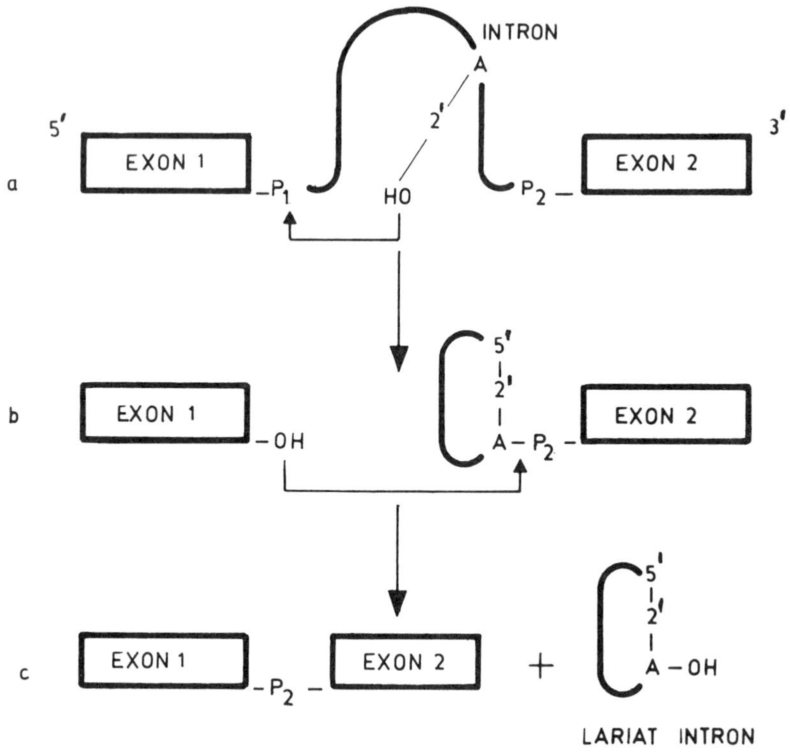

FIGURE 12. A general scheme by which the group II introns are removed. Some of the introns from this group are self-splicing, which does not require guanosine. One of the 2′-OH groups on the intron attacks the 5′ splice site (a); then the reaction joins the 5′ end of the intron, yielding a branched structure with a loop: the lariat (b); the subsequent reaction — ligation of the exons — frees the lariat of the DNA remnants (c).

introns are of two types: group I and group II. In both groups, splicing was shown to occur through two consecutive transesterification (phosphoester transfer) steps,[108] a reaction during which the substrate G is covalently joined to the 5′ end of the intron.

The mechanistic pathway for group II introns is characterized by two steps (Figure 12):

- **Step 1** — A 5′ splice site cleavage results from nucleophilic attack by the 2′-end hydroxyl of an internal A residue located upstream of the 3′ splice site. This causes the release of the 5′ exon and the formation of a ''lariat'' intermediate.
- **Step 2** — The 3′-end hydroxyl of the upstream exon exerts a nucleophilic attack on the 3′ splice site; this attack removes the intron and joins the two exons together. During this process the number of phos-

phodiester bonds is conserved and no exogenous energy source is required. This scheme of intron removal is also characteristic for nuclear pre-mRNA splicing, which proceeds through a lariat intermediate in a two-step reaction.[109]

The main differences between group II autocatalytic splicing and nuclear intron removal might be classified into three groups:

Conserved structural elements — In contrast to the highly conserved elements that reside within group II introns, there are short region motifs at or near the splice junctions. In yeasts these motifs are (1) a conserved hexanucleotide at the 5′ splice; (2) an invariant heptanucleotide UACUAAC box, surrounding the branchpoint A (underlined), called the branchsite; and (3) a generally conserved enrichment for pyrimidine residues adjacent to the invariant AG dinucleotide at the 3′ splice site.

Energy requirement — While group II autocatalytic splicing does not require an exogenous energy source, nuclear pre-mRNA splicing depends on the presence of adenosine triphosphate (ATP).[110]

Splicing machinery — Another characteristic of nuclear mRNA splicing that distinguishes it from autocatalytic splicing is the dependence on some components that are necessary for the reactions. Among the most thoroughly studied of these components are the highly conserved small nuclear ribonucleoproteins (snRNPs or "snurps"), a complex consisting of small nuclear RNAs (snRNAs) designated U1, U2, U4, U5, U6, and U7[111] and proteins (SC-35, essential for assembly of the spliceosome) known as splicing factors. During the splicing reaction, the snRNPs and pre-mRNA form a macromolecular complex known as a spliceosome. Assembly of the spliceosome involves the binding of snRNPs and auxiliary protein factors[112] to the pre-mRNA through a poorly understood mechanism which consists of specific RNA-RNA, RNA-protein, and protein-protein interactions. U1 snRNPs interacts with the 5′ splice site and U2 snRNPs with the branchsite of pre-mRNA; both of these interactions involve Watson-Crick base pairing.[113]

Very few things are known about the way the other factors, such as U4/U6 and U5 snRNPs, function in splicing. Recent studies have shown that U6 snRNA interacts directly with U2 snRNA by a mechanism involving base pairing.[114] This interaction seems to be necessary for the splicing of a pre-mRNA *in vivo*. The U2-U6 interaction is the fourth example of RNA-RNA base pairing involved in spliceosome assembly and/or splicing, the others being the U1-5′ splice site,[113] the U2 branchsite, and U4-U6 interactions.[113-116]

This mechanism shows at least a superficial similarity with two highly conserved and critical stem-loop structures found in group II self-splicing introns and is also an argument for the idea that sRNA is essential for pre-mRNA splicing.[117] These studies suggest that nuclear pre-mRNA splicing may have evolved from group II splicing, snRNA having a catalytic role in this process.[114]

According to this view, the proteic components of the spliceosome do not have a catalytic capacity. They may serve only to modulate the fidelity of the process.

One of the multiple questions regarding the nuclear pre-mRNA splicing is how and where the spliceosomes are formed. Recent studies have shown that spliceosome assembly could be compartmentalized within the nucleus.[112] This compartmentalization of spliceosomes within the nucleus could be important to the relationship between transcription and pre-mRNA processing. The electron micrographic and immunofluorescence data obtained from *Drosophila* genes suggest that the splicing machinery can function even during transcription. Studies of the amphibian oocyte nucleus have shown that snRNP occurs in at least three morphologically and biochemically distinct granules. These granules have been called A, B, and C snurposomes.[118] The composition of these granules differs: (1) the B snurposomes are more complex in composition, containing all five splicing snRNAs in roughly equivalent amounts; (2) the A snurposomes contain only U1 snRNA and U1-associated proteins; and (3) the C snurposomes present one or more B snurposomes embedded in their surfaces; they have a remarkably spherical shape, being known in the cytological literature as spheres; it seems that the C snurposomes don't contain splicing snRNAs.

The snurposome functions are a matter of speculation, at present. Their discovery promises new insights into the cell biology of RNA processing.

Another interesting aspect of snRNA is represented by the RNA-protein interactions. Recent studies of a diverse group of RNA binding proteins, including some U snRNPs, tried to elucidate the nature and the function of these interactions.[119] The RNA binding proteins present a region of 80 to 90 amino acids in length, called the "RNA binding domain" which includes a short stretch of 8 amino acids that is much more tightly conserved, called the "RNP consensus". This RNA binding domain may be singular or repeated within the protein molecule.

Query[120] studied in detail the sequences responsible for the binding activity of the U1 70K protein. The human U1 snRNP is composed of the 165 nt U1 RNA and two classes of proteins: U1-specific proteins, 70K, A, and C; and the Sm complex, consisting of six U snRNP-common proteins.[121,122] The 70K protein contacts the 5′ half of U1 RNA. The results of the study indicate that a 111-amino acid segment of this protein, including the entire RNA binding domain, is required for efficient specific interaction with U1 RNA. The RNA binding domain encompasses an 80-amino acid motif that is common to many other RNA associated proteins. This motif has a central sequence representing the RNP consensus octamer.[123,124] The presence of a RNA binding domain, including the RNP consensus, might have a particular function in regulating alternative splicing.[125] U1 snRNP plays a fundamental role in the recognition of 5′ splice sites. U1 RNP binds to the 5′ splice sites by direct base pairing between nucleotides of the 5′ splice site and the 5′ end of U1 RNA which is in contact with the 70K protein.

2. Alternative RNA Splicing

Alternative splicing of eukaryotic pre-mRNA represent a mechanism occurring naturally *in vivo* for generating multiple transcripts from a single gene. To ensure fidelity in RNA splicing, short consensus sequences are found at the intron boundaries (splice sites) and even within the introns (branchsites). These sequences are important signals recognized in splicing. However, these sites are not by themselves sufficient to explain the accuracy of splicing, especially in the case of multi-intron genes, or in genes containing very long introns, which present many sequences that resemble the consensus elements, called cryptic splice sites. Cryptic splice sites can be activated upon mutation of the authentic splice sites.[3] Except perhaps for the U1 snRNP,[125] it is not known whether the same set of factors that recognizes the authentic splice sites is also capable of recognizing the cryptic sites when the natural sites are mutated. Little is known about the proper matching of cognate 5′ and 3′ splice sites which avoids exons skipping. This specific recognition of splice sites and the match of exons can accommodate gene-specific control signals.[126]

Recent studies on splicing factors showed that SF2 — a protein factor required for the first cleavage-ligation step of splicing — plays an important role in *in vivo* regulation of alternative splicing. In addition to its more general role in the splicing reaction,[127] SF2 strongly influences the 5′ splice site selection, discriminating properly between authentic and cryptic splice sites. The molecular mechanism responsible for this capability of SF2 remains unknown, but its elucidation should help clarify the understanding of splice site selection.

Another aspect of the site splice selection is represented by scrambled exons. Using an assay for RNA expression, Nigro and his collaborators[128] identified several abnormally spliced transcripts in which exons from a candidate tumor suppressor gene (DCC) were scrambled during the splicing process *in vivo*. Their studies showed that the exons were joined accurately at consensus splice sites, but in a different order from that present in the precursor transcript. The factors responsible for generating the scrambled transcripts are not known. Exon scrambling could be the result of one of two possible mechanisms: (1) an intermolecular splicing between two different DCC transcripts (trans-splicing), a process characteristic for trypanosomes and nematodes[129] and less common for the higher eukaryotes, or (2) an intramolecular process influenced by the snRNP particles. Although the mechanism of this process is yet unknown, the scrambled splicing of mRNA precursors has implications for the fidelity of the splicing process and for the creation of unexpected RNA products.

D. MECHANISMS OF THE TRANSLATIONAL PROCESS

1. Translation of mRNA Codons

The translation of mRNA codons is conditioned by the first two bases; most of the degeneracy of the genetic code refers to the third base — many amino acids are encoded by four codons with the first base being unimportant

TABLE 3
Genetic Code

Second Position

	U	C	A	G	
U	UUU UUC } Phe UUA UUG } Leu	UCU UCC UCA UCG } Ser	UAU UAC } Tyr UAA[a] UAG[a] } Stop	UGU UGC } Cys UGA[a] Stop UGG Trp	U C A G
C	CUU CUC CUA CUG } Leu	CCU CCC CCA CCG } Pro	CAU CAC } His CAA CAG } Gln	CGU CGC CGA CGG } Arg	U C A G
A	AUU AUC AUA } Ile AUG[b] Met	ACU ACC ACA ACG } Thr	AAU AAC } Asn AAA AAG } Lys	AGU AGC } Ser AGA AGG } Arg	U C A G
G	GUU GUC GUA GUG } Val	GCU GCC GAA GCG } Ala	GAU GAC } Asp GAA GAG } Glu	GCU GCC GGA GGG } Gly	U C A G

[a] End chain codons (or nonsense codons).
[b] The codons are used for formyl Met-tRNA initiator.

First Position 5′-End

Third Position 3′-End

(Table 3). In order to explain this, Crick[130] advanced his ''wobble'' hypothesis, according to which only the recognition of the first two codon bases occurs by the usual complementary base pairing. The third position has some ''freedom'' — several bases might pair with a certain base of the tRNA anticodon. Thus, a G from the anticodon could pair to a U or C of the third position of mRNA codons; I (inosine) with U, C, or A; C with G; U with A or G; and the modified uridine from the anticodon with U, A, or G.

The RNA anticodon is located on a loop at the opposite end to the portion which binds the amino acid (see Figure 13). The polysome codon-tRNA anticodon association constant is larger by far than that of the association codon trinucleotide-tRNA or codon trinucleotide-anticodon trinucleotide. Entropy effects explain these differences: in the two latter associations there is a strong entropy decrease ($\Delta S << 0$) for fixing the structure in the complex compared to the relative freedom of internal motions in the free trinucleotide, while the codon in the polysome and the anticodon in the tRNA have their spatial conformations already fixed.[131]

The specificity of initiation and termination in protein synthesis cannot be explained by complementary base pairing only. The establishment of the

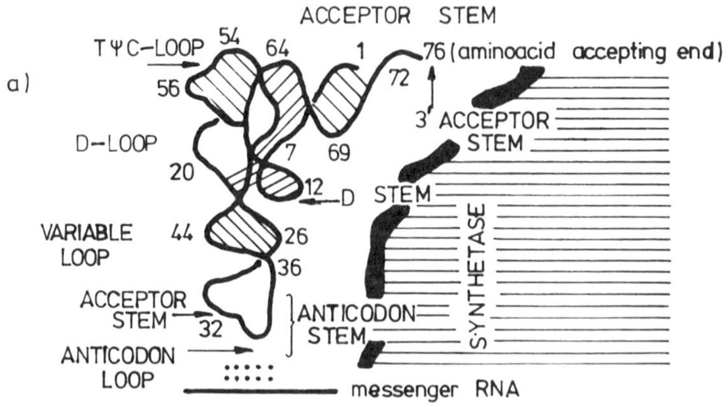

b) $pGCCGGAsUUUAm^2G_{10}CUAG_{15}DDGGG_{20}$

$AGAGC_{25}m^2GCCAG_{30}ACmUGmA_{35}AYA\Psi$

$m^5C_{40}UGGAG_{45}m^7GUCm^5CU_{50}GUGT\Psi_{55}CG$

$mAUC_{60}CACAG_{65}AAUUC_{70}GCACC_{75}A$

FIGURE 13. Transfer RNA folds into a compact L-shaped tertiary structure — a: A schematic drawing of aminoacyl-tRNA sythetases which may contact tRNA along the inside of the L structure. Various regions of the molecule are depicted (acceptor stem: ACC; anticodon: GAA). Nucleotides are numbered in the standard way. b: The sequence of the molecule which, in fact, forms a cloverleaf.

primary structure of phage R17 RNA allows us to ellucidate this specificity. This RNA has a length of 3300 nucleotides, but the very first translated gene only appears at about the 50th base from the 3′ terminal of the molecule. It is likely that initiation depends upon the secondary structure of RNA: in R17 RNA there is a double helical region just before the first AUG-initiating codon, which is presumably exposed at the binding of a loop.[132] Otherwise, the AUG and GUG codons within the gene are read as Met. For initiation, certain other protein factors are also required (see Chapter 4, Section V.B). For termination, next to the sequence, on R17 RNA, encoding for the envelope protein there are two ending codons (UAA and UAG) and a further 30-nucleotide sequence before the AUG codon which initiates the next protein.[133] The two tandem terminating codons may increase the accuracy of termination. Special protein factors are also required for recognition of the termination codons.

2. The Genetic Code

The genetic code is considered universal[3-5] in the sense that the trinucleotide codon-amino acid relationship is the same throughout the living world. From one class of organisms to another the primary structures of tRNA and aminoacyl-tRNA synthetases differ, so that systems with tRNA from one organism, and ribosomes or synthetases from another, function with a higher sensibility of error rate than systems with all components originating from the same organism. Nevertheless, there are also exceptions from the universality of the genetic code — in the yeast *Candida cilindracea* the codon CUG codes for Ser, not for Leu.[128]

The code is degenerated (see Table 3) in the sense that most of the amino acids have more than a single corresponding tRNA; the relative concentrations of various tRNA are different in various tissues of the same organism.[134] This seems to be a result of selection from a code most resistant to mutations. An analysis of monosubstitutions (point mutation) in codons shows that in 25% of cases they do not produce amino acid replacements and in another 50% they do not change the polar or apolar character of the amino acid. It is interesting that the frequency of various amino acids, averaged over the proteins of known primary structure, is approximately proportional to the number of codons corresponding to the amino acid.

3. Specificity in mRNA Translation

In the usual sense, the genetic code is the relation between the trinucleotide codon on the translated mRNA (or on the DNA strand complementary to the transcribed one) and the amino acid coded by the trinucleotide. This genetic code is listed in Table 3. In fact, this code implies subcodes in following processes:

1. DNA transcription that is based, at least in part, on the formation of complementary base pair triplets. The transcribed DNA strand is transcribed in the 5′-3′ direction;
2. mRNA translation on polysome, based on the mRNA codon-aminoacyl tRNA anticodon interaction and based also on complementary base pair triplets; and
3. Aminoacyl-tRNA formation out of the amino acid and the corresponding tRNA, a process catalyzed by the corresponding aminoacyl-tRNA synthetase (E). This process consists of two reactions:
    a. adenylation of the amino acid: $AA_i + ATP + E_i \rightarrow AA_i \cdot E_i +$ pyrophosphate; and
    b. tRNA aminoacylation: $tRNA_i + AA_i - AMP \cdot E_i \rightarrow AA_i - tRNA_i + AMP + E_i$.

The recognition between the amino acids, tRNA, and synthetase is often termed the second genetic code.[135] There are two different synthetases in living cells, one for each amino acid, and 45 to different tRNA corresponding to one or more codon trinucleotides. The tRNA is relatively constant in size (a sequence of 74 to 93 nucleotide bases), coiled like a tangled "garden hose" into a crude "L" shape having a three-dimensional structure. The spatial structure of yeast phenyl-tRNA, according to Quigley and Rich,[131] is depicted in Figure 13. The characteristic structural domains are the same for all tRNAs, and three of them are involved in contacting the synthetase: (1) the acceptor stem, which must be charged on the 3′ terminus with amino acid; (2) the D stem, involved by reactions in which enzymes are cross-linked photochemically to their tRNA; and (3) the anticodon stem (arm).

The frequency of errors in the primary structure of natural proteins must be small enough so that most protein molecules are functional. This requires an error of at most 10^{-3} to 10^{-4} per codon and for each of the processes from DNA transcription to mRNA translation and tRNA charging with amino acids.

The fidelity and efficiency of translation by eukaryotic ribosomes depend mainly on the particular structure of the mRNA,[136] namely: (1) the presence of a m7G cap;[137] (2) the position of the AUG codon and the context around this codon; (3) the length of the leader sequence;[138] and (4) the types of secondary structures both upstream and downstream.

The scanning mechanism[139] for initiation was proposed to explain how the above features of mRNA structure could work to ensure the fidelity of the initiation of translation. This mechanism considers that a 40S ribosomal subunit, carrying Met tRNA and an imperfectly defined number of initiation factors, binds at the 5′ end of the mRNA and slides along till it reaches the AUG codon, whereupon a 60S ribosomal subunit joins it, leading to the first peptide bond. Although most mammalian mRNAs have the ability to ensure the fidelity of initiation of translation (by a correct selection of the AUG codon), many do not achieve a very efficient translation.

4. Charging of tRNAs. The Second Genetic Code

The interaction of tRNAs with synthetases and amino acids has been studied by cleavage of tRNA and by testing the affinity of various fragments towards the enzyme, or by chemical modifications of the tRNAs at various positions.[140] Each of the 20 synthetases aminoacylates all tRNAs corresponding to a given amino acid. In turn, deacylation of tRNA Met does not interfere with the binding of the latter to the corresponding AUG codon; the dissociation constant, $K = 1.5 \times 10,^{-4}$ remains unchanged. There is also a specificity in the interaction of the amino acid with the synthetase alone, which is increased by the presence of the tRNA.[141]

The interactions of the tRNA with synthetase and amino acid are very intricate, with several contact points along the internal part of the L-shaped tRNA (see Figure 13)[135,142,143] For example, the tip of the long L-arm fits into a deep little pocket of the protein. This tip region contains the anticodon. The enzyme interacts mainly with two bases of the anticodon and much less strongly with the third, a fact which explains how one synthetase recognizes two different $tRNA^{Gln}$ molecules which differ by the third anticodon base. On the opposite end of the tRNA, the tip of the short leg of the L is inserted into a gaping cavern of the enzyme. This cavern is the active enzyme site catalyzing the formation of the link between the tRNA and the amino acid. The ATP molecule is bound to the lower surface of this cavern. The cavernous active site appears, with minor variations, in all the synthetases whose structures are known and in several other enzymes.

Some specific elements for $tRNA^{Gln}$ recognition by the synthetase are as follows: (1) the enzyme makes a contact with the $2\text{-}NH_2$ groups of the guanine G2 and G3 of the minor tRNA groove; (2) there is interaction of the enzyme with the ACC anticodon nucleotides; and (3) the U1.A72G73 portion of $tRNA^{Gln}$ is forced into a conformation stabilized by the Gln synthetase at lower energy cost than other noncognate tRNA.[142,143] Another recognition element is the hydrogen bond of the co-carboxyl group of Asp^{235} of the synthetase with the G3.U70 base pair of $tRNA^{Gln}$ in the minor groove of the acceptor stem. A certain mutation in position 235 of *E. coli* glutaminyl synthetase produces misacylation of $tRNA^{Tyr}$. The Ile^{129} of glutaminyl synthetase which is positioned on the face of the phosphate of the C74 nucleotide part of a hairpin structure of $tRNA^{Gln}$ also seems involved in the specific recognition.

The contribution of the anticodon to the discrimination between cognate and noncognate tRNAs by the arginine synthetase of *E. coli* was demonstrated in studies with tRNA mutants. If the Arg anticodon CCG is substituted for the CAU anticodon of methionine in $tRNA^{Met}$, there is a dramatic increase in the presence of Arg synthetase, in Arg acceptance by this modified $tRNA^{Met}$. This false acceptance is increased 4×10^4 times, as compared to the correct $tRNA^{Met}$. If position 20 in $tRNA^{Met}$ is also occupied by A, this mutated $tRNA^{Met}$ (CCG)A20 has an even 5×10^6 increased acceptance for Arg, only 2 times lower than the Arg acceptance of $tRNA^{Arg}$ in presence of the Arg synthetase.[144]

Less is known about structural features related to the specific amino acid acceptance. The G3:U70 base pair on the acceptor stem (implicated in acceptor helix recognition) is important for binding of alanine by $tRNA^{Ala}$ or by other tRNA synthetases in which the corresponding 3 to 70 base pair is introduced (glutamine and serine).[145]

V. ERRORS IN PROTEIN SYNTHESIS

A review of errors in protein synthesis was given by Nirenberg.[146] Although most codons must be translated with a high accuracy-error probability of at most 10^{-3} to 10^{-4}, for certain codons an error probability of 0.5 seems possible. According to von Ehrenstein,[147] at least 6 positions out of the 141 on the rabbit hemoglobin α chain contain more than 1 amino acid; in these multiplications only neutral amino acids seem to be involved. Defective substitutions of Val by Ile in rabbit reticulocyte hemoglobin has a frequency of 3×10^{-4} per residue,[148] while from the percentage of thermolabile glucophosphate dehydrogenase in cultured young fibroblasts, a mean error level of 2×10^{-4} per residue is inferred.[149] In ovalbumin synthesis in hen oviduct (experiments with labeled amino acids),[150] Leu, Ile, and Val do not substitute for each other with a probability higher than 3×10^{-4}.

Protein synthesis in various *in vitro* systems has a more reduced precision; errors of 2.5×10^{-3} for Leu instead of Phe incorporation templated by poly-U[151], or only 5% for the correct satellite tobacco necrosis virus coat protein if viral RNA is translated by a cell-free *E. coli* system.[152] Antibiotics, such as streptomycin, cause at the ribosome level an erroneous reading of pyrimidine codons (U and C read as A).[153]

As regards aminoacyl-tRNA interactions with mRNA codons in polysomes in a system derived entirely from *E. coli,* Arg tRNA and Ile tRNA recognize the false UUU (Phe) codon with a relative rate lower than 1%, but Ser $tRNA^{Ser}$ recognizes the false UGU (Cys) codon at a relative rate of 3%.[154]

Relatively abundant data exist concerning the precision of the tRNA aminoacylation process. In systems with all components from the same organism, the specific Ile synthetase distinguishes (in the AMP coupling reaction) between Ile and Val with an error of 4.5% in the absence of the specific $tRNA^{Ile}$, but the error level drops to 0.5% in its presence. The transfer reaction of an aminoacyl residue from AMP into $tRNA^{Ile}$, catalyzed by the same Ile synthetase, falsely transfers a Val instead of an Ile residue at an error level of, at most, 0.02%.[155] This demonstrates the requirement of the tRNA-aminoacylation reaction for a high precision. A series of data concerning the precision of this reaction are listed in Table 4, according to Lotfield and Eigner[155] and to other data.[156-158] Concerning the recognition of tRNA by synthetases of various specificities from data about the inhibition of $tRNA^{Val}$ recognition by the Val synthetase through alien tRNA, the relative affinities are for the $tRNA^{x}$ of various specificities; x: Met, 0.07; Arg, 0.05; His, 0.04; Tyr, 0.03; Asp, 0.02; Lys and Asp, 0.01; and Leu, Phe, Thr, Gly, Ile, 0.002.

A. MOLECULAR INTERACTIONS AND ERRORS

Concerning the error level of at most 10^{-3} to 10^{-4} per transcribed nucleotide or translated codon, the Watson-Crick mechanism of base pairing could hardly ensure the $\Delta G = 5$ to 6.5 kcal/mol difference in affinity. A mechanism of enzymic "touching-feeling" of the type advanced by Kubitscheck and Henderson[78] (see Section IV), easily solves the specificity requirements.

Problems appear when one considers the false aminoacylation of $tRNA^{Ala}$ by Gly in the presence of the Ala synthetase, or of $tRNA^{Ile}$ by Val in the presence of the Ile synthetase.[150] The difference between Gly and Ala and, respectively, Val and Ile, is the lack of a methyl group. The G affinity difference between the false and the correct amino acid consists in the unfilled portion of the synthetase-tRNA receptor site, occupied by the CH_3 group for the correct amino acid (Ala and Ile, respectively). According to Chapter 2, Section III.E, the maximal ΔG difference would be 1.5 kcal/mol, and the error probability in a single-step recognition would be as high as 0.1. Even with some supplementary control mechanisms (regarding the cascade regulation, Chapter 1, Section I.A) one may expect higher error rates for Ala by

TABLE 4
Precision of Transfer RNA Aminoacylation Process

Process	Enzyme	Substrate	K_M(*M*)	k_2(%)	k_2/K_M(rel)
Phosphate exchange	Val synthetase + $tRNA^{Val}$	L-Val	6.7×10^{-5}	100	1
		L-Thr	7.0×10^{-5}	21	2.0×10^{-3}
		D,L-Amino butyric acid	1.5×10^{-3}	6	2.8×10^{-3}
		L-Ile	0.1	22	1.5×10^{-6}
Competitive substitution	Phe synthetase +	Phe	1.0×10^{-5}		
		Phe-trityl ester	8.0×10^{-5}		
		$C_6H_5CH_2OH$	0.6×10^{-5}		
		Gly	>1		
		Tyr	$>1.10^{-2}$		
Amino acylation	Ile synthetase	Ile + $tRNA^{Ile}$	7.0×10^{-9}		1
		Ile + $tRNA^{Phe}$	$>1.4 \times 10^{-4}$		8.0×10^{4}
Amino acylation[152]	Tyr synthetase + $tRNA^{Tyr}$	L-Tyr	2.7×10^{-5}		1
		D-Tyr	1.1×10^{-4}		0.04
Amino acylation[139]	Arg synthetase	Arg + $tRNA^{Arg}$	1.1×10^{-6}	100	1
		Arg + $tRNA^{Met}$ (CCG)A20	1.2×10^{-6}	45	0.2
		Arg + $tRNA^{Met}$ (CCG)	5.2×10^{-6}	1.5	4×10^{-3}
		Arg + $tRNA^{Met}$ A20	4.4×10^{-4}	0.035	1×10^{-4}
Amino acylation[139]	Met synthetase	Met + $tRNA^{Met}$	1.1×10^{-6}	100	1
		Met + $tRNA^{Met}$ A20	1.6×10^{-6}	80	0.5
		Met + $tRNA^{Met}$ (CCG)	2.9×10^{-6}	0.01	4×10^{-5}
		Met + $tRNA^{Met}$ (CCG)A20			1.2×10^{-5}

[a] $tRNA^{Met}$ (CCG)A20 means mutant $tRNA^{Met}$ with the CAU codon substituted by the CCG codon and A substituted in position 20.

Gly substitutions and Ile by Val substitutions. Possibly, such substitutions do not change the shape and function of proteins.

B. ERRORS IN PROTEIN SYNTHESIS AND ORGEL'S THEORY OF AGING

Errors in protein synthesis play a crucial role in Orgel's theory of aging by error accumulation[1,2] (producing increasing quantities of defective DNA polymerase and thus increasing mutation frequency in the following generation, in absence of a cell selection mechanism — the "error catastrophy"). As we discussed in Section III, DNA does not replicate with 100% fidelity and there is much endogenous DNA damage. For instance, in a normal rodent, cells have about 10^6 oxidative adducts at any time and this increases with age.[160] In addition, the new oxidative adducts (about 10^5 per cell) are formed every day, but most are repaired. This DNA damage induced by endo- or exogenous agents could be considered as a major factor in aging and in the degenerative diseases of aging (e.g., cancer). These DNA lesions increase the likelihood of DNA damage, either by altering the DNA itself, or by an increase in DNA replication, then finally cell proliferation or hyperfunction.[161] Prolonged hyperfunction produces degeneration of the respective organ.

It would be of interest to see how hyperfunction conditions could enhance errors in protein synthesis. The cell apparatus in hyperfunction will have abilities to produce highly endoenergetic molecules that absorb, by active transport, species not required in the hyperfunction process. The intracellular concentration of amino acids which are relatively rare in blood may thus decrease, and the probability of false aminoacylation of the corresponding tRNA by more common acids may increase. Recent evidence in molecular biology has demonstrated that cancer and aging share many molecular pathways. The genes involved in tumor initiation and progression could be considered as the regulators of cellular senescence.

VI. CONCLUSIONS

Nucleic acids and proteins represent the biopolymers whose shape is determined by the noncovalent bonding between various groups of atoms. These biopolymers contain simple bonds which confer upon them the ability to preferably take an energetic configuration. Some of the molecules present a helicoidal structure that is maintained in a regular shape by the secondary structure formed between different groups on the backbone.

The interaction between biopolymers, and especially between DNA and proteins, represents an essential problem in replication, transcription, and translation processes. These interactions play a key role not only in the synthesis of these biopolymers, but also in regulation and control of gene expression and fidelity in the perpetuation of these molecules.

Three major enzymatic systems are responsible for the high precision of DNA replication. These systems are controlled by a cascade regulation mechanism. The first system is DNA polymerase, which selects the correct nucleotides at the level of the triphosphate to monophosphate transformation by alignment with the "old" bases. The error probability of this system is about 10^{-5}. The second error-avoidance mechanism involves a proofreading activity. There are other post replicative mechanisms, such as the mismatch repair and the sister chromatid exchange.

The problem of the mismatch mechanism is the recognition by the enzyme of the new strand (that contains an error) from the old one. The distinction is based on hypomethylation of the new strand that could contain the error.

The initiation of RNA transcription is a complex, multistep process which involves interactions of RNA polymerases, promoter sequence, and multiple accessory transcriptional factors. Each step of the intricate process is subject to intense scrutiny. The error in the entire process of protein synthesis is less than 10^{-4} per amino acid residue and the error per nucleotide in DNA transcription is not higher than this figure.

Posttranscriptional control by RNA splicing plays an important role in the fidelity of RNA maturation. The fidelity and efficiency of translation by eukaryotic ribosomes depends mainly on the particular structure of the mRNA. These consist of: the presence of m7G cap, the position of the AUG codon, the length of the leader sequence, and the types of secondary structures both upstream and downstream. The rules of the genetic code were determined by specific aminoacylation of tRNAs by aminoacyl tRNA synthetases. The specificity of aminoacylation may be determined by a few nucleotides and can be converted from one acid to another by changing only a few bases.

REFERENCES

1. **Simon, Z.,** *Quantum Biochemistry and Specfic Interactions,* Abacus Press, Tunbridge Wells, Kent, U.K., 1976, chap. 7.
2. **Kirkwood, T. B. L., Rosen, R. F., and Galas, D. J., Eds.,** *Accuracy in Molecular Processing. Its Control and Relevance to Living Systems,* Chapman & Hall, London, 1986.
3. **Lewin, B.,** *Genes,* 3rd ed., John Wiley & Sons, New York, 1987.
4. **Alberts, B., Bray, D., Lewis, L., Raff, M., Roberts, K., and Watson, J. D.,** *Molecular Biology of the Cell,* 2nd ed., Garland Publ., New York, 1989.
5. **Watson, G. D., Hookins, N. H., Roberts, J. W., Steitz, J. A., and Weiner, A. M.,** *Biologie Moleculaire du Gene,* 4th ed., Intereditions, Paris, 1989.
6. **Stryer, L.,** *Biochemistry,* 3rd ed., W. H. Freeman, New York, 1988.
7. **Clore, M. G. and Gronenborn, A. M.,** Determination of three-dimensional structure of proteins and nucleic acids in solution by nuclear magnetic resonance spectroscopy, *Crit. Rev. Biochem. Mol. Biol.,* 24, 479, 1989.
8. **Liquori, A. M.,** The stereochemical code and the logic of a protein molecule, *Quart. Rev. Biophys.,* 2, 65, 1969.
9. **Dayhoff. M. O. and Eck, R. N., Eds.,** *Atlas of Protein Sequence and Structure,* National Biomedical Research Foundation, Washington, D. C., 1968.

10. **Cohen, N. C., Blancy, G. M., Humblet, C., Gund, P., and Barry, D. C.,** Molecular modelling software and methods for medicinal chemistry, *J. Med. Chem.*, 33, 88, 1990.
11. **Sundaralingam, O. M.,** Stereochemistry of nucleic acids and their constituents. IV. Allowed and preferred conformations of nucleosides and nucleotides, *Biopolymers,* 7, 821, 1969.
12. **Felsenfeld, G., Davies, D. R., and Rich, A.,** Formation of three - stranded polynucleotide molecule, *J. Am. Chem. Soc.*, 79, 2023, 1957.
13. **Nielsen, P. E.,** Sequence selective DNA recognition by synthetic ligands, *Bioconjugate Chem.*, 2(1), 1, 1951.
14. **Scheraga, H. A.,** Theoretical and experimental studies of conformations of polypeptides, *Chem. Rev.*, 71, 193, 1971.
15. **Skvortsov, A. M., Birshtein, T. M., and Zalevski, A. O.,** Calculation of thermodynamic characteristics of α- helixes in solution by the Monte-Carle method, *Molek. Biol. (Moscow),* 5, 390, 1971.
16. **Ramachandran, G. N.,** *Symmetry Function in Biological Systems at Macromolecular Level,* W. Almqwist, Stockholm, 1969, 79.
17. **Wendelski, J. J., Gardner, K. H., Hirshinger, J., Miura, A., and English, A. D.,** Molecular dynamics in ordered structures. Computer simulations and experimental results for nylon 66 crystals, *Science,* 247, 431, 1990.
18. **Chou, K. C. and Carlacci, L.,** Energetic approach to the folding of α/β barrels, *Proteins Struct. Funct. Genet.*, 9, 280, 1991.
19. **Carlacci, L. and Chou, K. C.,** New developments on energetic approach to the packing of proteins, *J. Computat. Chem.*, 12, 410, 1991.
20. **Howard, A. E. and Kollman, P. A.,** An analysis of currents methodologies for conformational searching of complex molecules, *J. Med. Chem.*, 31, 16, 1988.
21. **Shiraton, Y., Nakagawa, S., Hori, H., Murakami, K., and Umeyama, H.,** Protein modelling of human prorenin using the molecular dynamics method, *J. Mol. Graphics,* 8, 163, 1990.
22. **Prothero, J. W.,** Secondary structure prediction from primary structure, *Biophys. J.*, 6, 367, 1966.
23. **Kotelchuck, D. and Scheraga, H. A.,** Influence of small range interactions on protein conformation. II. Mode for prediction of α-helical regions in proteins, *Proc. Natl. Acad. Sci. U.S.A.*, 62, 14, 1969.
24. **Ptitsyn, O. B. and Finkelstein, A. V.,** Relation of secondary structure of globular proteins to their primary structure, *Biofisika (Moscow),* 15, 757, 1970.
25. **Pachamabhan, S., Margusee, S., Ridgeway, T., Lane, T. M., and Baldwin, R. L.,** Relative helix forming tendencies of nonpolar aminoacids, *Nature,* 344, 268, 1990.
26. **Cohen, F. E., Cohen, B. I., Collach, N., and Presnell, S. R.,** Protein secondary and tertiary structure predictions: new methods and old problems, *J. Mol. Graphics,* 8, 154, 1990.
27. **Chou, P. Y. and Fasman, G. D.,** Conformational parameters for aminoacids in helical β-sheet and random coil regions calculated from proteins, *Biochemistry,* 13, 222, 1974.
28. **Maelicke, A.,** Predictions of protein structure, *Nachr. Chem. Tech. Lab.*, 36, 1119, 1988.
29. **Zimm, B. H. and Kallenbach, N. R.,** Aspects of chemical physics of polynucleotides and nucleic acids, *Annu. Rev. Phys. Chem.*, 13, 171, 1962.
30. **Von Hippel, P. A. and Mcghee, J. D.,** DNA protein interactions, *Annu. Rev. Biochem.*, 41, 232, 1972.
31. **DeCohen, J. L.,** Folding of polypeptide chain during biosynthesis, *J. Mol. Biol.*, 49, 405, 1970.
32. **Lewis, P. N., Momany, F. A., and Scheraga, H. A.,** Folding of polypeptide chains in proteins: a proposed mechanism for folding, *Proc. Natl. Acad. Sci. U.S.A.*, 68, 2293, 1971.

33. **Ikai, A. and Tanford, C.,** Kinetic evidence for incorrectly intermediate states in refolding denaturated proteins, *Nature,* 230, 100, 1971.
34. **Ellis, R.-J.,** Molecular chaperons as auxiliary factors in protein folding, *Biochem. Soc. Symp.,* 55, 145, 1989.
35. **Phickthun, A.,** Protein folding, *Nach. Chem. Tech. Labor,* 38, 219, 1990.
36. **Mitraki, A., Fane, B., House-Pettingel, C., Sturtewant, J., and King, J.,** Global suppression of protein folding defects and inclusion body formation, *Science,* 235, 54, 1991.
37. **Clarck, D. A., Barton, G. C., and Rawlings, C. J.,** A knowledge based architecture for protein sequence analysis and structure prediction, *J. Mol. Graphics,* 8, 94, 1990.
38. **Ycas, M.,** Computing protein tertiary structure, *J. Protein Chem.,* 9, 177, 1990.
39. **Sanders, C.,** Database of homology derived protein structure, *J. Mol. Graphics,* 8, 104, 1990.
40. **Welscher, H.,** Correlations between aminoacidic sequence and conformation. Hydrophobicity and fractional charge, *Int. J. Protein Res.,* 1, 253, 1969.
41. **Schulz, G.,** Rules for structure of globular proteins, *Angew. Chem.,* 89, 24, 1977.
42. **Jenin, J.,** Surface and inside volumes for globular proteins, *Nature,* 277, 491, 1979.
43. **Zamiatnin, A. A.,** Volume of proteins in solution [Obem belka v rastvore], *Biofizika (Moscow),* 16, 163, 1971.
44. **Neuberg, R. R. and Thomson, W.-J.,** How does a key fit a flexible lock? Structure and dynamics of receptor function, *BioEssays,* 11, 136, 1989.
45. **Pluckthun, A.,** Sequenzmotive: Hits'89, *Nach. Chem. Tech. Labor.,* 38, 222, 1990.
46. **Jones, N. C.,** Transcriptional regulation by dimerisation: two sides of an incestuous relationship, *Cell,* 61, 9, 1990.
47. **Kouzarides, T. and Ziff, E.,** The role of the leucine zipper in the Fos-Jun interaction, *Nature,* 336, 646, 1989.
48. **Turner, R. and Tjian, R.,** Leucine repeats and an adjacent DNA binding domain mediate the formation of functional c-Fos-c-Jun heterodimers, *Science,* 243, 1689, 1989.
49. **Hunter, T.,** Cooperation between oncogenes, *Cell,* 64, 249, 1991.
50. **Benbrook, D. M. and Jones, N. C.,** Heterodimer formation between CREB and Jun proteins, *Oncogenes,* 5, 295, 1990.
51. **Pavlevich, N. P. and Pabo, C. O.,** Zinc finger-DNA recognition: Crystal structure of a Zif 268-DNA complex at 21 Å, *Science,* 253, 809, 1991.
52. **Luizi, B. F., Xu, W. X., Otwinowski, Z., Freedman, L. P., Yamamoto, K. R., and Sigler, P. B.,** Crystallographic analysis of the interaction of the glucocorticoid with DNA, *Nature,* 352, 497, 1991.
53. **Green, S. and Chambon, P.,** Nuclear receptors enhance our understanding of transcriptional regulation, *Trends Genet.,* 4, 309, 1988.
54. **Blackwell, T. K., Kretzner, L., Blackword, E. M., Eisenman, R. N., and Weintraub, H.,** Sequence specific DNA-binding by the c-myc protein, *Science,* 250, 1149, 1990.
55. **Ma, J. and Ptashne, M.,** Converting an eukaryotic transcriptional inhibitor into an activator, *Cell,* 55, 443, 1988.
56. **Horikoshi, M., Carey, M. F., Kakidani, M., and Roeder, R. G.,** Transcription factor ATF interacts with the TATA factor to facilitate establishment of a preinitiation complex, *Cell,* 54, 1033, 1980.
57. **Lewin, B.,** Commitment and activation at pol II promoters: a tail of protein-protein interactions, *Cell,* 61, 1161, 1990.
58. **Courey, A. J., Holtzman, D. H., Jackson, S. P., and Tjian, R.,** Synergistic activation by the glutamine-rich domains of human transcription factor Spl, *Cell,* 59, 827, 1989.
59. **Monrod, N., O'Neil, E. A., Kelly, T. J., and Tjian, R.,** The proline-rich transcriptional activator of CTF/NF-1 is distinct from the replication and DNA binding domain, *Cell,* 58, 741, 1989.

60. **Pugh, B. F. and Tjian, R.,** Mechanisms of transcriptional activation by Spl: evidence for coactivators, *Cell,* 61, 1187, 1990.
61. **Ptashne, M. and Cann, A. A. F.,** Activators and Targets, *Nature,* 346, 329, 1990.
62. **Han, M. and Grunstein, M.,** Nucleosome loos activates yeast downstream promoters *in vivo, Cell,* 55, 1137, 1989.
63. **Stringer, K. F., Ingles, C. J., and Greenblott, J.,** Direct and selective binding of an acidic activation domain to the TATA-box factor TEIID, *Nature,* 345, 783, 1990.
64. **Tinoco, T., Jr., Uhlenbeck, O. C., and Levine, M. D.,** Estimation of secondary structure in ribonucleic acids, *Nature,* 230, 260, 1971.
65. **Delsi, Ch. and Crothers, M.,** Prediction of RNA secondary structure, *Proc. Natl. Acad. Sci. U.S.A.,* 68, 2682, 1971.
66. **Zucker, M.,** An algorithm for folding in ribonucleotide acids, *Science,* 244, 48, 1989.
67. **Delbrück, M. and Stent, G.,** On the mechanism of DNA replication, in *The Chemical Basis of Heredity,* Elory, W. D. and Class, B., Eds., John Hopkins Press, Baltimore, 1975.
68. **Kowalski, D. and Laiskowski, M. Dr.,** Functional characterisation of nucleophoposodiesterases, in *Practical Handbook of Biochemistry and Molecular Biology,* Fasman, G. D., Ed., CRC Press, Boca Raton, FL, 1990.
69. **Gefter, M. L., Hihota, Y., Kornberg, T., Wechsler, J. A., and Barnoux, C.,** Analysis of DNA-Pol-2 and DNA-Pol-3, in mutants of *E. coli* thermosensitive for DNA synthesis, *Proc. Natl. Acad. Sci. U.S.A.,* 68, 3150, 1971.
70. **Darnell, J., Lodish, H., and Baltimore, D.,** *Molecular Cell Biology,* 2nd ed., Scientific American Books, New York, 1990.
71. **Weinberg, D. H., Collins, K. L., Simancek, P., Russo, A., Wold, M. S., Virshup, D. M., and Kelly, T. J.,** Reconstitution of simian virus 40 DNA replication with purified proteins, *Proc. Natl. Acad. Sci. U.S.A.,* 87, 8692, 1990.
72. **Prelich, G. and Stillman, B.,** Coordinate leading and lagging DNA strand synthesis during SV 40 DNA replication requires PCNA, *Cell,* 53, 117, 1988.
73. **Rawn, J. D.,** *Biochemistry,* Carolina Biological Supply Co., Burlington, NC, 1989.
74. **Linn, S.,** How many pols does it take to replicate nuclear DNA?, *Cell,* 66, 185, 1991.
75. **So, A. G. and Downey, K. M.,** Mammalian DNA polymerases and current status in DNA replication, *Biochemistry,* 27, 4591, 1988.
76. **Wolfendin, R. V.,** Tautomeric equilibria in inosine and adanine, *J. Mol. Biol.,* 40, 307, 1969.
77. **Sukhorukov, B. I., Poltew, V. I., and Knobdi, Y. K.,** Tautomeric transformations in complementary DNA base pairs, *Dokl. Akad. Nauk. S.S.S.R.,* 175, 948, 1967.
78. **Kubitschek, H. E. and Henderson, T. R.,** DNA replication, *Proc. Natl. Acad. Sci. U.S.A.,* 55, 512, 1966.
79. **Kornberg, A.,** Regulation of replication, in *DNA Replication,* W. H. Freeman, San Francisco, 1980, chap. 13.
80. **Brown, T., Kennard, O., Kneale, G., and Rabinovici, D.,** High resolution structure of a DNA helix containing mismached base-pairs, *Nature,* 315, 604, 1985.
81. **Cox, E. C.,** Bacterial mutator genes and the control of spontaneous mutation, *Annu. Rev. Genet.,* 10, 135, 1976.
82. **Modrich, P.,** DNA mismatch and correction, *Annu. Rev. Biochem.,* 56, 435, 1987.
83. **Lindahl, T.,** DNA repair enzyme, *Annu. Rev. Biochem.,* 51, 61, 1982.
84. **Radman, M. and Wagner, R.,** The high fidelity of DNA implication, *Sci. Am.,* 8, 24, 1988.
85. **Lu, A. L., Clarck, S., and Modrich, P.,** Methyl-directed repair of DNA base-pair mismatches *in vitro, Proc. Natl. Acad. Sci. U.S.A.,* 80, 4639, 1983.
86. **Loeb, L. A. and Kunkel, T. A.,** Fidelity and DNA synthesis, *Annu. Rev. Biochem.,* 52, 429, 1982.

87. **Murray, R. K., Granner, D. K., Mayes, A. P., and Rodwell, V. W.,** *Harper's Biochemistry,* 22nd ed., Prentice-Hall, Englewood Cliffs, NJ, 1990, chap. 21.
88. **Friedberg, E. C., Bonura, T., Cone, R., Simmons, R., and Anderson, C.,** in *DNA Repair Mechanisms,* Hanawalt, P. C., Friedberg, E. C., and Fox, C. F., Eds., Academic Press, Orlando, FL, 1978, 163.
89. **Varshney, U. and Van de Sande, J. H.,** Specificity and kinetics of uracil excision from uracil containing DNA oligomers by *E. coli* uracil DNA glycosilase, *Biochemistry,* 30, 4055, 1991.
90. **Lin-Chao, S. and Cohen, S. N.,** The role of processing and degradation of antisense RNAI regulates the replication of ColEl-type plasmids *in vivo, Cell,* 65, 1233, 1991.
91. **Beebee, T. J. C.,** Quantitative regulation of transcription in eukaryotes: theoretical considerations of RNA polymerase involvement, *J. Theor. Biol.,* 86, 803, 1980.
92. **Conaway, J. W., Haxley, J. P., Garrett, K. P., and Conaway, R. C.,** Transcription initiated by RNA polymerase II and transcription factors from liver, *J. Biol. Chem.,* 266, 7804, 1991.
93. **Sumimato, M., Ohkuma, Y., Yamamoto, T., Horokishi, M., and Roeder, R. G.,** Factors involved in specific transcription by mammalian RNA polymerase II: Identification of general transcription factor TF IIG, *Proc. Natl. Acad. Sci. U.S.A.,* 87, 9158, 1990.
94. **Ruvkun, G. and Finney, M.,** Regulation of transcription and cell identity by POU domain proteins, *Cell,* 64, 475, 1991.
95. **Felsenfeld, C.,** Chromatin as essential part of the transcriptional mechanism, *Nature,* 355, 219, 1992.
96. **Greenblatt, B.,** Roles of TFIID in transcriptional initiation by RNA polymerase II, *Cell,* 66, 1067, 1991.
97. **Buratowski, S., Hahn, S., Guarante, L., and Sharp, P. A.,** Five intermediate complexes in transcription initiation by RNA polymerase II, *Cell,* 56, 549, 1989.
98. **Kao, C. C., Lieberman, M. P., Schmidt, M. C., Zhou, Q., Pei, R., and Berk, A. J.,** Cloning of a transcriptionally active human TATA binding factor, *Science,* 248, 1646, 1990.
99. **Mitchel, P. J. and Tjian, R.,** Transcriptional regulation in mammalian cells by sequence-specific DNA binding proteins, *Science,* 245, 371, 1989.
100. **Flores, D., Ma, I., and Reinberg, D.,** Factors involved in the transcription by mammalian RNA polymerase II: purification and subunit composition of transcription factor IIF, *J. Biol. Chem.,* 265, 5629, 1990.
101. **Mullis, K. B.,** The unusual origin of the polymerase chain reaction, *Sci. Am.,* 4, 56, 1990.
102. **Morales, V. M. and Bagdasarian, M.,** Cloning in bacteria: vectors and techniques, in *Recombinant DNA Technology and Applications,* Prokop, A., Bajpai, R. K., and Ho, C., Eds., McGraw-Hill, New York, 1991, chap. 1.
103. **Piccirilli, J. A., Kransch, T., Moroney, S. E., and Benner, S. A.,** Enzymatic incorporation of a new base pair into DNA and RNA extents the genetic alphabet, *Nature,* 343, 33, 1990.
104. **Springate, C. F. and Loeb, L. A.,** Leukemic cells-mutagenic polymerase, *Proc. Natl. Acad. Sci. U.S.A.,* 70, 245, 1973.
105. **Preston, B. D., Poiesz, B. J., and Loeb, L. A.,** Fidelity of HIV-1 reverse transcriptase, *Science,* 242, 1168, 1988.
106. **Roberts, J. D., Bebenek, K., and Kunkel, T. A.,** The accuracy of reverse transcriptase from HIV-1 *Science,* 242, 1171, 1988.
107. **Krainer, A. R. and Maniatis, T.,** Frontiers in molecular biology, in *Transcription and Splicing,* Hames, B. D. and Glover, D. M., Eds., IRL Press, Washington, D. C., 1988, 131.

108. **Cech, Th. R.,** RNA as an enzyme, *Sci. Am.*, 1, 255, 1986.
109. **Berget, S. M., Moore, C., and Sharp, P. A.,** Spliced segments at the 5′terminus of Adenovirus 2 late mRNA, *Proc. Natl. Acad. Sci. U.S.A.*, 74, 3171, 1977.
110. **Maniatis, T. and Reed, R.,** The role of the small nuclear ribonucleo-protein particles in pre-mRNA splicing, *Nature,* 325, 673, 1987.
111. **Mowry, K. L. and Steitz, J. A.,** Identification of the human U7 snRNP as one of several factors involved in the 3′ end maturation of histone pre-mRNAs, *Science,* 238, 1682, 1987.
112. **Fu, X. D. and Maniatis, T.,** Factor required for mammalian spliceosome assembly is localised to discrete regions in the nucleus, *Nature,* 343, 437, 1990.
113. **Zhuoeng, I. and Weiner, A. M.,** A compensatory base change in U1 and RNA suppresses a 5′ splice site mutation, *Cell,* 46, 827, 1986.
114. **Wu, J. and Manley, J. L.,** Base pairing between U2 and U6 mRNAs is necessary for splicing of a mammalian pre mRNA, *Nature,* 352, 818, 1991.
115. **Parker, R., Siliciano, P. G., and Guthrie, C.,** Recognition of the TACTAAC box during mRNA splicing in yeast involves base pairing to the U2-like mRNA, *Cell,* 49, 229, 1987.
116. **Vankan, P., McGuigan, C., and Mattai, I. W.,** Domains of U4 and U6 mRNAs required for mRMP assembly and splicing complementation in *Xenopus* oocytes, *EMBO J.*, 9, 3397, 1990.
117. **Steitz, J. A., Black, D. L., Gerke, V., and Parker, K. A.,** Functions of the abundant UsnRNPs, in *Structure and Function of Major and Minor Small Nuclear Ribonucleo-protein Particles,* Bimsteil, M. L., Ed., Springer-Verlag, New York, 1988, 115.
118. **Gall, J. G.,** Spliceosomes and snurposomes, *Science,* 252, 1499, 1991.
119. **Bell, L. R. and Maine, E. M.,** Sex-lethal a *Drosophila* sex determination switch gene, exhibits specific RNA splicing and sequence similarity to RNA binding proteins, *Cell,* 55, 1037, 1988.
120. **Query, Ch. C.,** A common RNA recognition motif identified within a defined U1 RNA binding domain of the 70 k U1 RNP protein, *Cell,* 57, 89, 1989.
121. **Mattaj, J. W.,** A binding consensus: RNA-protein interactions in splicing, snRNPs, and sex, *Cell,* 57, 1, 1989.
122. **Petterson, I.,** The structure of mammalian small nuclear ribonuclearproteins, *J. Biol. Chem.*, 259, 5907, 1984.
123. **Adam, S. A.,** mRNA polyadenilate-binding protein: gene isolation and sequencing and identification of a ribonucleoprotein consensus sequence, *Mol. Cell Biol.*, 6, 2932, 1986.
124. **Dreyfuss, G.,** Heterogenes nuclear ribonucleoprotein particles and the pathway of RNA formation, *Trends Biochem. Sci.*, 13, 86, 1988.
125. **Kuo, H. Ch., Nasim, F. H., and Grabowski, P. J.,** Control of alternative splicing by the differential binding of U1 small nuclear ribonuclear particle, *Science,* 251, 1045, 1991.
126. **Leff, S. E., Rosenfeld, M. G., and Evans, R. M.,** Complex transcriptional units: diversity in gene expression by alternative RNA processing, *Annu. Rev. Biochem.*, 55, 1091, 1986.
127. **Krainec, A. R., Conway, G. C., and Kozak, D.,** The essential pre-mRNA splicing factor SF2 influences 5′ splice site selection by activating proximal sites, *Cell,* 62(1), 35, 1990.
128. **Nigro, J. M., Cho, K. R., and Fearon, E. R.,** Scrambled exons, *Cell,* 64, 607, 1991.
129. **Agapia, Nina,** Trans splicing of nuclear pre-mRNAs, *Cell,* 61, 1157, 1990.
130. **Crick, F. C. H.,** Codon-anticodon pairing: The Wobble hypotheses, *J. Mol. Biol.*, 19, 548, 1966.
131. **Quigley, G. J. and Rich, A.,** Structural domains of transfer RNA molecules, *Science,* 194, 796, 1976.

132. **Kozak, M.,** Comparison of initiation of protein synthesis in prokaryotes, eukaryotes and organelles, *Microbiol. Rev.,* 47, 1, 1983.
133. **Kohli, J. and Grosjean, R.,** Usage of the three termination codons: compilation and analysis of the known eukaryotic and prokaryotic translation termination sequences, *Mol. Gen. Genet.,* 182, 430, 1981.
134. **Capecchi, M. R. and Gussin, G.,** Suppression *in vitro:* identification of sex-sRNA as a non-sense suppressor, *Science,* 149, 417, 1965.
135. **Waldorp, M. M.,** The structure of the second genetic code, *Science,* 246, 1122, 1989.
136. **Kozak, M.,** An analysis of vertebrate mRNA sequences: initiations of translational control, *J. Cell Biol.,* 13, 887, 1991.
137. **Shatkin, A. J.,** Gapping of eukaryotic mRNAs, *Cell,* 9, 645, 1976.
138. **Kozak, M.,** Context effects and (inefficient) initiation at non-AUG codons in eukaryotic cell free translation systems, *Mol. Cell Biol.,* 108, 229, 1989.
139. **Kozak, M.,** The scanning model for transcription: an update, *J. Cell Biol.,* 108, 229, 1989.
140. **Högenauer, G.,** Stability of a codon-tRNA complex, *Eur. J. Biochem.,* 12, 527, 1970.
141. **Ramachandran, G. N. and Lakshminurayanan, A. V.,** Macromolecular storage and transfer, *Biol. Inform. Pra. Symp. Bombay,* 17–20, 1969.
142. **Perona, J. J., Swanson, R. N., Rould, M. A., Stetz, T. A., and Soll, D.,** Structural basis for misaminoacylation by mutant *E. coli* glutamine tRNA synthetase enzymes, *Science,* 246, 1152, 1989.
143. **Rould, M. A., Perona, J. J., Soll, D., and Stetz, T. A.,** Structure of *E. coli* glutaminyl tRNA synthetase complexed with tRNA Gln and ATP at 2.8 Å Resolution, *Science,* 246, 1135, 1989.
144. **Schulman, L. D. H. and Pelba, H.,** The anticodon contains a major element of the identity of arginine transfer RNAs, *Science,* 246, 1595, 1989.
145. **Wright, K.,** Language of translation, *Sci. Am.,* 1, 14, 1988.
146. **Nirenberg, M.,** The genetic code, *Angew. Chemie,* 87, 1017, 1969.
147. **Von Ehrenstein, G.,** Translational variances in the aminoacid sequence of rabbit hemoglobin α chain, *Cold Spring Harbor Symp. Quant. Biol.,* 31, 705, 1966.
148. **Lotfield, R. B. and Vanderyagt, D.,** The frequency of errors in protein biosynthesis, *Biochem. J.,* 128, 1353, 1972.
149. **Holliday, R. and Tarrant, G. M.,** Evaluation of false aminoacid incorporation in protein synthesis, *Nature,* 238, 26, 1972.
150. **Lotfield, L. B.,** Error frequency in protein biosynthesis, *Biochem. J.,* 89, 82, 1963.
151. **Gynewskaia, V. A., Skarlat, I. V., Kalinina, N. O., and Agol, V. I.,** On precision of translation in acellular systems from mammalian cells, *Molek. Biol. (Moskow),* 5, 132, 1971.
152. **Rice, R. and Fraenkel-Conrat, H.,** Fidelity of translation of sattelite tobako necrosis virus RNA in cell free *E. coli* system, *Biochemistry,* 12, 181, 1973.
153. **Davies, J.,** Streptomycin and the genetic code, *Cold Spring Harbor Symp. Quant. Biol.,* 31, 665, 1966.
154. **Ozaki, H., Inokuchi, H., Yamo, F., and Kodairo, Y.,** Genetics of nonsense suppressor of tRNAs in *E. coli,* in *Transfer RNA, Biological Aspects,* Soll, D., Abelson, J. N., and Schimmel, P. R., Eds., Cold Spring Harbor Laboratory, Cold Spring Harbor, N.Y. 1980, 341.
155. **Lotfield, R. B. and Eigner, E. A.,** Increase of specificity of the Ile activating enzyme of *E. coli* induced by tRNA, *J. Biol. Chem.,* 240, 1482, 1965.
156. **Mulivar, R. and Rappaport, H. P.,** Analysis of phenylalanine binding to phenylalanine-tRNA-synthetase, *J. Mol. Biol.,* 76, 123, 1973.
157. **Jarus, M.,** Binding of Ile-tRNA by Ile-tRNA-synthetase: solvents, the strengths of interaction and a proposed source of specificity, *Biochemistry,* 11, 2050, 1972.

158. **Calendar, H. and Berg, P.,** D-tyrosyl-DNA: formation, hydrolysis and usage in protein synthesis, *J. Mol. Biol.*, 26, 39, 1967.
159. **Bonnet, J., Kern, D., and Ebel, J. P.,** Recognition of tRNA's by tRNA synthetases of different specificities, *Molek. Biol. (Moscow),* 9, 48, 1975.
160. **Ames, B. N.,** Carcinogenesis mechanisms: the debate continues, *Science,* 252, 902, 1991.
161. **Cohen, S. M. and Ellewein, L. B.,** Carcinogenesis mechanisms: the debate continues, *Science,* 252, 902, 1991.

Chapter 6

SPECIFICITY IN IMMUNE PHENOMENA

Zeno Simon and Geza Szegli

TABLE OF CONTENTS

0-8493-5398-X/93/$0.00 + $.50

I. INTRODUCTION

The immunological defense system is the natural system with the highest specificity. Regardless of the intricate mechanism of the immune response, the immunocytes which detect (and finally elicit a response to) a foreign, nonself antigen do not detect a very large number, N, of the self antigens with which they come into contact. Each of these immunocytes possesses on its surface an antibody-like combining site, with the role of a biological receptor, R_0, which detects an alien antigenic determinant, E_0 (with the role of a ligand or effector), at a concentration c_0, but none of the self antigenic determinants, E_i, at their concentrations, c_i.

The number of different antibody specificities elicited by a mammalian organism, polysaccharidic or short, overlapping peptidic sequences of the self protein, can be evaluated from the length of the genome, to $N = 10^9$, as an order of magnitude.[2] The minimal immunogenic antigen concentrations quoted for soluble antigens in mice and rats are in the range of 10^{-5} to 10^{-12} *M* (mol/l).[3,4] The concentration of nonrecognized self antigens may reach 10^{-3} *M;* the concentration of serum globulins, with much sequence homology, is 10^{-3} to 10^{-4} *M*.

We shall discuss the mechanism of these highly specific phenomena based upon structural features of antibody recognition sites and antigenic epitopes, as well as upon chemical equilibrium considerations for antibody-antigen interactions. Equilibrium considerations for antigen-antibody interactions have already been introduced by Svante Arrhenius[5-7] and are throughly discussed in a series of excellent immunochemistry textbooks and in more recent publications.[3,4,8-11] Nevertheless, in the main line of contemporary research in the field of self-nonself discrimination, rather little attention is paid to peculiarities of equilibria in the role of antigen and antibody concentrations in immune defense.

Let us first review some basic phenomena which take place in the immune response[12,13] and to which special attention will be paid here. A macromolecular, viral, or cellular antigen usually presents a large number of different epitopes, i.e., regions which are recognized per se by the antibody sites in a ligand-receptor type interaction. These antibodies are either in solution (circulating antibodies) or on the surface of various types of immunocytes (B cells, T_H cells, etc.). The recognition of an antigen epitope by the antibodies on the surface of an immunocyte results in a transformation of the cell (activation or preactivation). The elicitation of an immune response, humoral or other type, requires some additional processes: the antigen is recognized by two types of cells, a clone of B cells, T-killer cells on one hand, a clone of T-helper (T_H) cells on the other. The two types of cells are preactivated and interact (B with T_H), and thereby the first type of cells (B, T-killer) is activated. Differentiation and proliferation occur and the immune response against the recognized antigen is elicited (antibody secretion, etc.).

II. STRUCTURAL FEATURES OF ANTIGEN-ANTIBODY INTERACTIONS

A. ANTIGENS AND ANTIBODIES

The immune system contains two interception mechanisms which recognize foreign antigens. The first, based upon B cells, recognizes intact proteins and other large antigens. The second, based upon T-helper (T_H) cells, involves processing of the protein antigen and presentation of some of the resulting peptides (in association with a major histocompatible coded protein), by another cell, in order to be recognized. For the elicitation of an antibody response, the antigen must be recognized by both mechanisms, although some antigens seem to elicit response only by B cell recognition.[13]

Antigens are macromolecules, usually polypeptides or polysaccharides, but one can elicit antibodies to several micromolecules, the so-called haptens, if they are chemically coupled to polypeptides. The minimal molecular weight for antigenicity (of the whole antigen) is of about 4000 to 10,000 Da, although some exceptions have been noted, such as angiotensin (1031 Da) or arsanyl-*N*-acetyl-tyrosine (451 Da) which can possibly be explained by *in vivo* coupling to some protein.[12,13] Some proteins have very poor antigenicity, such as gelatin, poly-Ala, poly-Lys, and poly-Glu; by adding aromatic amino acids to them their antigenicity is increased.[14] Nucleic acids are rather rarely adduced as antigens; they are usually confined to the interior of the cell. Of the known antigens against nucleic acids, the one elicited against phenylalanine tRNA recognizes the odd nucleoside Y with a strongly modified tricyclic imidazopurine base.[15]

The size of epitopes (antigenic determinants), previous to the development of X-ray crystallography, was studied by inhibition of the reaction of the oligomer-coupled-to-protein-antigen with the antibody (elicited by the oligomer-coupled-to-protein) by the added oligomer, as a function of the number n of monomeric units (within the oligomer). The size of haptens determined by this method is in the range of 6 monosaccharidic units for polysaccharides[3] and 10 to 12 amino acid residues for peptides.[1] The smallest known determinant is trialanine (as hapten).[16] For such small haptens, the epitope may also comprise part of the carrier protein; thus, while gelatin does not elicit antibody production, when it is coupled with 2% tyrosine it elicits antibodies against both poly-Tyr and gelatin.[14]

Recent X-ray crystallographic studies of complexes of antibody sites (Fab) with their proteinic antigen indicate, for epitopes of lysozyme, surface areas of about 750 $Å^2$,[2] comprising 15 amino acid residues.[2,17] The epitope of a dodecasaccharide of the cell surface of a pathogenic *Salmonella* (studied as a complex with the Fab antibody fragment) is trisaccharidic.[18] Peptidic epitopes recognized by MHC class I molecules consist of 9 amino acid residues,[19] those recognized by MHC class II molecules have 13 to 17 residues.[20] The MHC class II recognized epitopes are often of the type Hph-X-Hph-Hph-X-

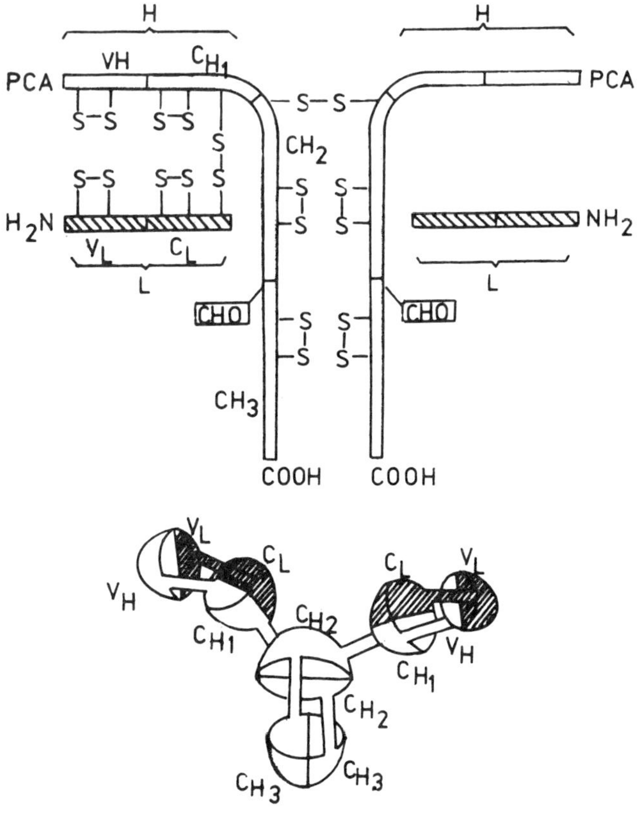

FIGURE 1. Spatial structure of immunoglobulin. IgG(H_2L_2) V_H, V_L — variable regions; C_L, C_{H1}, C_{H2}, C_{H3} — constant regions in the L and H chain; PCA — pyrrolidine carboxylic acid; CHO — carbohydrate. (From Simon, Z., *Quantum Biochemistry and Specific Interactions*, Abacus Press, Tunbridge Wells, Kent, 1976. With permission.)

Hph or B-Y-B-Y-B (Hph — hydrophobic amino acid residue).[21] Smaller epitopes have also been found recently, for example, a heptapeptide as an epitope recognized by the murine monoclonal antibody mAb 1/1.[22]

Antibody molecules are flexible associations of light and heavy chains, with each chain folded into several characteristic globular domain structures, having approximately 110 amino acid residues and two combining sites (Figure 1).[2,12,13] The domains are tightly joined in pairs to form functional modules, such as VH:VL and CH1:CL. The antigen binding site is at the top of the Fab fragment; the six hypervariable regions of VH and VL (variable heavy and light chain regions) are brought together so as to form a continuous hypervariable region. For the monoclonal anti-lysozyme antibody there are 17 amino acid residues in the combining site in direct contact with the lysozyme epitope.[17] The domain structure of the antibody molecule is a β barrel consisting of seven antiparallel β strands in constant domains and the other nine in variable domains.

The above-described structure of immunoglobulin molecules is similar to that of other functional surface proteins, such as the human leukocyte antigen molecule, certain domains of β_2 microglobulin, class II major histocompatibility antigens, Thy-I antigens, T cell receptors (TCR), polymeric immunoglobulin receptors, the CD-4 and CD-8 antigens of T cells, and the neural cell adhesion molecule.[17]

B. ANTIGEN-ANTIBODY COMPLEXES

X-ray diffraction studies helped to decipher the structure of proteic antigen-antibody complexes and to obtain answers to questions concerning antigen-antibody function. Three complexes of egg white lysozyme with three different antibodies were studied relatively recently.[2,17,23] The three antibodies, D1.3, HyHEL-5, and HyHEL-10 bind to different epitopes of the lysozyme molecule. The three epitopes present a small overlap and cover approximately half the total surface area of the lysozyme molecule. The lysozyme epitope which bind the monoclonal D1.3 antibody and the amino acid contacts are presented in Table 1. The corresponding dissociation constant for the complex with D1.3 is $K = 2.2 \cdot 10^{-8}\ M$.

Analyses of these three lysozyme epitope-antigen complexes, as well as of a fourth complex, NC41 influenza virus neuraminidase,[24] demonstrate that the epitope is located in several discrete segments of the polypeptide chain, contiguous on the surface as a result of folding. The residues from the antibody that are in contact with the antigen come from all six complementarity determining regions. All accessible regions of proteins seem to be potentially antigenic and there is a good correlation between antigenicity and accessibility, but not with mobility of the protein regions. The conformational changes range from a rigid "lock-and-key" fit, for the backbone of the lysozyme antigen in the D1.3 Fab-lysozyme complex, to a locally induced fit in the complexes with HyHEL-5, HyHEL-10, and NC41. The locally induced fit corresponds, for the lysozyme-HyHEL-5-complex, in a C_α of the Pro^{70} residue moving 1.7 Å and a rotation by 180° of the C_β-C_γ side chain of Trp^{63} and, for the lysozyme-HyHEL-10 complex, in a carbonyl oxygen atom (of Asp^{101}) moving by 2.5 Å. No correlation was found between antigen or hapten binding and the "elbow bend" of Fab.

X-ray crystallographic studies were also performed for the MHC class I molecule peptidic fragment complexes[19] and for an oligosaccharide-Fab (dodecasaccharide from the surface of a pathogenic *Salmonella*) fragment complex.[18] For the oligosaccharide-Fab complex the dissociation constant is $K = 5 \cdot 10^{-6}\ M$ and binding is entropy assisted ($T \cdot \Delta S = +2.4$ kcal/mol). The recognition site of this Fab is unusual in that it is defined by aromatic amino acids and structured water molecules. This site is a hydrophobic pocket,

TABLE 1
Amino Acid Sequence of a Lysozyme Epitope and Amino Acid-Amino Acid Contacts in Binding to the D1.3 Recognition Site[2,17]

Lysozyme Epitope

Asn^{19} Arg^{21} Gly^{22} Tyr^{23} Ser^{24} Leu^{25} Asn^{27} Lys^{116} Gly^{117} Thr^{118} Asp^{119} Val^{120} Gln^{121} Ile^{124} Leu^{129}

Amino Acidic Contacts

	Antibody	Lysozyme
	Light Chain Regions	
CDR1	His^{30}(NH-ϵ)	Leu^{129}(O)
	Tyr^{32}(OH)	Leu^{25}, Gln^{121}, Ile^{124}
FR2	Tyr^{49}	Gly^{22}
CDR2	Phe^{91}(O)	Gln^{121}(NH-ϵ)
	Trp^{92}	Gln^{121}, Ile^{124}
	Ser^{93}	Gln^{121}
	Heavy Chain Regions	
FR1	Thr^{30}(OH-ι)	Lys^{116}(O), Gly^{117}
CDR1	Gly^{31}(NH)	Lys^{116}(O), Gly^{117}
	Tyr^{32}	Lys^{116}, Gly^{117}
CDR2	Trp^{52}	Gly^{117}, Thr^{118}, Asp^{119}
	Gly^{53}(NH)	Gly^{117}(O)
	Asp^{54}	Gly^{117}
CDR3	Arg^{99}(NH-η)	Gly^{22}(O)
	Asp^{100}(O-δ1, O-δ2)	Asn^{27}(NH-δ), Ser^{24}(OH-ι), Gly^{22}, Tyr^{23}
	Tyr^{101}(OHη)	Asp^{119}(O-δ), Val^{120}(N), Gln^{121}(N), Thr^{118} Asn^{19},
	Arg^{102}	Gly^{22}

Note: For hydrogen bounding contacts the implied atoms are also indicated.

8 Å deep and 7 Å wide, filled with the trisaccharidic epitope. Saccharide-saccharide and saccharideprotein hydrogen bonds are also implied in this complex.

1. Structural Features in Antigen-Antibody Interactions

Structural features in antigen-antibody interactions were studied by competition experiments, for the antibody binding site, between the eliciting haptens and various other molecules. More recently, X-ray crystallographic examination of antigen-antibody complexes have also contributed to such studies.

Table 2 presents affinity decreases (Δ log 1/K) produced by substitution in haptens, as revealed by competition experiments between the hapten and

TABLE 2
Correlation of Affinity Decrease Produced by Amino Acid Substitutions in Bradykinin toward Antibodies Elicited by Bradykinin Nonapeptide Coupled to Different Macromolecular Carriers

	−Δ log (1/K)					
	aBK-TC-PL		aBK-TC-OA			
Substitution	**BK**	**Ac-BK**	**T24**	**T36**	**aKin**	**MSD**
Arg^9/Ala	—	0.20	0.52	0.62	2.95	6
Phe^8/Ala	—	0.50	2.85	—	2.19	6
Pro^7/Ala	1.00	0.80	1.61	—	0.76	2
Ser^6/Ala	0.38	0.35	0.75	1.62	1.79	1
Phe^5/Ala	1.57	1.85	3.77	3.94	1.93	6
Gly^4/Ala	1.58	1.96	1.05	1.42	0.56	1
Pro^3/Ala	2.58	2.63	0.76	1.13	0.08	2
Pro^2/Ala	—	1.35	0.11	0.83	0.15	2

Note: −Δ log (1/K) is the decrease of affinity (dissociation constant, K, log units) in competition with the substitution derivative with bradykinin for the antibody; aBK-TC-PL is the antibody elicited by bradykinin coupled to polylysine for bradykinin (BK) and acetylated bradykinin (Ac-BK) derivatives; aBK-TC-OA is the same, but ovalbumin is the carrier, in two different rabbits (T24 and T36); aKin is the antibody elicited against akininogen; and MSD is the minimal steric difference corresponding to the amino acid substitution.

Data from Reference 26.

amino acid substitution derivatives of the hapten. As a first observation, one and the same amino acid substitution in the hapten produces substantially different effects for the complexes with the antibody elicited by bradykinin coupled to polylysine (a BK-TC-PL) or to ovalbumin (a BK-TC-OA) and even for antibodies elicited in two different animals.

Amino acid substitution effects are sometimes dramatic.[25] Two peptides, corresponding to residues 315 to 329 of the HIV-1 gp 160 virus envelope protein were synthesized and 9 of the 15 amino acid residues were identical. The two peptides react with two different populations of cytotoxic lymphocytes. If the amino acids in position 325 — Val in the first peptide, Tyr in the second — are inverted, this single substitution produces an inversion of the reactivity (of the two peptides) with respect to the two lymphocyte populations.

a. Steric Fit

A high degree of complementarity between the interacting surfaces is perhaps the most striking feature of the three lysozyme-antibody complexes

TABLE 3
Inhibition of Precipitin Reaction between Purine-6-Oyl-Human Serum Albumin and the E22–27 Bovine Antiserum by Nitrogen Base Derivatives

Nitrogen base derivative	A_i	$j(x_{ij} = 1)$	MTD
N-Purine-6-oyl-ϵ-aminocaproic acid	2.00	1–9,11–22	0
N-Purine-6-oyl-glycine	1.30	1–9,11–17	2
Purine-6-carboxamide	0.88	1–9,11–13	5
Hypoxanthine	−0.17	1–9,11	7
6-Methylpurine	−0.55	1–9,11	7
Adenine	−0.71	1–9,11	7
Deoxyadenosine	−1.08	1–9,11,23,24,26–31	7
Uracil	−1.14	1–6,9,11	9
Adenosine	−1.23	1–9,11,23–31	7
9-Methylpurine	−1.44	1–9,23	8
Thymine	−1.55	1–7,10,11	9
Cytosine	−1.61	1–7,11	9
ϵ-Aminocaproic acid	−2.57	13–22	11

Note: Here $A_i = -\log C_{rel}$, with C_{rel} the relative value of the nitrogen base concentration producing 50% inhibition of the precipitin reaction.

Data from Reference 27.

described in the previous paragraph.[2] This complementarity extends over almost all the molecular contact surface area and implies that any cavity between the two surfaces is too small to accommodate even one water molecule.

The minimal steric difference, MSD (see Chapter 2, Section IV.C) corresponding to each pair of substituted amino acids, is indicated in Table 2. There is a certain correlation between MSD (as a measure for stereochemical differences between the substituted pair) and the affinity decrease in the interactions with the antibody elicited against akininogen ($r = 0.775$) and against ovalbumin-carried bradykinin ($r = 0.610$), but none for the antibody elicited against polylysine-carried bradykinin.

The role of steric fit also results in two QSAR studies for affinity decreases as obtained from competition studies (for the antibody) by the eliciting hapten and various other molecules. Table 3 lists results for a QSAR based upon the MTD method (see Chapter 2, Section IV.C) for inhibition by various nitrogen based derivatives of the precipitation reaction between purine-6-oyl human serum albumin and the E22-27 bovine antiserum, elicited against this antigen.[28] The hypermolecule for this MTD study is depicted in Figure 2. The optimized receptor map (here part of the antibody binding site) and the correlational equation are

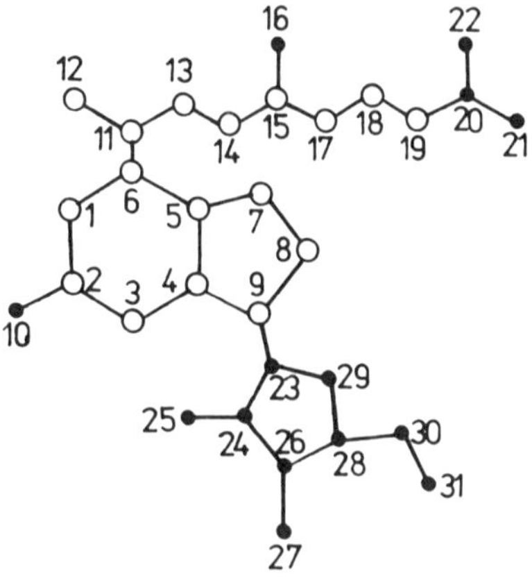

FIGURE 2. Hypermolecule for correlation of inhibition of precipitin reaction of purine-6-oyl human serum albumin with its corresponding antibody by nucleic acid derivatives. Numerotation of j vertices. Cavity ($\epsilon_j = -1$) vertices — open circles (○); wall ($\epsilon_j = +1$) vertices — black circles (●); irrelevant ($\epsilon_j = 0$) vertices — dots (·). See Table 3.

$$S_1^* \begin{cases} j(\epsilon = -1)\text{: } 1\text{–}9,\ 11\text{–}15,\ 17\text{–}19 \\ j(\epsilon = \ \ 0)\text{: } 10,\ 16,\ 20\text{–}31 \\ j(\epsilon = +1)\text{: } - \end{cases}$$

$$\hat{A}_1 = 2.24 - 0.42\ \text{MTD}\ ;\ r = 0.0959\ ;\ N = 13 \qquad (1)$$

The correlation also holds for the precipitin reaction and the complement fixation reaction for two other bovine antisera.

A QSAR study based on the SIBIS method (see Chapter 2, Section IV.C) was performed for a series of substituted benzoate haptens competing for the anti-*p*(*p*′-azophenylazo) benzoate antibody.[29] The haptens are listed in Table 4, the hypermolecule in Figure 3. The optimized receptor map and correlational equation are

$$S_2^* \begin{cases} j(\epsilon = -1)\text{: } 1\text{–}6,\ 9\text{–}11,\ 14\text{–}22 \\ j(\epsilon = \ \ 0)\text{: } 23 \\ j(\epsilon = +1)\text{: } 8,\ 12,\ 23 \end{cases}$$

$$\hat{A}_2 = -1.049 - 1.671q_4 + 0.179\text{SMD}_c - 0.516\text{SMD}_w$$

$$r = 0.990\ ;\ N = 18 \qquad (2)$$

$\hat{A}_2 = -1.049 - 1.671q_4 + 0.179\ \text{SMD}_c - 0.516\ \text{SMD}_w$; $r = 0.990$, $N = 18$ (2) SMD_c is the antibody binding site cavity volume occupied by the

TABLE 4
Benzoate haptens in the Anti-*p*-(*p*′Azophenylazo) Benzoate System — Equilibrium Constants and Structural Parameters

No	A	q_4	SMD_C	SMD_W
1. Benzoate	0.0000	−0.0554	6.0	0.0
2. 3-Cl-B	−0.3665	−0.0191	6.0	1.2
3. 3-NO_2-B	−0.9208	−0.0034	7.0	2.0
4. 3-NO_2,4-Cl-B	−0.3872	−0.1794	8.2	2.0
5. 3NO_2,5-Cl-B	−1.1549	0.0158	7.0	3.2
6. 3,5-diCl-B	−1.6198	0.0005	6.0	2.4
7. 4-F-B	0.5563	−0.2825	6.8	0.0
8. 4-Cl-B	0.7324	−0.2414	7.2	0.0
9. 4-Me-B	0.2553	−0.0570	7.0	0.0
10. 4-NH_2-B	0.3222	−0.1084	7.0	0.0
11. 4-OH-B	0.6721	−0.1488	7.0	0.0
12. 4-NO_2-B	0.2553	−0.2600	8.0	1.0
13. PAB	1.8261	−0.1622	14.0	0.0
14. 4′-NH_2-PAB	1.8451	−0.1666	15.0	0.0
15. 4′-Me-PAB	1.8129	−0.1631	15.0	0.0
16. 4′-OH-PAB	2.0453	−0.1645	15.0	0.0
17. 2′-Me,4′-OH-PAB	1.9085	−0.1674	16.0	0.0
18. 3′-Me,4′-OH-PAB	2.0969	−0.1643	17.0	0.0

Note: A = log K_{rel}, with K_{rel} being the hapten antibody relative binding constant; K_{rel} = 0 for benzoate; q_4 is net charge of the atom occupying vertex j = 4, calculated by the CNDO/2-method; SMD_C and SMD_W approximate (relative) volumes of the hapten molecule overlapping.

Data from Reference 4.

hapten molecule; SMD_w is the cavity wall volume overlapping with the hapten. The net electronic charge, q_4, of the atom situated in the vertex j = 4 of the hypermolecule, was calculated by the CNDO/2 method. The charge q_4 of the j = 4 vertex atom gives a partial correlation coefficient of r = 0.612.

Conclusively, steric complementarity of the epitope and the antibody site is a certain requirement for antigen-antibody interactions.

b. Intermolecular Forces

Substantive studies of the importance of different types of intermolecular forces in the above three lysozyme antibody complexes and for other antigen-antibody complexes, performed by a new technique, are described in a review by Davies et al.[2] In this technique, protein libraries are used to study the interactions of an antibody with different proteic antigens having similar three-dimensional structures with the eliciting antigen, but differing by only one or a few amino acid substitutions. The contribution of each amino acid to binding can be determined by this technique.

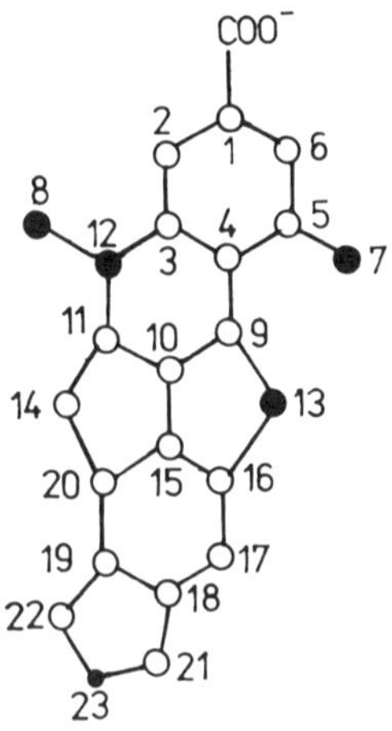

FIGURE 3. Hypermolecule for benzoate haptens in the anti-*p*-(*p*′-azophenylazo) benzoate system. Legend — as for Figure 2, see Table 4.

Charge complementarity is important in the binding of the HyHEL-5 antibody with the Glu^{35} and Glu^{50} residues to two Arg residues of the lysozyme epitope and in the binding of HyHEL-10 with Arg^{32} to Lys^{97} of the corresponding lysozyme epitope. Also, in the McPC 603 binding of phosphocholine, the Arg^{91} and Glu^{35} residues of the binding pocket interact with the positively charged trimethylammonium group, while Arg^{52} interacts with the negatively charged phosphate group.

Hydrogen bonds also play a key role in the specificity of antigen-antibody interactions, although not all theoretically possible hydrogen bonds between donors and acceptors are actually formed. Substitution of Arg^{68} by Lys in bobwhite quail lysozyme lowers the affinity to the HyHEL-5 antibody by about 4 kcal/mol, despite the like charges, probably due to some specific hydrogen bond arrangement.[2]

Charge transfer complexes seem important in specific binding of nitrophenyl derivatives and the aromatic ketone, menadione, to the MOPC-460 protein of a mouse plasmacytoma.[30] The dissociation constants, K (at 4°C), for ϵ-dinitrophenyl-Lys, other aromatic nitrophenyl derivatives, and menadione are between $2.5 \cdot 10^{-5}$ and $3.9 \cdot 10^{-4}$ *M,* while purine-6-oyl-aminocaproate and various aromatic amine and sulfonamide derivatives all have, $K > 10^{-3}$. All molecules of the first group have an electron-acceptor character for charge transfer complexes, while those of the second group do not present this character. Large steric dissimilarities (MSD values) exist between the molecules of the first group with only moderate effects upon dissociation constants.

C. Conformational Entropy

Conformational flexibility is expected to decrease upon antigen-antibody combination, producing a negative $T \cdot \Delta S$ term which decreases binding affinity. This may be related to the fact that the highest binding affinities (in the 10^{-12} *M* range) were observed for antibodies against small rigid organic molecules, such as haptens (fluorescein, digoxin), while for flexible molecules (oligosaccharides, peptides), affinities are typically in the 10^{-6} *M* range.[31]

C. SPECIFICITY AND SIZE OF BINDING SITES

The relation between the specificity and size of binding sites was discussed at length in Chapter 1. An antibody, circulating or on the surface of an immunocyte, must recognize a foreign antigen, but not the approximately $N = 10^9$ epitopes of the self antigen, as discussed in Section I. This evaluation is based upon the total length of the human genome, but as only a few percent of the genome code proteins and several other proteins and macromolecules do not come into contact with the immune system (intracellular proteins, anatomically shielded regions), $N = 10^7$ is probably a more realistic estimate of the number of self epitopes which must escape recognition.

According to the type of reasoning of Thomas[32] and Conaughy and McCarthy,[33] if the size of the recognition site and of the epitope is n monomeric units, with α types of monomers, the probability that such a self epitope should be recognized by chance by an antibody recognition site must be small (see also Chapter 1, Section II.B, Equation 12), i.e.,

$$\alpha^n \gg N \tag{3}$$

With $\alpha = 20$ amino acid types and $N = 10^7$ to 10^9, the size must be at least $n = 6$ to 7 amino acid residues.

As discussed in Chapter 1, Section II.B, a single amino acid difference in certain self epitopes may not be sufficient. Concentrations of certain self antigens may reach $c_i = 10^{-3}$ *M* while soluble antigens are recognized at much lower concentrations, in the range $c_0 = 10^{-5}$ *M* (weak antigens) to 10^{-12} *M* (strong antigens; see Section I). As demonstrated in Chapter 1, Section II.B, Equation 14, the affinity difference for the antibody, $\Delta\Delta G$, between the recognized low concentration antigen and the unrecognized high concentration self antigen must be

$$\Delta\Delta G \geq 2.3\ RT\ \lg \frac{c_i}{c_0} \tag{4}$$

For the weak antigen limit one obtains $\Delta\Delta G \geq 3$ kcal/mol; for the strong antigen limit, $\Delta\Delta G \geq 12.5$ kcal/mol. Competition studies for the antibody between the eliciting oligopeptides hapten and modified haptens (by amino acid substitution) indicate a mean affinity decrease of about Δg = 2.0 kcal/mol per amino acid substitution.[34,35] For peptidic antigens, the unrecognized self epitopes must differ from the recognized alien epitope by at least $\nu = 2$ to 6 amino acid substitutions (weak and strong antigen limit, respectively). The final formula for an upper limit for the size of the combining site (Chapter 1, Equation 18) is

$$n > \nu - 1 + \frac{\lg N + (\nu - 1)\lg \dfrac{n}{\nu - 1}}{\lg \alpha} \qquad (5)$$

The upper size for epitopes for the weak and strong antigen limit, respectively, results in the range n = 9 to 16 for $N = 10^9$, and n = 7 to 14 for $N = 10^7$.

Comparison with the experimental figures for the size of peptidic epitopes, between 7 and 15 amino acid residues (Section II.A) indicate that these epitopes are certainly large enough for the required specificity. The same can be said for polysaccharidic epitopes. The stated figure of 3 saccharidic residues[18] (if one considers mean molecular weights) corresponds to about 5 amino acid residues, and there are many more than 20 types of monosaccharidic units in antigenic polysaccharides, especially if various acetylated, aminated, and sulfated monosaccharidic derivatives are also considered.

III. ANTIGEN-ANTIBODY INTERACTIONS AS CHEMICAL EQUILIBRIUM

A. THRESHOLD CONCENTRATION CONDITION IN ANTIGEN RECOGNITION

A cell able to recognize antigens presents a certain number of antibody receptor sites (of a given type), Ab_i, able to interact with an antigen, Ag_j, and to form a complex. C_{ij}. It is a natural supposition that the cell recognizes the Ag_j if the degree, θ, of saturation of its Ab_1 sites is higher than a "critical" threshold, θ_{cr}. In terms of chemical equilibria, this means that Ag_j is recognized by the cell clone "i" if its concentration $[Ag_j]$ is higher than a threshold concentration, c^0_{ij}. If this Ag-Ab (epitope-antibody receptor) interaction takes place independently, Ab site by Ab site:

$$Ab_i + Ag_j \rightleftharpoons C_{ij} \qquad (6)$$

the degree θ of surface site saturation is given by a Langmuir isotherm:

$$\Theta = \frac{[Ag_j]}{[Ag_j] + [K_{ij}]} \quad K_{ij} = e^{-\frac{\Delta G_{ij}}{RT}} \qquad (7)$$

where K_{ij} is the dissociation constant of the c_{ij} complex and ΔG_{ij} is the affinity of Ag_j for Ab_i (more exactly, the highest affinity of an epitope of Ag_j for the receptor site of Ab_i). As proven in Appendix 1, the condition for the recognition of Ag_j by clone "i", $\theta \geq \theta_{cr}$, yields a critical concentration requirement:[36]

$$[Ag_j] \geq c^0_{ij} \cong e^{-\frac{\Delta G_{ij}}{RT}} \tag{8}$$

It is possible that the adsorbtion of Ag molecules on the cell surface Ab sites has a cooperational character with interactions between the adsorbed Ag molecules, as in the case of multiple ligand binding by a protein molecule.[11] Equation 8 remains valid in this case also, but ΔG_{ij} has the significance of a mean affinity per Ag-Ab pair (Appendix 1).

The threshold concentration, c^0_{ij}, thus depends upon the affinity of the pair Ag_j-Ab_i (the most favorable epitope). The higher the affinity, the lower the threshold concentration. If the minimal immunogenic antigen concentrations are equated to this threshold concentration, the range of 10^{-5} to 10^{-12} *M* (as determined for different soluble antigens in mice and rats[3,4]) corresponds to a ΔG_{ij} affinity range of 7 to 17 kcal/mol.

The neutralization of soluble antigens by antibodies corresponds to a precipitation reaction. A solubility product requirement is to be used in this case, i.e., the reaction takes place if

$$[Ab_i] \cdot [Ag_j] \geq P_{ij} = e^{-\frac{\Delta G_{ij}}{RT}} \tag{9}$$

ΔG_{ij} as used here is also the Ag-Ab affinity (per pair) but it must not be identical to the affinity for the Ag recognition, i.e., for the adsorbtion of Ag molecules upon the Ab surface sites.

Equation 8 is deduced for soluble antigens. For cell surface antigens the concentration ($[Ag_j]$ in Equation 8) is constant and does not depend upon the number of antigen-bearing cells introduced in the immune system. For such antigens, their surface density assumes the role of concentration in Equation 8.

Recognition of an antigen or its precipitation by antibodies is also a problem of antigen and antibody concentrations in the immune system.[36] For example, the $H\text{-}2^K$ antigen is recognized by an IgM antibody in a certain immune system. The quotient of surface densities of $H\text{-}2^K$ antigens on the surface of C11D and Y2C cells is 1:0.55; the first cells are sensitive, but the second are relatively resistant to IgM mediated cytotoxicity.[37]

The recognition of an antigen introduced into an immune system should take place as follows. The antigen (bacteria, virus, or macromolecule) usually presents several epitopes ("j"). The immune system contains several B and

T cell clones, each with a specific antibody combining site ("i"). A clone (i) will recognize an antigen (j) if it is found at a concentration $[Ag_j]$ larger than the threshold c^o_{ij} corresponding to the ΔG_{ij} affinity. An antigen, at a certain concentration, can be recognized by all clones with sufficient affinity for one of its epitopes and, in principle, one clone can recognize several epitopes (antigens), but at different threshold concentrations. If the condition of Equation 8 is not satisfied for either clone (i) or epitope (j) of the antigen pair, no recognition of the antigen will take place.

B. KINETICS OF ANTIGEN-ANTIBODY INTERACTIONS AND RECOGNITION

Some biological responses may be determined by the rate of antigen-antibody combination or by the mean life of the complex before its dissociation. Kinetics, rather than equilibrium may be the determinant in this case.[38] In the simple case of bimolecular association-dissociation, the equilibrium constant is equal to the quotient of the two rate constants, $K_{dis} = k_{off}/k_{on}$. For the interaction of an antibody with a hapten, $k_{on} = 10^8\ M^{-1}s^{-1}$ is a typical value; for the interaction with a protein, $k_{on} = 10^5\ M^{-1}s^{-1}$. The mean lifetime of the complex before its dissociation, for $K_{dis} = 10^{-9}\ M$, is 7 s for the complex with the micromolecular hapten and 115 min for the one with the protein. Such differences are critical for understanding serological reactions.[38]

Foot and Milstein[39] point out the importance of the combining rate, $v = k_{on}[Ag] \cdot [Ab]$, in the maturation of the antibody response against an oxazolone derivative as a hapten. In a certain immunologic response, one group of antibodies predominates in the primary response, while in the second and tertiary response a new set of antibodies appears, together with clones of the original group. Nevertheless, there are no ΔG differences between the two sets of antibodies, but kinetic studies demonstrate that the new antibodies have a k_{on} rate 10 times larger than for the secondary antibodies of the first set of clones.

C. CLONAL SELECTION THEORY AND THRESHOLD CONCENTRATION-CONDITION

According to Burnet's clonal selection theory,[40] explanations for tolerance and autoimmunity were attractively simple in the early 1960s. During phylogeny, natural selection will favor organisms with clones of lymphocytes producing antibodies with high affinity against the usual pathogenic agents. Autoreactive lymphocytes are deleted or inactivated during embryonic life, when a clonal diversification process take place, and self tolerance is provided.[41,42] In the animal with a mature immunologic system, for each pair of self antigen (epitope) - lymphocyte clones the threshold antigen concentration for recognition (Equation 8) exceeds the normal, physiologic concentration of the self antigen. Hence, no self antigen is recognized and no autoimmune process is elicited.

According to this type of reasoning, autoaggression could appear if, due to some pathologic process, there is a strong increase in the concentration of a self antigen, exceeding the threshold (Equation 8), or if a strong antibody response is elicited against an alien antigen (with an epitope similar to some self epitope) and the solubility product requirement (Equation 9) for antigen neutralzation is satisfied for both the alien, eliciting antigen and for some self antigen. Both conditions are often connected with autoimmune diseases — the idea of molecular mimicry[43] or autoaggression following an intense, prolonged infection.[12,13]

Each antibody of the immune system must have a sufficiently high affinity against one (or several) foreign antigen, but sufficiently low affinities against all epitopes of self antigens, in order not to recognize them. This requires a suitable size for antigen epitopes and antibody combining sites, a requirement which is satisfied, as demonstrated in Section II.C above.

One may ask, then, why the elicitation of an immune response also usually requires recognition of the antigen by a second partner, the T_H (helper) cell? The response may be found in the high variability of pathogenic agents, which also requires a continuous diversification of the antibody repertoire during mature life. Clonal diversification mechanisms (mutational hot spots, exon transpositions[44]) also had to appear after birth, possibly during elicitation of secondary immune response. While producing "better" antibodies, this diversification mechanism may accidentally produce antibodies with a sufficiently high affinity for some self antigen to be able to recognize it. A mechanism had to evolve in order to prevent elicitation of immune response against an accidentally recognized self antigen.[36] Recognition of the antigen by the second partner, the T_H cell, "validates" the primary recognition, which is performed by B cells or T cells implied in cytotoxic mechanisms.[45] Most self antigens will remain unrecognized by B cells, despite antibody diversification after birth. Small structural changes in the antibody site or the pathogenic agent-antigen generally produce small changes in their reciprocal affinities (point mutation, amino acid substitution, see Section II.B.1 and Table 2). The probability of a self antigen to be recognized accidentally by both a B clone and a T_H clone will remain sufficiently small (product of two probabilities).[46]

The threshold concentration-condition must be valid for antigen recognition both in eliciting immune response and in the mechanism of deletion of inactivation of autoreactive lymphocyte clones in certain embryonic stages. Older theories explain such deletion by immunologic paralysis — inactivation by an excess of antigen;[47] experimental facts seem to contradict these theories although there are some indications of tolerance induced in the B cells of mature mice by endogenous self antigens.[48] Insufficient antigen (humoral or surface) concentrations may explain some puzzles, for example the presence of T cells, autoreactive *in vitro,* without autoimmune disease in the recipient (mouse) organism.[41]

Another concept to consider is Jerne's theory of idiotype-antiidiotype networks for regulating immune response.[49,50] Such theories require complicated nets of antibodies, anti-antibodies, etc., and there is no general acceptance for them.[51] Nevertheless, according to the threshold concentration condition, one may expect elicitation of anti-antibodies toward an antibody whose concentration is greatly increased. There were some recent successes in using antiidiotypic antibodies as vaccines.[52] Recent results favor the role of immune networks in preventing autoimmune disease.[53-55]

IV. ROLE OF T CELLS IN SELF-NONSELF DISCRIMINATION

The primary antigen recognition for elicitation of humoral response is performed by a B cell, which thereby becomes preactivated. The B cell processes this antigen and presents a certain fragment of the antigen in a complex with a MHC class II molecule (MHCII) to a T-helper (T_H) cell able to recognize the complex. This B-T_H interaction, mediated by the antigen fragment-MHC II complex, activates the preactivated B cell, which proliferates, maturates to plasmocytes, and elicits large quantities of antibodies against the antigen. It is possible that the antigen is processed by a macrophage, presented in the complex with MHC-II to a T_H cell, which becomes preactivated and thereafter interacts with the preactivated B cell. In cytotoxic response there is a similar cooperation between T_H cells and another type of T cell.[12,45]

The antigens for the recognition by T_H cells are usually proteins from the exterior of the cells, recognized as alien (by the B cell or a macrophage), and are first processed to rather short peptides.[21] Some of these peptides bind to MHC-II and the complex is then presented on the surface of the processing cell to the T_H cell. These peptides and antigen fragments bind to MHC-II proteins with rather large dissociation constants, $K = 10^{-6}$ *M,* or even higher.[56,57] There are 10^4 to 10^5 MHC-II molecules in a B cell or macrophage, already complexed with endogenous self-peptides, 50% of them with only 5 to 7 predominant peptides. Access of various proteins to peptidases and the number of MHC-II molecules per cell are suggested to have a primary importance for recognition or absence of recognition by T_H cells. The minimal number of peptide-MHC-II complexes on the surface of the B cell of the macrophage required to be recognized (to stimulate) by the T_H cell is evaluated to be 100 or even as few as 10 such complexes.[58,59]

The specificity of interaction between B and T_H cells, preactivated by the same antigen, is ascertained by the interaction of antigenic fragment peptide-

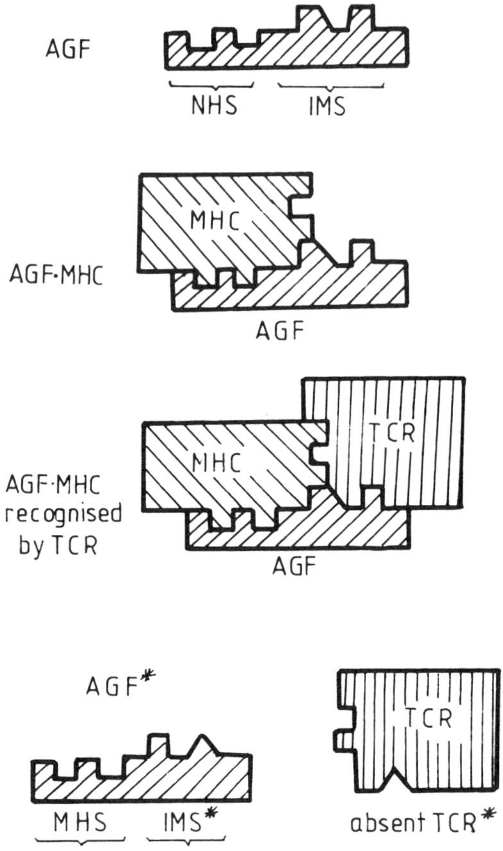

FIGURE 4. Schematic representation of recognition by T cell receptor (TCR) of MHC class II molecule (MHC)-antigen fragment (AGF) complexes. The individual mark sequence (IMS) on the AGF and the MHC recognized sequence (MHS) on the AGF. Self antigen fragment (AGF), self individual mark sequence (IMS), and the corresponding T cell receptor (TCR) which is absent. (AGF, IMS, and TCR derived from Simon, Z. *Quantum Biochemistry and Specific Interactions,* Abacus Press, Tunbridge Wells, Kent, 1976. With permission.)

MHC-II complexes on the surface of the B cell and the specific T cell receptors (TCR) on the T_H surface. Interactions occur between pairs of receptors on the B and T_H surfaces, respectively, and interleukins intervene in the subsequent transformations of B and T_H cells.[60]

The sites recognized in the peptide-MHC-II complex-TCR interaction are depicted in Figure 4, according to References 20, 21, and 45. The MHC-II molecule recognizes a site (denoted MHS in Figure 4) of the peptide (antigen fragment, AGF · MHC in Figure 4). The T_H receptor, TCR, recognizes a site on the MHC-II molecule and another site (IMS in Figure 4) on the peptide.

Tolerance to self is induced during thymus maturation of T_H cells when clones (TCR*) able to recognize complexes between MHC-II and peptides derived from self antigens (AGF* in Figure 4) are deleted or inactivated. Thus, B cells preactivated by accidental recognition of a self antigen will have no T_H partner to recognize the same self antigen and to induce the transformation of this preactivated B clone to plasmocytes.[36,46]

One may ask, why the requirement for complexation with molecules of the major histocompatibility complex (MHC-II) for antigen recognition by T_H cells? The reason could be the reduction of the number of T_H clones required for self-nonself discrimination. The number of B clones in an immune system is very large, although only a few clones are used in defense against an infection. Comparatively, the number of genes of the major histocompatibility complex is much lower. MHC class II molecules probably bind to common and functionally important (therefore, strongly conserved during evolution) amino acid sequences. The set of MHC-II types within the immune system must cover practically all proteins and biomacromolecules with which the system may come into contact, in the sense that there is at least one MHC-II type able to bind a fragment of any such biomacromolecule. These functionally important sequences (MHS of Figure 4) have attached sequences also required for TCR recognition (IMS of Figure 4) which play the role of "individual markers". T_H clones able to recognize complexes with self individual markers are deleted (inactivated) out of the immune system, and no recognition takes place even if such complexes are presented accidentally by some B clone. If a foreign cell, virus, or a sufficiently large macromolecular complex is recognized by a B cell and processed, some of the complexes presented on the cell surface will have "individual markers", different from the self markers, and there will be at least one T_H clone able to recognize the complex.

The rather large dissociation constants (low affinities) for binding peptidic fragments to MHC class II molecules could be related to the prevention of recognition of self proteins coded by genes modified by somatic mutations. Such proteins may contain modified "individual markers" which could be recognized by T_H cells, but each such mutated protein will be found in a reduced concentration. These rather low affinities may also prevent accidental recognition of short sequences of self protein fragments, differing from the fragment "devised" for recognition by MHC-II; such accidentally recognized fragments could also contain sub-sequences that allow recognition by a T_H clone of the organism (Appendix 2).

A. SELF-NONSELF RECOGNITION IN T CELL MEDIATED CYTOTOXICITY

T cells implicit in cytotoxic mechanisms have on their surface a CD-8 antigen recognition site, instead of the CD4 site of T_H cells. The mechanism of antigen recognition by cytotoxic T cells seems rather similar to that of T_H cells.[12,20,21,45,47,61]

The MHC class I molecules, also encoded by the major histocompatibility gene complex, are more ubiquitous than MHC-II. These MHC-I molecules bind peptidic fragments of about eight to nine amino acids in length, resulting from proteolysis of cytoplasmic proteins. The total number of MHC-I molecules per cell is in the order of 10^5, and about 500 peptidic fragments of a given sequence can be bound by them.[20,51] The MHC-peptide complexes presented on the cell surface are recognized by the TCR of cytotoxic T cells; the dissociation constant for the fragment is about 10^{-7} *M*, but only 10^{-4} *M* for the soluble antigen fragment peptide alone.[62] Cytolysis of the recognized cell implies the porphyrin gene of cytotoxic T lymphocytes and natural killer cells, a gene which encodes for the pore forming protein.

Cytotoxic T cells are involved in destruction of cells infected by viruses. The antigen fragment recognized by MHC class I molecules should also contain highly conservative sequences (as in the case of MHC-II), found in proteins synthesized by both cells and viruses. A second sequence of the fragment, near the first one, should also be involved in the T cell recognition of the complex, similar to the mechanism depicted in Figure 4. This second sequence should be the one which differs in self and nonself antigenic fragments bound to the one described for recognition by T_H cells.

The transport system of the MHC-I-peptide complex from the cytosol to the cell surface also takes part in the recognition process mediated by class I molecules. MHC-I molecules are composed of two subunits, a membrane-spanning heavy chain and a small exoplasmic protein, the first being polymorphic and encoded by genes in the MHC. ATP and the chaperon BiP are required for folding and assembling the heavy chain and the small exoplasmic protein (β_2 microglobulin) so that the peptide can bind.[63] Alloreactive T cell recognition by rat MHC-I molecules was dramatically affected by the cim (class I modifier) locus which lies within the class II region of the major histocompatibility complex and is related to the multidrug-resistant family of transmembrane transporters. The cim locus genes are thought to encode a pump (a peptidic channel) which delivers peptides to class I MHC molecules in the endoplasmic reticulum.[64] Polymorphism localized in side-chain binding pockets of MHC-I molecules does alter the requirements for peptide binding, but it alone is not a sufficient explanation for the specificity of peptide uptake by class I molecules. Interactions of the peptides with this pump are probably involved in the specificity of peptide uptake.[65]

B. IMMUNO DEFENSE AND CANCER

Immuno defense is considered to have an important role in cancer prevention. The invasiveness of cancer cells is determined largely by their behavior with respect to the immunologic defense system in blood and with respect to T-killer lymphocytes.[66] Cancer cells should present some special surface antigens, different from normal cells, in order to be recognized as alien. Fetal antigens, normally found in large quantities only in fetal tissues, are often found on cancer cells;[67] such antigens should normally be present at a stage before elimination of self-recognizing lymphocyte clones (in the ontogeny of the immune system). Some antigens favoring escape of growth control mechanisms or invasiveness could represent the point of attack for the immune system. A large fraction from a great number of breast cancer and other tumoral cell lines present the so-called Gal epitope (Gal-α 1-3 Gal-β 1-4 GlcNAc-R) not found in a normal human cell line. The presence of the anti-Gal antibody prevents the attachment of tumoral cells on the surface of blood vessels.[68]

V. CONCLUSIONS

Immunologic defense implies, as a first step, recognition of epitopes from nonself antigens at concentrations in the 10^{-5} to 10^{-12} *M* range (for *in vitro* experiments) but no aggression against self antigens, although there are 10^7 to 10^9 epitopes on them, and their concentrations can reach 10^{-3} *M*. Antibody recognition sites for epitopes do not differ in principle from other biologic receptors, except by a rather reduced conformational flexibility that imposes more stringent steric fit requirements. The size of antibody sites and epitopes, about 6 to 17 amino acid residues in direct contact, ascertains sufficient specificity for discrimination between the recognized nonself and self epitopes, as demonstrated by use of equilibrium and probabilistic considerations.

Antigen recognition by formation of antigen-antibody complexes is an equilibrium phenomenon, although kinetic criteria may also be important in certain cases. This recognition requires not only the existence of an antibody site with significant affinity for one epitope of the antigen, but also an antigen concentration (cell surface density) above a threshold which depends upon this affinity. B cell clones must have antigen sites with a sufficiently low affinity for the self antigens in order to not recognize them at their physiological concentrations. Higher attention to such concentration requirements could probably explain various types of autoaggression and other puzzlements in self-nonself discrimination by the immune system.

Diversification of B clones during mature life produces new antibodies which could, accidentally, recognize a self antigen. The additional requirement of antigen recognition by a T-helper clone and interaction between the two types of recognition cells makes a double accidental recognition of a self antigen and the subsequent autoaggression quite improbable. Recognition of antigen fragments by molecules coded by the major histocompatibility complex (the MHC class II molecules) and fragments probably representing some functionally important amino acid sequences, directs T-helper recognition to a reduced set of epitopes, thus reducing the required number of T_H clones for validation of antigen recognition by B cells. Immune networks are also likely to intervene in preventing autoimmune aggression.

APPENDIX A
ANTIGEN-ANTIBODY RECOGNITION AS CHEMICAL EQUILIBRIUM

The recognition of an antigen, Ag, by antibody receptors on the surface of immune cells:

$$Ag + Ab \rightleftharpoons C \tag{A1.1}$$

can be considered as a surface adsorption process from an Ag solution. If each Ag-Ab interaction takes place separately, the saturation degree Θ of surface receptors will be given by a Langmuir isotherm:

$$\Theta = \frac{[Ag]}{[Ag] + K} \; ; \; K = e^{-\frac{\Delta G}{RT}} \; ; \; 0 \leq \Theta \leq 1 \tag{A1.2}$$

in which [Ag] is the antigen concentration, K is the dissociation constant for the Ag-Ab complex C, and ΔG is the Ag-Ab affinity. We assume recognition to take place if Θ exceeds a critical saturation degree, Θ_{cr}, i.e., $\Theta \geq \Theta_{cr}$. Expressing the concentration [Ag] as function of Θ, this condition implies:

$$\Theta \geq \Theta_{cr} \rightarrow [Ag] \geq K \frac{\Theta_{cr}}{1 - \Theta_{cr}} \tag{A1.3}$$

With $\Theta_{cr} = 0.5$, Equation 3 of paragraph 3 results.

If the AgAb interaction has a cooperational character:

$$Ag + R \rightleftharpoons C_1; C_1 + Ag \rightleftharpoons C_2; \ldots; C_{n-1} + Ag \rightleftharpoons C_n \tag{A1.4}$$

i.e., an $Ag_nR = C_n$ complex results, (where R is a n-tuplet of Ab receptors) the mean number, r, of Ag molecules complexed with R is in analogy with the ligand-protein interaction model of Klotz and Hunston:[11]

$$r = \frac{k_1[Ag] + 2k_1k_2[Ag]^2 + \ldots + nk_1k_2 \ldots k_n[Ag]^n}{1 + k_1[Ag] + k_1k_2[Ag]^2 + \ldots + k_1k_2 \ldots k_n[Ag]^n} \tag{A1.5}$$

with k_1, k_2, . . . the association constants of equilibria (Equation 4). The increase of r with [Ag] is more abrupt than the one corresponding to Equation 2 due to the terms with higher [Ag] potencies in Equation 5. In an extreme case with:

$$k_1 \leq k_2 \leq \ldots \leq k_{n-1} \ll k_n \; ; \; k_n[Ag]^n \gg 1 \qquad (A1.6)$$

$$r \cong \frac{nk_1k_2\ldots k_n[Ag]^n}{1 + k_1k_2\ldots k_n[Ag]^n}$$

$$\Theta = \frac{r}{n} \cong \frac{[Ag]^n}{K_n + [Ag]^n}$$

$$K = \frac{1}{k_1k_2\ldots k_n} \qquad (A1.7)$$

The critical [Ag] concentration (for the $\Theta \geq \Theta_{cr}$ condition) is roughly equal to $K^{1/n}$. With ΔG_1, ΔG_2, . . . , ΔG_n the Ag-Ab affinities for the successive steps (Equation 4), one obtains, finally:

$$[Ag]_{cr} \cong K^{1/n} = e^{-\frac{\Delta G_1 + \Delta G_2 + \ldots + \Delta G_n}{nRT}} \qquad (A1.8)$$

APPENDIX B MEAN STATISTICAL CONCENTRATIONS OF AMINO ACID n-TOUPLETS

The probability for a certain n-tuplet of amino acids (as a sequence or at a certain position n, within a protein) is 20^{-n}. In a protein solution with a 10% concentration, the total molar concentration of overlapping sequences of n amino acid residue is 100 n/125 n $\cong$ 1 *M*. The mean molecular weight for an amino acid residue is 125 Da. If, as assumed here, all random sequences have the same probability to appear, the mean concentration of an n-tuplet, in the 10% protein solution is about 20^{-n} *M*. For n = 5, as an example, the mean concentration is $3 \cdot 10^{-7}$ *M*, already lower than the dissociation constants, $K_d = 10^{-6}$ *M*, for binding to MHC class II molecules (Section IV).

REFERENCES

1. **Burnet, M. F.,** *Cellular Immunology,* Melbourne University Press, Melbourne, 1969.
2. **Davies, D. R., Sheriff, S., and Padlan, E. A.,** Antibody — antigen complexes, *J. Biol. Chem.,* 263, 10541, 1988.
3. **Kabat, E. L.,** *Structural Concepts in Immunology and Immunochemistry,* Holt, Reinhart & Winston, New York, 1968.
4. **Pressman, D. and Grossberg, A. L.,** *Structural Basis of Antibody Specificity,* Benjamin/Cummings, Menlo Park, CA, 1968.

5. **Ehrlich, P.,** *Studies in Immunity,* John Wiley & Sons, New York, 1906.
6. **Arrhenius, S.,** *Immunochemistry,* Macmillan, New York, 1907.
7. **Albert, A.,** *Selective Toxicity,* John Wiley & Sons, New York, 1951.
8. **Gandolfi, A. and Ciovenco, N. A.,** Reversible binding of multivalent antigen in control of B lymphocyte activation, *J. Theor. Biol.,* 74, 513, 1975.
9. **Berzovsky, J. A. and Schechter, A. N.,** The concepts of cross reactivity and specificity in immunology, *Mol. Immunol.,* 18, 751, 1981.
10. **Munson, P. J.,** A computerised analysis of ligand binding data, in *Immunoassay of Antigens and Antibodies,* Munson, P. J., Ed., Academic Press, New York, 1987.
11. **Klotz, I. M. and Hunston, D. L.,** Protein affinities for small molecules: conceptions and misconceptions, *Arch. Biochem. Biophys.,* 193, 314, 1979.
12. **Male, B., Champion, A., and Cooke, A.,** *Advanced Immunology,* Gower Medical Publishing, New York, 1987.
13. **Moraru, I.,** *Immunology,* Edit. Medicala, Bucharest, 1984.
14. **Sela, M.,** Antigens and antigenicity, *Naturwissenschaften,* 56, 206, 1969.
15. **Fuchs, S., Aharanov, A., Sela, M., v.d. Haar, F., and Cramer, F.,** Antibodies to yeast phenylalanine transfer ribonucleic acid are specific for the odd nucleoside Y in the anticodon loop, *Proc. Natl. Acad. Sci. U.S.A.,* 71, 2800, 1974.
16. **Schaechter, I.,** Mapping of the combining sites of antibodies specific for poly-L-alanine determinants, *Nature,* 228, 639, 1970.
17. **Amit, A. G., Mariuzza, R. A., Phillips, S. E., and Poljack, R. J.,** Three dimensional structure of antigen-antibody complex at 2.8 Å resolution, *Science,* 233, 747, 1986.
18. **Cygler, M., Rose, D. R., and Bundle, D. R.,** Recognition of a cell surface oligosaccharide of pathogenic *Salmonella* by an antibody Fab fragment, *Science,* 253, 442, 1991.
19. **Nikolic-Zugic, J. and Bevan, M. J.,** Role of peptides in positively selecting T-cell repertoire, *Nature,* 344, 65, 1990.
20. **Germain, R. N.,** Antigen presentation. The second class story, *Nature,* 353, 605, 1991.
21. **Marx, J. L.,** What T cells see and how they see it, *Science,* 242, 863, 1988.
22. **Brown, L. E.,McInerney, T., Anders, E. M., White, D. O., and Jackson, D. C.,** Probing the idiotype/anti-idiotype antibody interaction with a set of synthetic peptide homologues, *Immunol. Lett.,* 28, 213, 1991.
23. **Huber, R.,** Structural basis for antigen-antibody recognition, *Science,* 233, 702, 1986.
24. **Colman, P. M., Lawer, W. G., Varghese, J. N., Backer, A. T., Tulloch, P. A., Air, G. M., and Webster, R. G.,** Three dimensional structure of a complex of antibody with influenza virus neuraminidase, *Nature,* 326, 358, 1987.
25. **Takahashi, H., Merli, S., Putney, S. D., Houghten, R., Moss, B., Germain, R. N., and Berzowsky, J. A.,** A single aminoacidic interchange yields reciprocal CTL-specificities for HIV-1 gp 160, *Science,* 246, 118, 1989.
26. **Haber, E., Richards, F. F., Spragg, J., Austen, K. F., Valloton, M., and Page, L. D.,** Modifications in the heterogeneity of the antibody response, *Cold Spring Harbor Symp. Quant. Biol.,* 32, 299, 1967.
27. **Butler, V. P., Tanenbaum, S. W., and Beiser, S. M.,** A study of the cross reactivity of antipurin-6-oyl serum with desoxyribonucleic acid, *J. Exp. Med.,* 121, 9, 1965.
28. **Miklos, J. and Simon, Z.,** Structural characteristics in the affinity of antibodies towards antigenic sites. Antinucleic acid antibodies. Affinity-minimal steric difference correlation, *Rev. Roum. Biochim.,* 13, 283, 1976.
29. **Motoc, I.,** Biological receptor maps. 2. Steric maps of benzoate antibody and carbonic anhydrase, *Quant. Struct.-Act. Relat.,* 3, 47, 1984.
30. **Jafee, B. M., Simms, E. S., and Eisen, H. N.,** Specificity and structure of the myeloma protein produced by mouse plasmacytoma MOPC-460, *Biochemistry,* 10, 1693, 1971.
31. **Novotny, J.,** A static accessibility model for protein antigenicity, *Int. Rev. Immunol.,* 2, 379, 1987.

32. **Thomas, C. A., Jr.,** Oligonucleotide size and specificity in DNA-RNA hybridisation, *Prog. Nucl. Acid Res. Mol. Biol.,* 5, 315, 1966.
33. **Conaughy, B. L. and McCarthy, B. J.,** Interaction of oligonucleotides with denaturated DNA, *Biochim. Biophys. Acta,* 149, 180, 1967.
34. **Simon, Z. and Ghetie, V.,** Molecular size and specificity in immunology, *Rev. Roum. Biochim.,* 8, 261, 1971.
35. **Simon, Z.,** Molecular size and specificity of interactions, *Rev. Roum. Biochim.,* 9, 337, 1972.
36. **Simon, Z., Margineanu, I., Vlad, A., and Kerek, F.,** Clonal selection theory, antigen concentration threshold and validation by T-cells. Hypothesis for self-nonself discrimination, *Rev. Roum. Biochim.,* 27, 245, 1990.
37. **Rubio, N.,** Surface H-2 antigen concentration requirement of somatic hybrid cells for IgM-mediated cytotoxicity, *Nature,* 249, 461, 1974.
38. **Williams, A. F.,** Cellular interactions. Out of equilibrium, *Nature,* 352, 473, 1991.
39. **Foote, J. and Milstein, C.,** Kinetic maturation of an immune response, *Nature,* 352, 530, 1991.
40. **Burnet, M. F.,** *The Clonal Selection Theory of Acquired Immunity,* Canbridge University Press, London, 1959.
41. **McDonald, H. R.,** Mechanism of immunological tolerance, *Science,* 246, 982, 1989.
42. **Appenzeller, T.,** Living with your self, *Sci. Am.,* 259(1), 18, 1988.
43. **Bjork, R. L.,** HIV-1, seven facets of functional molecular mimicry, *Immunol. Lett.,* 28, 91, 1991.
44. **Tonegawa, S.,** The molecules of the immune system, *Sci. Am.,* 253(2), 104, 1985.
45. **Unanue, E. R. and Allen, P. M.,** The basis for the immunoregulation role of macrophages and other accessory cells, *Science,* 236, 551, 1987.
46. **Simon, Z.,** Chemical equilibrium considerations in self-nonself discrimination. Role of B and T cells, *Universitatii Timisoara Ser. Fiz.-Chim.,* in press.
47. **Forsdyke, D. R.,** A theory of immunity, *J. Theor. Biol.,* 25, 173, 1969.
48. **Manser, T.,** The B cell grows up, *New Biologist,* 2, 605, 1990.
49. **Jerne, N.,** Towards a network theory of immune response, *Ann. Immunol.,* (Paris), 125C, 373, 1974.
50. **Victor-Kobrin, C. and Bona, C. A.,** The molecular origin of regulatory idiotypes, *Intern. Rev. immunol.,* 5, 117, 1989.
51. **Nossal, G. J.,** Immunologic tolerance: collaboration between antigen and lymphokine, *Science,* 245, 147, 1989.
52. **Bona, C. A.,** HIV-Genetic structure, immunodeficiency syndrome, therapeutical and vaccine approaches, *Universitatii Timisoara Ser. Fiz.-Chim.,* in press.
53. **Bona, C. A.,** Recognition of antigens by T cells, *Universitatii Timisoara Ser. Fiz.-Chim.,* in press.
54. **Zanetti, M.,** Immune network and autoimmunity, *Universitatii Timisoara Ser. Fiz.-Chim.,* in press.
55. **Bona, C. A.,** Postulates to define pathogenic autoantibodies and T cells, *Universitatii Timisoara Ser. Fiz.-Chim.,* in press.
56. **Marrack, Ph.,** New insights into antigen recognition, *Science,* 235, 1311, 1987.
57. **Guillet, J. G., Lai, M. Z., Briner, T. J., Buus, S., Sette, A., Gray, H. M., Smith, J. A., and Efter, M. L.,** Immunological self-nonself discrimination, *Science,* 235, 865, 1987.
58. **Rudensky, A. Yu., Preston-Hurlburt, P., Hong, S. C., Barlow, A., and Janeway, C. A., Jr.,** Sequence analysis of peptides bond to MHC class II molecules, *Nature,* 353, 622, 1991.
59. **Rudensky, A. Yu., Rath, S., Preston-Hurlburt, P., Murphy, O. B., and Janeway, C. A., Jr.,** On the complexity of self, *Nature,* 353, 660, 1991.

60. **De Franco, A. L.,** Between B cells and T cells, *Nature,* 351, 603, 1991.
61. **O'Shea, J. J. and Ortaldo, J. R.,** Killer T's, macrophagues and NK's, *New Biol.,* 2, 779, 1990.
62. **Schneck, J., Lee Maloy, W., Colligan, J. E., and Margulies, D. H.,** Inhibition of an allospecific T cell hybridoma by soluble class I proteins and peptides. Estimation of the affinity of a T cell receptor for MHC, *Cell,* 56, 47, 1989.
63. **Dobberstein, B.,** Who needs peptide transporters?, *Nature,* 355, 109, 1992.
64. **Purham, P.,** Flying the first class flag, *Nature,* 357, 193, 1992.
65. **Powis, S. J., Deverson, E. V., Coadwell, W. J., Ciruella, A., Huskinson, N. S., Smith, H., Butcher, G. W., and Howard, J. C.,** Effect of polymorphism of an MHC-linked transporter on the peptides assembled in a class I molecule, *Nature,* 357, 211, 1992.
66. **Felman, M. and Eisenbach, L.,** What makes a tumor cell metastatic? *Sci. Am.,* 259(5), 40, 1988.
67. **Castro, J. E., Lance, E. M., Medawar, P. B., Zanelli, J., and Hunt, R.,** Foetal antigens and cancer, *Nature,* 243, 225, 1973.
68. **Galilis, U. and Macher, A.,** Interaction between anti-Gal and human tumor cells: a natural defense mechanisms?, *J. Natl. Cancer Inst.,* 81, 178, 1989.

Chapter 7

SPECIFIC INTERACTIONS IN IONIC TRANSPORT THROUGH BIOLOGICAL MEMBRANES

Maria-Luiza Flonta

TABLE OF CONTENTS

0-8493-5398-X/93/$0.00 + $.50

I. INTRODUCTION

Biological membranes are recognized as the site of many energy conversions (photochemical, chemoosmotic, and chemoelectric) which proceed in living systems, and also as the site of information reception and transmission in biological systems. Membranes are involved in these functions due to their ability to act as selective barriers for water, ions, and biomolecules.

The flows of various molecules and ions are produced either by physical forces, such as concentration or electrical potential gradients (passive fluxes), or are energized from different energy sources such as light, redox processes, or hydrolysis of macroergic compounds (active transport). The active antientropic transport of ions or other compounds keeps the steady state far from thermodynamic equilibrium, a situation characteristic of biological systems.

In the following we will discuss transport specificity for ions, which are transferred by membranar integral proteins (channels, carriers, and pumps), not involving major reorganization of the cell membrane as in the case of exo- or endocytosis.

One can distinguish two types of transporters: (1) channels or carriers which allow passive fluxes produced by transmembrane electrochemical gradients, and (2) ionic pumps which perform active antientropic fluxes by coupling with an exergonic chemical or photochemical reaction.

The coherent organization of inflows and outflows in the cell, which is essential for maintaining the steady state, depends on the membrane selectivity and the capacity of the organism to modulate it.

II. SPECIFICITY OF ION CHANNELS

A. CHANNEL-ENZYME ANALOGY: A POSSIBLY FRUITFUL APPROACH TO EXPLAIN SPECIFICITY OF INTERACTION IN IONIC TRANSPORT

An analogue property to channel or ionic pump selectivity is enzyme specificity — a very well-studied and characterized property of biochemical systems. One can make an analogy between enzymes and channels, arguing that channels and enzymes have similar functions and therefore can be investigated with the same methods and concepts.

Ionic channels can be seen as catalysts of electric currents (ionic fluxes) through membranes.[1] Channels modify the flux of ions the same way enzymes modify the flux of reactants. They stabilize the transition state between substrate and product, if we define the transition state of a channel as the state with an ion in the pore. The channel protein stabilizes this state (compared to what would happen without the protein) because it provides charges produced by polarization to neutralize the permanent charge of the permeating ion. The channel protein has a high dielectric constant in the wall of its pore and thereby lowers the potential barrier to ion movement across the membrane.

Channels also modify the rate of a chemical reaction, just as enzymes do. Channels have substrates and products just as clearly defined as for enzymes. The substrate of a channel is the permeable ion on one side of the membrane; the product is the permeable ion on the other side of the membrane. The substrate and product of channels have different free energies, just as they do for enzymes. Substrates and products of enzymes are different chemical species with different free energies at the same location.

Substrates and products of channels are the same chemical species at different free energies and different locations. The spatial gradient of electrochemical potential drives a chemical reaction. In this sense, a channel is a catalyst for diffusion through membranes and also an enzyme, even if it does not mediate chemical reactions of the ordinary type.

This perspective suggests to the membranologist new types of experiments and interpretation of these.[2,3] It seems worthwhile to ask questions about channels like those that have been useful to enzymologists. Knowing some of the history of enzyme research we may be able to investigate channels more efficiently.[4]

The ability of membranes to select between ions that are chemically not too different (e.g., K^+ and Na^+) is critical to the life of cells: it permits membranes of animal cells to maintain a constant volume without having to support much hydrostatic pressure, thereby allowing animal cells to exist without the rigid retaining wall found in plant cells. In this way animal membranes allow motility and contain proteins and nucleic acids within the cell, while still allowing metabolites to flow in and out of the cell.

Selectivity is now known to be a property (for the most part) of the open channels: the open channel permits larger flows of some ions rather than others, while the opening process depends much less on the identity of the permeating ion, in most cases. Thus, selectivity is a property of the pore of a channel, much as some kinds of enzyme specificity are a property of the active site and not the conformational change of enzymes.[4]

Some specificity in enzymes, however, is supposed to result from a conformation change, from the fit of the enzyme to the substrate induced by the substrate's presence,[5] with the specificity depending on the plasticity as well as the structure of the active site.[6] Induced mechanisms for selectivity seem not to have been suggested for channels but are possible, and even likely. A channel may be something like a snake swallowing a rabbit; it may change shape significantly, stretching or shrinking while the ion passes through, and thus the fit of the ion may depend on the structural change induced by the ion. The pore may not have a definite invariant size, independent of the strength of the interactions between the channel protein and permeating ion.[7] From this point of view, channel permeation is not so rigidly separated from channel gating, because both involve significant conformational changes, albeit not necessarily the same ones.[1]

Here is a case in which the analogy between enzymes and channels can suggest a new idea to channologists: if gating and open channel permeation are related, they should vary in a qualitatively similar way with a range of experimental interventions. Thus it would be interesting to check whether pharmacological agents, divalent ions, permeating species, and so on, have similar effects on open channel conductance and on gating.

B. FACTORS WHICH DETERMINE IONIC SELECTIVITY

The channel selectivity, or in other words, the specific interaction between the channel protein and the transported solute, is supposed to be determined either geometrically, when the two surfaces which interact are precisely matched to each other, or by the capacity of the ion to lower or eliminate the interposed energetic barriers which hinder its passage through apparently open channels.[8]

Steric correspondence allows many noncovalent bonds to be formed by electrostatic or van der Waals interactions, which accounts for a lower free energy level of the ion included in a channel state or, more generally, for the specificity of biological recognition processes.

Although living systems are viewed energetically as spatial-temporal dissipative structures, the complementary idea that at much smaller space and time scales quasiequilibrium conditions exists, is valid.[9] These make the local organization of living matter at supramolecular level — to arise from the spontaneous tendency of molecular components to attain that state which minimizes free internal energy. The self organization of living matter on the basis of searching for the thermodynamically most favorable configurations can be best illustrated in the case of biological membranes.

In what follows we aim to discuss how selectivity of biological membranes is determined by the interaction specificity between the membrane transporter and the ions to be transported.

According to theories based on the assumption of independent movement of ions in a homogeneous membrane with a linear potential decrease,[10] selectivity in channels is governed by two factors: partitioning into the membrane, and mobility once inside. The product of these factors determines a single permeability coefficient P_s, which may be determined in absolute terms from the size of measured currents:

$$I_s = P_s z_s^2 \frac{EF^2}{RT} \frac{[S]_i - [S]_o \exp(-z_s FE/RT)}{1 - \exp(-z_s FE/RT)} \quad (1)$$

I_S = the current carried by ion S, z_s = the valence of the ion, E = the transmembrane voltage, and F = the Faraday number.

The Goldman-Hodgkin-Katz(GHK) current equation (Equation 1) allows one to calculate the absolute permeability (P_s) from a current measurement if the concentration and membrane potential are known. The permeability coefficient also may be determined in relative terms from zero-current (reversal) potential (E_{rev}):

$$E_{rev} = \frac{RT}{F} \ln \frac{P_k[K]_o + P_{Na}[Na]_o + P_{Cl}[Cl]_i}{P_k[K]_i + P_{Na}[Na]_i + P_{Cl}[Cl]_o} \quad (2)$$

The GHK voltage Equation 2, gives the membrane potential at which no net current flows. It allows one to calculate permeability ratios from measurements of reversal potential if ionic concentrations are known.

Even if the theoretical assumptions do not correspond closely to the mechanism of permeation in a pore, the resulting equations provide useful definitions of absolute and relative permeabilities. Ionic permeability ratios for many ions have been determined with Na^+, K^+, Ca^{2+} and endplate channels.[11] These measurements seek clues about the structures and forces that account for selectivity: ionic radius, hydration, pore radius, charge, and energies of interaction between ions, water, and pore, and motions of each of these components.

1. Ionic Channels Act as Molecular Sieves

We consider first how much of the ionic selectivity can be accounted for purely from the shape of the ions and the pore, neglecting the important energy changes that accompany permeation. If an ionic channel is highly ion selective, the pore must be narrow enough to force permeating ions into contact with the wall so that they can be sensed. Selection requires interaction. Similar ions hiding under a coat of water molecules could be almost impossible to distinguish. Consider K^+, Tl^+, and NH_4^+ ions whose crystal radii differ by, at most, 12%. In free solution their mobility difference is reduced to 2%. Only direct contact with the naked ion could give the degree of selectivity characteristic for Na or K channels. The narrow, ion-selective region of ionic channels is called the selectivity filter.[11]

A selectivity filter, only one or two atomic diameters wide, should cut off the permeation of larger particles. Although the filter might be soft and able to stretch a bit, in a very selective channel the flexibility cannot be large. Let us now attempt to deduce the pore size of ionic channels.

Figure 1 shows silhouettes of the permeant ions for Na and K channels. A water molecule has been drawn with some of the ions to provide the scale.

The K channel evidently does not allow passage of ions as large as those passed by the Na channel. The largest permeant metal in K channels is Rb^+ ($r = 1.48$ Å); Cs($r = 1.69$ Å) and methyl groups ($r = 2.0$ Å) do not pass. A circular selectivity filter with a diameter of 2.96 to 3.38 Å would account both for the ions that go through and for those that do not.[11] Recalling that a channel needs to provide oxygen dipoles as surrogates for water molecules whenever an ion is stripped of some of its contact water molecules, we can imagine that the selectivity filter of K channels is formed by a ring of oxygen atoms provided by the channel protein. A bracelet of five oxygens with centers 3 Å from the pore axis, as in Figure 2a, would explain the observations.

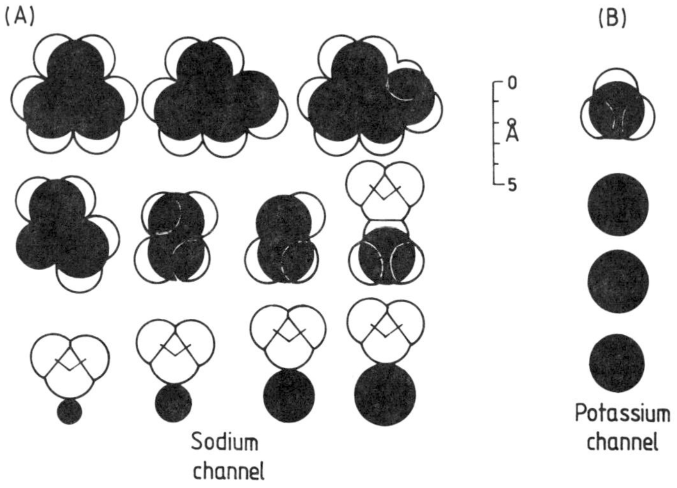

FIGURE 1. Silhouettes of permeant ions. Outline drawings of all known monovalent permeant ions, except H^+, for Na channels (A) and K channels (B) of frog nerve. Hydrogen atoms are made transparent to suggest the effective size of the permeant particle when probed by a hydrogen-bond acceptor. A transparent water molecule is drawn next to four of the ions. From left to right and bottom to top the ions are (A) Li^+, Na^+, K^+, Tl^+, formamidinium, hydrazinium, hydroxylammonium, ammonium, guanidinium, hydroxyguanidinium, aminoguanidinium; and (B) K^+, Tl^+, Rb^+, NH^+_4. (From Hille, B., *Ionic Channels of Excitable Membranes,* Sinauer Associates, Inc., Sunderland, MA, 1984. With permission.)

Can one make a similar, purely steric hypothesis concerning the selectivity filter of Na channels? At first glance it seems difficult, as the impermeant methylammonium appears much smaller than the permeant aminoguanidine. Nevertheless, there is a way. First, guanidine compounds are planar. As with a coin, the smallest hole that would pass a guanidine would be a slot as narrow as 3.2 Å. Such a slot would be too narrow for methyl groups to pass. However, can the tetrahedral ammonium and amino groups on many of the permeant ions be accommodated? Protonated amino groups, NH_3^+, are only marginally smaller than methyl groups, CH_3.[11] The problem can be solved by invoking hydrogen bonds between the amino groups and the channel walls. Groups that are hydrogen bond donors (-NH_2, NH_3^+, -OH) form links like -N-H . . . O with other oxygen- or nitrogen-containing groups. In these interactions, which take only 10^{-11} s to make or to break, the O and N atoms readily approach to a center-to-center distance of 2.8 Å. Subtracting the standard oxygen van der Waals radius of 1.4 Å leaves only 1.4 Å for the effective radius of the -NH_2 or -NH_3^+ group — when probed by an oxygen ligand. These elements suggest the following geometry for the selectivity filter of the Na channel (Figure 2b): a cluster of six oxygens defines a hole that is at least 3.1 Å wide and 5.1 Å high. These dimensions leave just enough room for permeant aminoguanidine to pass through, making hydrogen bonds with the selectivity filter *en passant,* while excluding the impermeant meth-

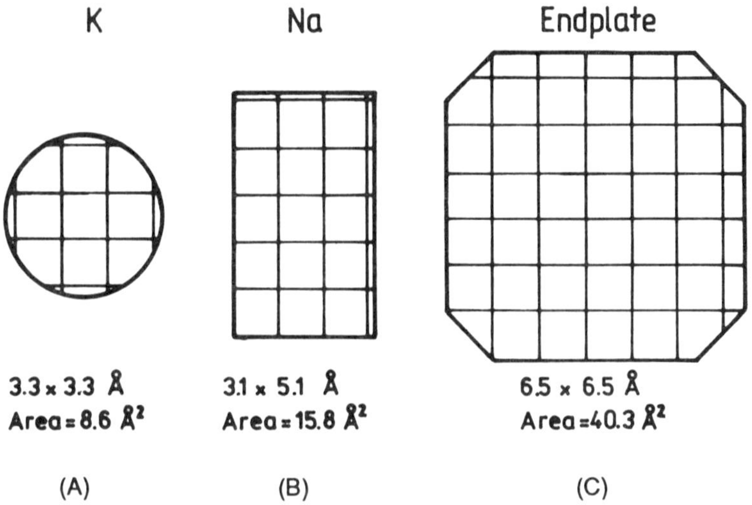

FIGURE 2. Dimensions of ionic selectivity filters. Outline of minimum pore size that will pass the known permeant ions in frog nerve and muscle. Grid marks in 1-Å steps. Sizes were evaluated from space-filling models of the permeant and impermeant ions. (From Hille, B., *Ionic Channels of Excitable Membranes,* Sinauer Associates, Inc., Sunderland, MA, 1984. With permission.)

ylated cations and triaminoguanidine. A Na^+ ion (r = 0.95 Å) would fit easily into one corner of the filter with a water molecule standing above it to fill the space. Note that the permeability of the K channel to NH^+_4 ion implies that hydrogen bonds are made to an oxygen-containing selectivity filter there, too.

For Ca channels we know that only small cations can pass: five alkali metals, three alkaline earths, hydrazine, and hydroxylamine cations. These results require a circular hole at least 3.40 Å in diameter.

In conclusion, the old idea that ionic channels act as molecular sieves explains one striking feature of ionic selectivity: the permeability cuts off at a definite ionic size. This may correspond to the size of the selectivity filter and provides an estimate of the caliber of the narrowest part of the pore. The remainder of the pore may be far wider. On the other hand, pore size alone is insufficient to explain the sequence of selectivity among ions small enough to enter the pore. Thus Na, K, and Ca channels are each large enough to pass Na^+ and K^+ ions, but with quite different selectivities. Such problems require a discussion of energy changes.

2. Na Channel Permeation Can Be Described by Rate-Theory Models

No perfectly selective channel is known. Several K channels pass at least 4 kinds of ions, Na channels pass at least 14, Ca channels pass at least 8, and endplate channels more than 50.[11] The permeability sequence in the wide endplate channel is largely accounted for by the relation of the ionic radius to friction in a hole. The sequence in narrower channels, such as the Na and K channels, cannot be explained in this way. For example, the sequence is $P_{Na} > P_K$ in one, and $P_K > P_{Na}$ in the other. Such selectivity must depend on the balance between the energy of removing water from the ion and the energy of allowing the partly dehydrated ion to interact with the stabilizing charges and dipoles of the selectivity filter.

When concentrations are raised sufficiently, all channels show saturation, competition, and blockage by small ions.[12-14] These phenomena cannot be described by theories based on ionic independence. Instead, they may be modeled by assuming that ions hop between discrete, saturable sites in the pore as they cross the membrane, that they pause at various sites within the channel while passing through, and that their presence in the channel has a profound effect on the passage of other ions.[15]

Deviations from independence and saturation of fluxes are readily explained by assuming that (1) ions must bind to certain sites in the pore as part of the permeation process, and (2) a site can bind only one ion at a time. The assumptions are fully analogous to those in the derivation of enzyme kinetics. Two classes of saturating barrier models, one-ion pores and multi-ion pores, can be distinguished.[16,17] The former may have several internal sites and be permeable to several kinds of ions, but can contain only one ion at a time in the pore. The latter may contain more than one ion in the pore simultaneously.

Let us describe the Na channel permeation using the rate-theory (RT) method. Figure 3 shows a proposed energy profile for Na^+ ions crossing amphibian Na channels.[18] The free energy levels are marked in RT units. Approaching from the outside, there is first a low barrier and then a well, -1.0 RT units deep. This is the external binding site that produces saturation with a dissociation constant, K_{Na}, of about 370 m*M*. Then there is a high barrier, the rate-limiting selectivity filter. Its height is chosen to give a reasonable single-channel conductance (11 pS) with physiological Na concentrations. Beyond that, there are several low inner barriers to make the predicted I-V curve more linear. The Na channel was postulated to be a one-ion pore in this model, with states of occupancy as in Figure 4. These assumptions lead to a predicted concentration dependence of currents that fit the observations better than the independence relation does.[18]

The RT model can be made to account for permeability ratios for other ions if one adjusts empirically the height of the selectivity filter, adding 0.1 RT unit for Li, 2.7 for K^+, and so on.

Different deviations from independence are accounted for if one varies the depth of the external binding site: the deeper the well, the stronger the binding and the lower the saturating concentration.

What could the barrier model mean in molecular terms? One interpretation is shown in Figure 3B. The diagram represents molecular events as a Na^+ ion passes through the narrow part of the pore. The events are intended to correspond to positions in the energy barrier profile. Hille[13] suggests as a

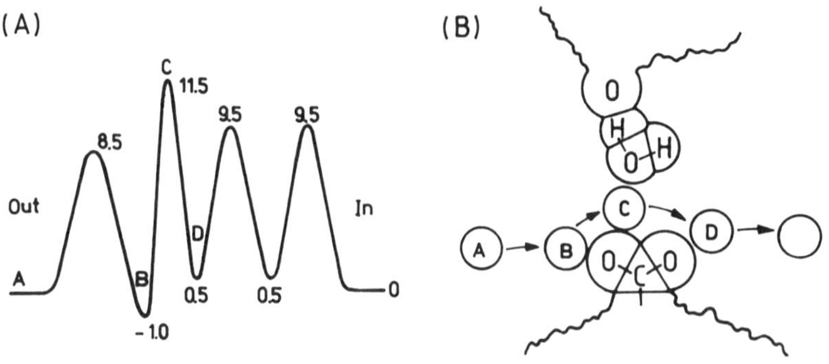

FIGURE 3. Barriers and binding sites in Na channels. (A): Representation of the diffusion path in an Na channel in terms of energy barriers and wells. The energy levels relative to bulk solution are labeled in multiples of RT with values appropriate for Na^+ ions. For less permeant ion such as K^+, rate-limiting energy peak C would be higher. For a more firmly bound ion, such as Tl^+ or Li^+, energy well B would be deeper. (B): Possible molecular interpretation of the barrier model in terms of a hydrated Na^+ ion moving through positions A, B, C, and D in the energy diagram. In position C the ion is at the narrow selectivity filter with a charged -COO group below and an oxygen above from the channel and one water molecule shown for scale. For the remainder of the membrane thickness, the Na channel may be a wide pore with little resistance to ion movement. (From Hille, B., *Ionic Channels of Excitable Membranes,* Sinauer Associates, Inc., Sunderland, MA, 1984. With permission.)

working hypothesis that low pH blocks Na channels by protonating an ionized carboxylic acid group in the selectivity filter, whose negative charge is needed to stabilize permeating cations as they pass through. This $-COO^-$ group is shown in cross section forming the bottom edge of the selectivity filter. The hydrated Na^+ ion diffuses over the first barrier and is drawn to the negative group, where it must lose a few H_2O molecules. Electrostatic attraction to the site more than compensates for the energy required to lose some water. This is a stable binding to the external binding site — the energy well. Next, the ion needs to squeeze through the narrow selectivity filter, retaining at most three H_2O molecules, one drawn on top, one in front, and one behind the ion as it passes through (not drawn). In the narrow region, the energy of the ion is raised considerably by the loss of H_2O ligands — the highest energy peak. When the ion emerges into the wider inner vestibule, it regains hydration and aqueous energy levels. In this view, the important binding and permeation steps occur as the ion moves only a few angstrom units.

How could selectivity arise in such a picture? Eisenman's theory of equilibrium of ion exchange[19] assumes that binding selectivity arises from the difference between hydration and site-interaction energies.

Here we need to explain why the free energy of a Na^+ ion sitting at the highest energy point in the selectivity filter is lower than that of a K^+ ion. Although not stably bound there, the ion is partly hydrated and interacts with a negative charge. Hence the elements of Eisenman's theory are applicable.[20] The negative charge in the filter would have to be equivalent to a high field-

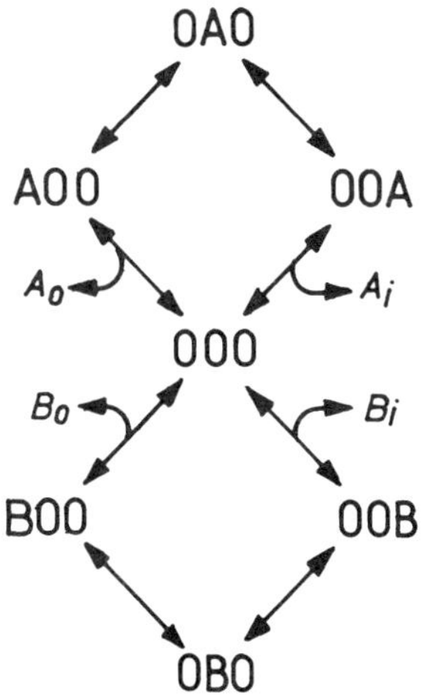

FIGURE 4. State diagram for a one-ion channel. The channel has three internal sites, but at most only one of them may be occupied. Two types of permeant ion, A and B are present. Successive occupancy states represented by triplets differ from each other by a single ionic jump. The top cycle transports one A ion per revolution and the bottom cycle, one B ion. (From Hille, B., *Ionic Channels of Excitable Membranes*, Sinauer Associates, Inc., Sunderland, MA, 1984. With permission.)

strength site so that when Na^+ ions approach 0.38 Å closer than K^+ ions, the electrostatic attraction overcompensates the larger energy consumption required to remove some water molecules from the Na^+.

The point of maximum free energy is to be considered, rather than a true binding site as in Eisenman's original ion exchange theory. As Benzanilla and Armstrong[21] pointed out, binding does not help an ion cross a one-ion channel. It slows permeation.

A similar idea could be applied to the K channel to explain why Na^+ and Li^+ are relatively impermeant. The K channel would have to be equivalent to a weak field-strength site so that neither Na^+ nor Li^+ would be stabilized enough to compensate for the even stronger dehydration required in the narrow K channel. While one ion pauses at a site, other ions have to wait their turn.[11]

Barrier models of the diffusion steps provide a guide to important questions, such as where the major energy barriers to diffusion lie with respect to the electric field in the pore, where the binding sites are located, or the minimum number of ions able to enter the pore.

3. Energetics of Ion Interaction with Channels and Channelogs

Despite their usefulness, barrier models of this form cannot be the ultimate description. If Nernst-Planck models can be criticized for treating a molecular pathway as a continuum, barrier models can be criticized for representing diffusion across the membrane as instantaneous jumps between only two or three vacant sites. There are no vacancies in a fluid medium. Solvent and solute particles remain permanently in contact. Descriptions of permeation might take into account motions of the channel macromolecules coupled to motions of water molecules. The barriers have then to be thought of as representing the combined effect of all these factors averaged over the period required for an ion to move from one major energy minimum to the next.[22] Läuger has given one method for calculating effective barriers from microscopic force constants and intermolecular potentials.[23]

Another promising technique is molecular dynamic simulation, which has been applied successfully to calculate molecular motions in water and proteins.[24] This procedure calculates the time course of motions by integrating Newton's law of motion, given the masses of all particles and the forces operating between them.

Eisenman and Alvarez[8] provide a method to calculate the energetics of ion interaction with any protein structure known at high resolution. This gives new insights into permeation, particularly as to the energetic factors controlling which species permeate and how well, what the effects of simple point mutations in the channel structure might be, and what effects helix dipoles and ionizable residues might have.

These calculations were made in the channel-like structures of icosahedral viruses (called channelogs), whose coordinates are known crystallographically at atomic resolution. The family of viral channelogs has a common pentameric structure formed from five beta barrels, which is different from the pentamers of parallel alpha helixes as in all proposed structures for membrane channels.[8] The viral pentameric architecture has narrow regions analogous to selectivity filters, which can be lined by negative or positive dipoles,[25] or even positively charged residues[26] where ion discrimination occurs. The channelogs also contain wider regions with a variety of geometries and polar linings which can provide models for channels "vestibules" in which interactions with ionizable, dipolar, and hydrophobic groups can modify the concentrations of permeant species from those in the bulk solution.

Because of their structural analogy to membrane channels, the channelogs therefore provide interesting models for cationic and anionic channels and their cationic and anionic binding sites.

The selectivity filter consists of a narrow ring of five backbone carbonyl oxygens surrounding a 2- to 4-Å-diameter aperture, or can be composed of different ligands, such as the side chain threonine -OH or arginine NH groups.

Calculations for the energy profile for a viral channelog of 2STV (satellite tobacco necrosis virus) whose selectivity filter is formed by backbone carbonyl oxygens, can answer the following questions: is this simple viral "channel" expected to bind cations, as observed experimentally[25,28] and is it more permeable to cations than anions, as expected intuitively for a carbonyl oxygen selectivity filter?

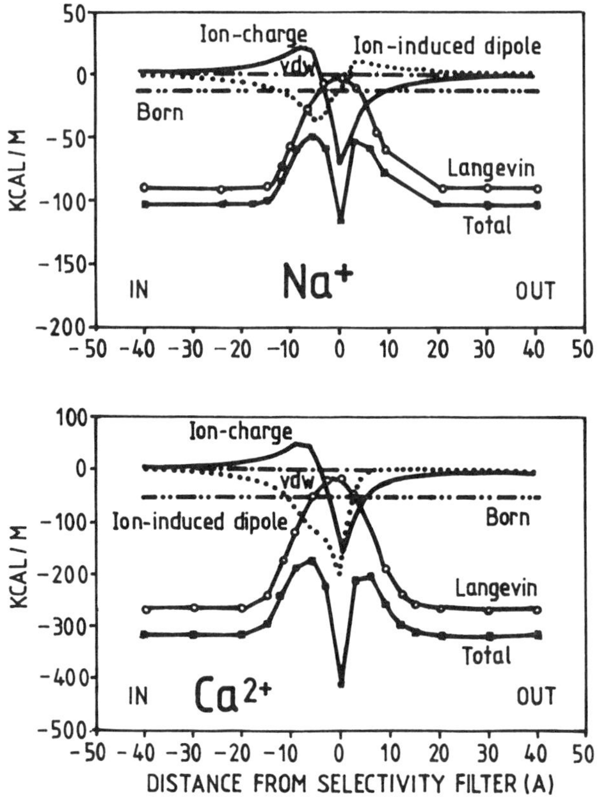

FIGURE 5. Energy profiles and extent of hydration as a function of distance from the selectivity filter for the satellite tobacco necrosis virus. Distances in Å. The total energy at any position is the sum of the following individual terms: the ion-charge energy, the ion-induced dipole energy, the van der Waals energy, and the Langevin dipole energy. (From Eisenman, G. and Alvarez, O., *J. Membr. Biol.*, 119, 109, 1991. With permission.)

The free energy profiles computed for Na,$^+$Cl$^-$, and Ca^{2+} permeation are presented in Figure 5. The energy profile for Na^+, given by the curve labeled "Total", is seen to consist of an energy minimum (i.e., a binding site) at the selectivity filter, separated from the fully hydrated reference levels by two relatively symmetrical energy maxima (i.e., barriers) to entering and leaving the site from or to the adjoining solutions. In channel parlance, this corresponds to the two-barrier one-site profile. In the case of Cl^-, a sharp maximum exists at the selectivity filter, corresponding to a strong repulsion. The "channel", therefore, is expected to bind Na^+ but not Cl^- and also to be much more permeable to Na^+ than to Cl^-. The channel prefers Na^+ over Cl^-, not only because the electrostatic interaction is unfavorable but also because the van der Waals repulsion energy for Cl^- is large. The hourglass shape of the channel correlates nicely with the number of water molecules found in it. This can be seen in the lower right part of Figure 5, which plots

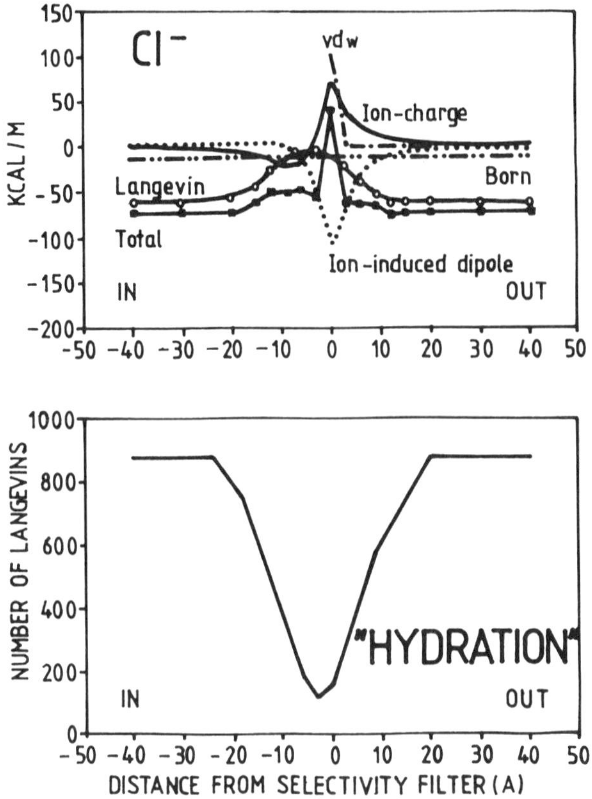

FIGURE 5. (continued)

the number of Langevin dipoles contained in the probing sphere as a function of distance from the selectivity filter. This number is a measure of the number of water molecules in the channel. Water is increasingly excluded as one gets nearer to the selectivity filter region.[8]

The profile for Ca^{2+} resembles that for Na^{+} except that the minimum for the total energy is deeper for Ca^{2+} than for Na^{+}. This implies that Ca^{2+} is more strongly bound than Na^{+} to this channel, in accord with experimental findings.[28]

The general features of the profile are consistent with the current views as to what is expected for a cation channel. However, the profile is not satisfactory quantitatively, for it implies an impossibly slow ion exchange for a ring of carbonyl oxygens. For example, the barrier for Na to leave the ring is about 60 kcal/mol and to enter is about 50 kcal/mol. Since a barrier of 50 kcal/mol implies a rate of motion slower than 1 ion per 10^{16} years, permeation through a structure with such an energy profile would be too slow for it to function as a channel. Indeed, such a channelog would not even be able to exchange an ion on the time scale of ion displacement studies.[28]

The unrealistically strong binding and impossibly slow permeation rate implied by the present energy profile are due to the inadequacies, discussed below, of the Langevin dipole procedure for modeling water in narrow regions. These results indicate that simple protein dipole/Langevin dipole (PDLD) modeling may not be appropriate for quantitative assessment of such an energy profile and indeed this energy profile can be reassessed by the more rigorous molecular dynamics and free energy perturbation (MD/FEP) procedure.

The pentameric viral channelog in 4SBV (southern bean mozaic virus) offers a model system in which to explore a selectivity filter formed by -OH groups.[8] Channels with -OH ligands in their selectivity filters are more complicated than those with backbone carbonyl ligands because the -OH groups can have different orientations of their dipoles relative to the channel axis. Such ligands are of considerable interest because side chain -OH groups (e.g., from Ser, Thr, or Tyr) are prime candidates for the polar lining of cell membrane channels, and synthetic serine-rich membrane-spanning alpha helixes, designed to form polar pores lined by -OH residues have, indeed, been found to produce cation-selective channels.[29,30] That these synthetic structures are cation selective seems paradoxical, since side chain -OH groups in polyene antibiotic channels are usually ligands for anions rather than cations,[31] owning to their ability to orient themselves with the H end of the -OH dipole facing the ion. However, Perutz has pointed out that the gamma OH of serine in an alpha helix donates a hydrogen bond to the carbonyl oxygen of the next helical turn, so that its lone-pair electrons point away from the helix.[32] He has suggested that a ring of serines might in this way form a cation-binding structure.

Computations made by Eisenman and Alvarez[8] show how different orientations of the hydroxyl group might enable a selectivity filter to be cation selective in certain situations, while being anion selective in others. In particular, they are intended to shed light on how a superfamily of architecturally and chemically similar transmitter-gated channels lined by hydroxyl groups[33] can contain members, some of which can be selective for anions,[34] while others are selective for cations,[35] a finding which till now was rationalized by invoking the conditioning effects of charged groups in the vestibules.[34]

The ''frozen'' level of modeling can be partially relaxed for the selectivity filter by allowing the side chain threonine hydroxyl ligands in the selectivity filter of 4SBV virus to rotate freely around their -C-OH single bond, in order to explore the effects on ionic selectivity of allowing hydroxyl ligands to orient with the positive end of their dipoles facing the middle of the channel or away from it.[8] Figure 6 shows the energy profile for K and Cl calculated for 4SBV in the same way as in Figure 5, but for the two different orientations of side chain hydroxyl ligands. The profiles are computed for two extreme conditions: one, when the H ends of the -OH dipoles are rotated so that they face maximally *in* towards the lumen of the channel; the other, when they are rotated so as to face maximally *away* from the lumen.

When the Thr-OH groups face *in* toward the lumen, the energy profile for K is seen to consist solely of an unfavorable energy barrier separating the solutions on the two sides. Such a profile implies that the channel, when its selectivity filter is in this conformation, is impermeant to K. In contrast, when the Thr-OH groups face *away* from the lumen, the energy profile exhibits a distinct energy minimum. Such a profile corresponds to a favorable binding site for K separated by barriers from the internal and external solutions, the barrier on the inside being somewhat higher than that on the outside. Therefore, one can conclude that this type of channelog can bind cations and that its external portion could be permeable to them, but only under conditions when -OH groups can face away from the fivefold axis.

In the case of Cl^- (Figure 6) the situation is essentially reversed. When the Thr-OH groups face *in,* a clear energy minimum corresponding to a favorable binding site is seen. When the Thr-OH groups face *away,* the site becomes less distinct, though it is still present. For all orientations of the -OH group the external portion of the channel should therefore be permeable to anions. The results imply that the ring of side chain -OH ligands can form a selectivity filter in the 4SBV channelog that can bind Cl and K equally well when the -OH ligands are oriented appropriately. Because the -OH group is free to rotate around the C-OH single bond, the -OH ligand is thus capable of being "ambidextrous".[8]

The various energies that go to make up the energy profile are now examined: the main differences between cationic and anionic behavior can be seen to originate in the "ion-charge" (i.e., coulomb) energy, which is attractive (negative) at the selectivity filter for Cl^- and repulsive for K when the positive end of the -OH dipoles are rotated inward. When the -OH dipoles are rotated away from the axis, the reciprocating behavior between cations vs. anions is quite apparent. For the frozen coordinate PDLD calculation the "ion-induced dipole" energies make a very important attractive contribution to the energy at the selectivity filter, for both cations and anions, particularly when the -OH dipoles face away. This type of calculation (PDLD) probably overstates the importance of ion-induced dipole interactions and understates the importance of ion-charge interactions because the "frozen" structure has not been allowed to rearrange to optimize ion-charge interactions. Aqvist and Warshel[36] have shown that when a structure has been energy minimized, it rearranges so that ion-charge interactions generally increase at the expense of ion-induced dipole interactions.

Similar calculations on a conjectured structure for a known channel (e.g., the acetylcholine receptor channel) show how parallel alpha helix dipoles can produce such a large electrostatic potential on the channel axis that, if uncompensated, a sterically quite open channel could be completely impermeable to ions.[8]

Though the helix dipole effect may be less severe in the native channel, where opposing dipoles from other antiparallel alpha helixes are present,[37]

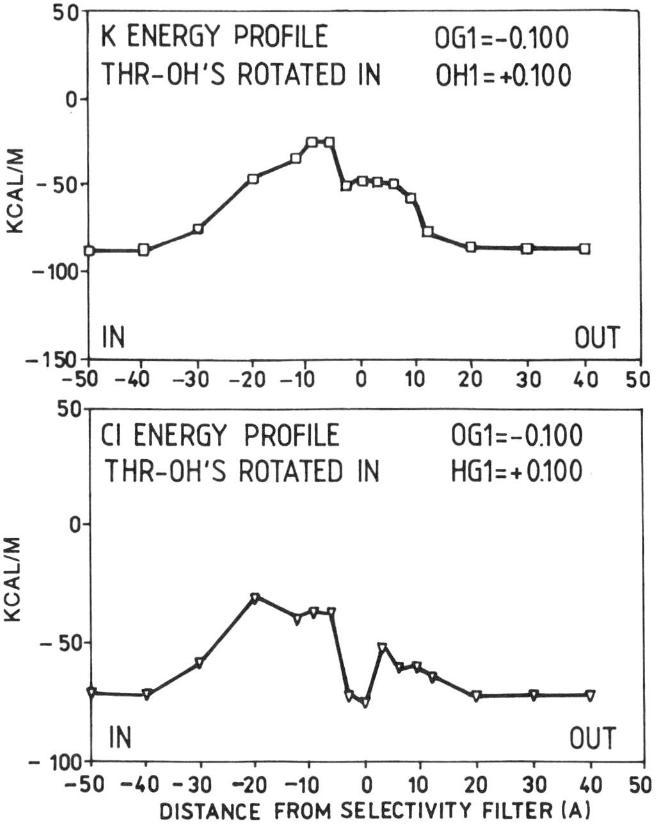

FIGURE 6. Energy profiles calculated for a selectivity filter composed of side chain hydroxyl ligands. The plotted energy profiles are for the southern bean mozaic virus computed for two extreme orientations of the Thr-OH ligands of the selectivity filter: one, when the H ends of the -OH dipoles are rotated so that they face maximally *in* toward the lumen of the channel, the other, when they are rotated so as to face maximally *away* from the lumen. A partial charge of +0.100 has been assigned to the H atom and of −0.100 to the 0 atom of the -OH group. The top half of the figure plots the profiles calculated for K, the bottom half for Cl^-. The left half of the figure shows the energy profiles when the -OH groups are rotated maximally *inward* toward the channel axis. The right half of the figure shows the profiles when they are rotated maximally *away* from the axis. (From Eisenman, G. and Alvarez, O., *J. Membr. Biol.*, 119, 109, 1991. With permission.)

the dipole effects are not likely to be completely neutralized in the channel if the five subunits lie closest to the channel axis. It is of considerable interest that differential ionization of the rings of carboxyl and amino residues known to be present at the ends of the channel can produce an almost perfect compensation of such a dipole barrier. The existence of rings of ionizable Glu and Lys side chains at the cytoplasmic end of the pentamer and of Arg side

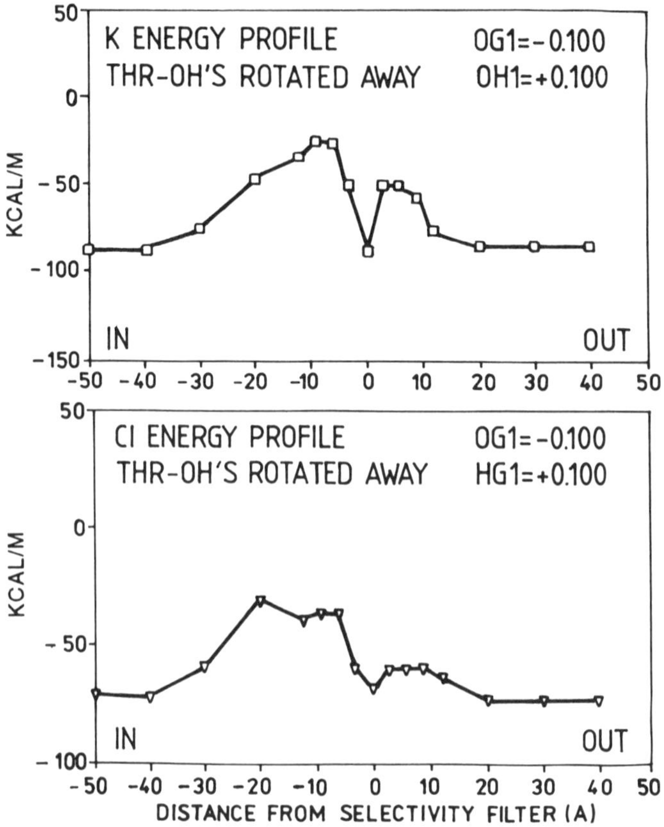

FIGURE 6. (continued)

chains at the extracellular end allows compensation of the helix dipole potential by alterations in their state of dissociation, which is regulated by the local H^+ concentration. Since the free H^+ concentration is decreased by a positive dipole potential and increased by a negative one, the inner, cytoplasmic end of the channel will tend to become negatively charged, while the outer, extracellular end will tend to become positively charged. These potentials provide a negative feedback compensation of the dipole which can be almost perfect.

At the cost of considerably more extensive computations, it is possible to remove the constraint to frozen coordinates in the channel by using a molecular dynamics simulation to relax the protein structure and optimize the free energy of the channel for various positions as an ion is moved through it.[8] The MD/FEP procedure calculates the energy of charging the ion inside the channel by changing the charge in small steps, from fully charged to zero, while allowing the neighboring atoms to continuously rearrange their positions so as to optimize their energy.

The system consists of the ion, the atoms of the protein, and a sphere of explicit water molecules centered at the position of the ion and filling the space not occupied by the ion or protein atoms. Two kinds of calculations can be made.[8] In the first, the energy profile is calculated for an ion moving along its optimum free energy path in a rearrangeable channel. In these calculations, the ion is constrained harmonically to a series of positions along the channel axis but allowed to move in a plane normal to the axis. In this way it finds a free energy minimum which need not lie on the channel axis. In the second, the selectivity is calculated for ions of differing size and charge constrained to lie in the plane of the carbonyl ligands. The energy profile for Na^+ shows that the barrier for an ion entering or leaving the site oriented to the external solution has now been reduced substantially (to about 20 kcal/mol instead of 60 kcal/mol). Such a site would exchange ions with the external solution with a time constant in the order of seconds. Extending the calculation to more atoms including also water molecules and allowing the protein atoms to rearrange, further reduces this barrier to about 14 kcal/mol. It therefore seems likely that the ring of carbonyl oxygens can actually load and release monovalent ions with sufficient rapidity that such a structure could function as the selectivity filter in an ion-selective channel.[8]

It is becoming clear from studying the structural rearrangements and the energy costs associated with them that the structures relax locally rather easily around a cation, so that van der Waals repulsion between the ion and its surrounding ligands is not as important in such an energy-minimized structure as it is in a truly rigid one.

One can anticipate that the ultimate description of selectivity in a protein will involve not only the local effects considered in the earliest modeling,[38] but also more distant effects, whose importance is already well recognized: a filter can rearrange itself under the kinds of constraints that actually occur in a protein, allowing the protein structure to rearrange.[39]

We can conclude, after examining all these findings, that the binding affinity selectivity represents a balance between the energy of hydration and energy of interaction with the filter, including the conformational energy changes that occur if the protein can rearrange. In a rigid selectivity filter, where ion exchange is strictly isomorphic, selectivity is expected to be inversely related to the hydration free energy if the size of the filter exceeds that of the largest ion. Only when the size of an ion exceeds the size of the filter do inversions of the selectivity sequence occur in which smaller ions become favored. In this situation it is the van der Waals repulsion energy that causes the selectivity reversal among ions of the same valence type, and not the partial charge (i.e., "field strength") of the ligands. In a rigid filter field strength is relevant only for selectivity between ions of differing valence and/

or sign. On the other hand, in a deformable selectivity filter, repulsion energy differences among differing species become negligible, and the major contributions to selectivity come from two factors. One is the "field strength" of the binding ligands, variations in which lead to species differences in electrostatic energies of the ions with the ligands locally. The other is the structural energy change of the protein itself, which is less local and can be quite widespread, even allosteric.

The different selectivity sequences obtained experimentally are probably due to structural rearrangements of the channel protein which achieve an energy minimum. An interesting point is that structural mutations, obtained by amino acid replacements in the protein, modify the calculated energetic profile and the channel capacity to bind ions.[8]

C. CHANNEL PROTEIN ENGINEERING

Another approach to the identification of molecular determinants of selectivity in ionic channels is channel protein engineering, that is, synthesis of functional peptides which have the ability to form ion channels by incorporating into a synthetic lipid bilayer. Montal[40] has obtained channels analogous to the Na channel of excitable cells and to the nicotinic acetylcholine receptor. The conductance selectivity ratio is comparable to that characteristic of the authentic channels in the case of the Ach receptor (Na^+: K^+: $Tris^+$ = 1.0 : 1.2 : 0.18); the synthetic Na channel is indeed cation selective, but lacks selectivity between Na^+ and K^+.

D. SELECTIVITY OF EPITHELIAL CHANNELS

1. The Epithelial Na^+ Channel

The epithelial Na^+ channel is highly selective for Na^+ over K^+ (Na^+/K^+ = 10:1 to 670:1).[41] Among the alkali metal cations, only Na^+ and Li^+ ions could enter the apical channels.[42] Palmer presents a geometric model for the epithelial Na^+ channel based upon analysis of a voltage-dependent block of Na^+ currents by impermeant ions.[43] In this model, the outer pore of the channel is envisaged as a negatively charged funnel that attracts cations indiscriminately. Further into the channel, the pore narrows to exclude all but the smallest cations, and a final narrowing excludes all ions except Na^+, Li^+, and H^+.

The expression of Na^+ channels in cultured A_6 epithelial cells seems to be related to cell-matrix interactions or to basolateral nutrient requirements.[44] A_6 cells cultured on impermeable supports possess Na^+ channels with low Na^+/K^+ selectivity, moderately high conductances (9 pS), and short mean open times (millisecond range). In contrast, A_6 cells grown on permeable supports have channels that are similar to those of the rat collecting tubules and toad urinary bladder, i.e., high Na^+/K^+ selectivity (19:1), low conductances (3 pS), and longer open times (seconds).[46]

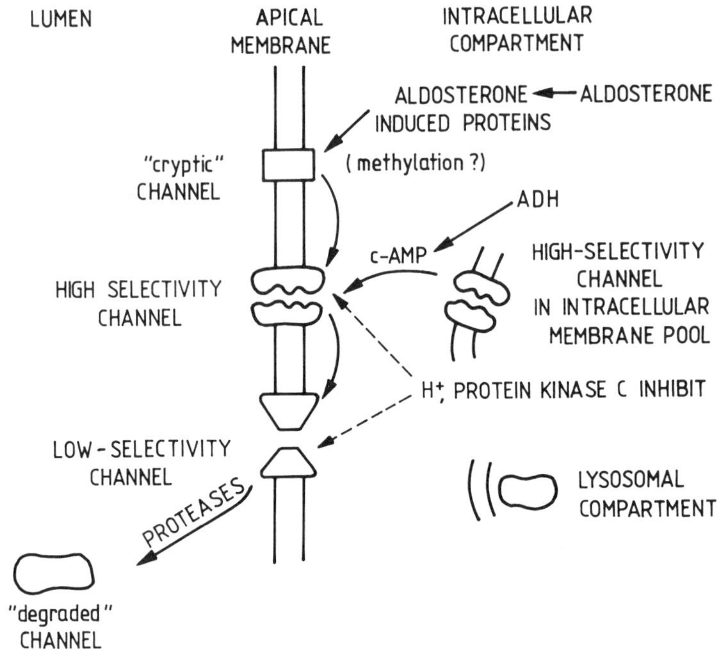

FIGURE 7. Hypothetical model of Na channel turnover. Cytoplasmic vesicles containing amiloride-sensitive and high-selectivity Na channels fuse with apical membrane, or alternatively, channels in vesicle wall and apical membrane are exchanged by brief contacts. Apical Na channels are degraded, possibly by kallikrein in urine, to low selectivity channels, which may be lost as degraded channels, or internalized into the cell. (From Eaton, D. and Hamilton, K. L., *Ion Channels,* Vol. 1., Narahashi, T., Ed., Plenum Press, New York, 1988. With permission.)

An interesting feature of channel selectivity is that it varies with the channel age: it was demonstrated that in epithelial cells where highly selective Na channels are present in their apical membrane, new high-selectivity channels can be inserted into the membrane by the fusion of intracellular vesicles with the apical membrane (the probable mode of action of ADH).[46]

Alternatively, aldosterone can ''unmask'' cryptic high-selectivity channels already present in the membrane. With time, in the absence of continued high levels of aldosterone or ADH, the high-selectivity channels gradually degrade into low-selectivity channels before finally becoming nonfunctional channels whose fate is unclear (Figure 7).

Loss of selectivity of the Na channels also is involved in the physiologic regulation of epithelial Na transport in the urinary bladder.[47] Newly inserted channels were more highly selective for Na^+ over K^+ (over 30:1) than ''older'' channels that were already present in the membrane (P_{Na}:$P_K \approx 2$:1). The loss of selectivity may be related to the presence of urokinase, kallikrein, and plasmin, which are trypsin-like enzymes secreted by the kidney and found in the urine.

The degradation process involves two steps: conversion of the Na^+ channel to a nonselective, amiloride-insensitive cationic channel, followed by a further proteolysis that leads to the loss of the channel into the urine. The number of channels can be regulated by this mechanism in which channel selectivity is related to the rate of channel turnover.[47]

It is conceivable that this cycle of channel synthesis, insertion, degradation, and removal may be tied directly or indirectly to the release of hormones and other endogenous factors which regulate Na^+ transport.

An understanding of the molecular interactions which condition the channel selectivity will provide the answer to specific questions concerning the regulation of the channels by hormones, other ions, intracellular messengers, or various pharmacological tools.[48-51]

2. The Epithelial K^+ Channel

Another well-characterized ion channel is the K^+ channel. In the excitable membranes, K^+ channels are generally thought to function mainly in determining membrane potential.[11] K^+ channels were also identified in transporting epithelia.[52] Both types of K^+ channels show the following selectivity sequence: $Tl^+ > K^+ > Rb^+ = NH_4^+$. It was proposed for the epithelial K^+ channel that ions penetrate the channel by single-file, multisite diffusion.[52] In other words, more than one cation is present in the channel at a time and transient binding of the cation occurs at one or more negatively charged polar groups along the channel (Figure 8). The evidence for this model is the finding of an ''anomalous mole-fraction'' effect when potassium is partially replaced by thallium. Specifically, low amounts of thallium inhibit the potassium current through the channel, because of the tight association of thallium within the channel. At high thallium levels, the current is increased because thallium now also strongly interacts at the entrance to the channel, overcoming the strong interaction within the channel. The model assumes that the channel gate is near the cytoplasmic side of the channel, and that a selectivity filter is located midway through the membrane.

This site and a ''modifier'' site at the mucosal entrance of the channel are the presumed sites for interaction between permeant cations, impermeant cations, and the channel gate. The latter is thought to arise from short-range electrical fields from the permeant cations or by charge screening. As for structural features of the channel, pH titration effects indicate that the channel possesses at least one acidic component, possibly a carboxyl group, that influences permeability. It is unclear whether this component might represent the site for blocker interactions (e.g., Ba^{2+}). Since the channel is not blocked by tetraethylammonium (TEA), Van Driesshe and Zeiske have suggested that the channel may possibly lack a hydrophobic site for TEA interaction, or alternatively, the entrance of the channel might be too small to admit this compound.[52]

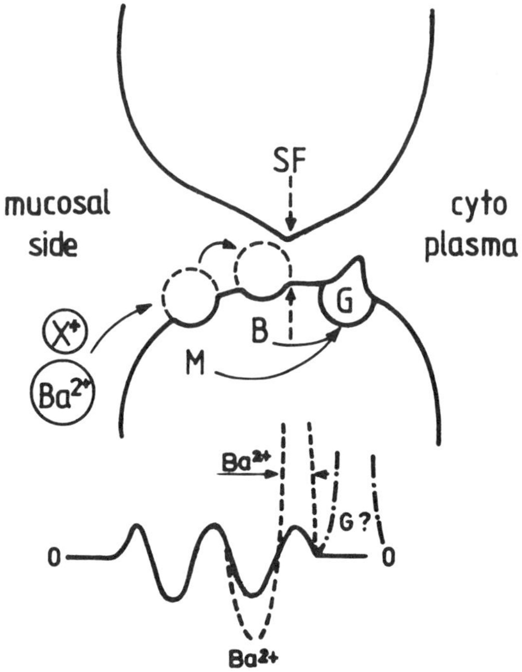

FIGURE 8. Hypothetical model of two sites and three barriers for frog skin apical K^+ channel. M and B denote sites for interaction with permeant ions (X^+) and Ba^{2+}, which "hop" from the solution and along single-file sites. Appropriate interaction of X^+ and/or Ba^{2+} at M (modifier site) and/or B (blocker site) influences kinetics of channel gate (G). Selectivity filter (SF) allows passage of only X^+(NH_4^+, Rb^+, K^+, Tl^+) but not Ba^{2+} or Ca^{2+} to cytoplasmic channel side. Lower panel: tentative potential profile (not including electrical potentials), with barrier heights and well depths assumed to be uniform. (From Van Driesshe, W. and Zeiske, W., *Physiol. Rev.*, 65, 833, 1985. With permission.)

Electrophysiological studies of the Ca channel have provided evidence that the channel is a single-file pore that contains at least two high affinity ($\sim$1 μM) binding sites for Ca.[53] These high affinity Ca binding sites are believed to impart the selectivity of the Ca channel for Ca and to play an intimate role in the permeation of Ca through the channel.[54]

III. SPECIFICITY OF ION PUMPS

A. TYPES OF ION PUMPS

All cells maintain, at the expenditure of energy, differences in the electrochemical potential of ions (H^+, Na^+, K^+, Ca^{2+}) between the cytoplasm and extracellular medium. This task is performed by active transport systems, namely, the ion pumps, which are capable of moving ions across cell membranes against electrochemical gradients. These transmembrane ion gradients are essential for cellular functions such as energy storage, excitability, or volume regulation.

Ion pumps may be classified according to the nature of the energy source utilized for uphill transport. Bacteriorhodopsin and halorhodopsin are light-activated translocators of H^+ and Cl^-, respectively, which are found in halophilic archaebacteria.[55,56] They represent a separate and probably ancient class of ion pump in which the transport cycle is driven by light-induced conformational changes in a retinal moiety covalently attached to the protein. With a single polypeptide chain of only 248 amino acids, bacteriorhodopsin is structurally the simplest ion pump known so far.

Cytochrome oxidase and NADH oxidase are redox-linked ion pumps.[57] Cytochrome oxidase in the inner mitochondrial membrane and in the plasma membrane of bacteria couples electron transfer from cytochrome *c* to O_2, to proton pumping.

Ion motive ATPases represent the largest and most diverse class of ion pumps. They may be subdivided into three distinct groups.[58] F-type (or F_oF_1) proton ATPases are found in the inner mitochondrial membrane, in the thylakoid membrane of chloroplasts, and in the plasma membrane of bacteria. In mitochondria and chloroplasts, the enzyme seems to operate under physiological conditions exclusively in the downhill-flux mode carrying out ATP synthesis driven by an electrochemical H^+ gradient. Bacterial F_oF_1-ATPases appear to run in the reverse direction, also as a normal function, hydrolyzing ATP and extruding protons from the cell.[59] In contrast to the P-type ATPases to be discussed below, the reaction cycle of F-type ATPases does not involve phosphorylated intermediates. Instead, the enzyme is thought to function by performing a cycle of transitions between states of high- and low-affinity nucleotide binding that are coupled to ion translocation steps.[60,61]

Vacuolar (or V-) type H^+-ATPases represent a ubiquitous class of proton pumps that are found in cellular organelles such as lysosomes, endosomes, and secretory and storage granules, as well as in vacuoles of fungi and higher plants.[62] Their main function is acidification of the aqueous inner space of the organelle.

P-type ATPases differ fundamentally from the other ion motive ATPases in their reaction mechanism.[63,64] They form a phosphorylated intermediate in which phosphate is bound to an aspartyl residue. P-ATPases have a comparatively simple structure. The H-ATPases in the plasma membrane of fungi and higher plants and the Ca-ATPases consist only of a single polypeptide chain with a molecular weight of approximately 100 kDa.[65] The Na,K-ATPase and also the H,K-ATPase contain an additional, smaller polypeptide (the beta chain) of unknown function.[66,67]

P-ATPases, which are sometimes referred to as E_1E_2-ATPases, are thought to assume, in the course of the pumping cycle, two principal conformations, E_1 and E_2, with inward- and outward-facing ion-binding sites.

P-type ATPases are found in virtually all eukaryotic cells and also in bacteria. The Na,K-ATPase is the principal ion pump in the plasma membrane of animal cells; it is responsible for the maintenance of electrochemical potential differences $\Delta\tilde{\mu}$ of Na^+ and K^+. $\Delta\tilde{\mu}_{Na}$ serves as an energy source for secondary active transport of organic solutes and is essential for electrical excitability of nerve and muscle cells. The counterpart of the Na,K-ATPase is the H-ATPase in the plasma membrane of fungi and higher plants. Secondary active transport of solutes in fungi, algae, and plants is driven by the inward flow of protons and thus depends on the presence of an H^+ pump. ATP-driven Ca pumps are found in the plasma membrane of animal cells, as well as in the membrane of cellular organelles. The Ca pump in the plasma membrane extrudes Ca^{2+} from the cell keeping the cytoplasmic Ca^{2+} concentration at the low levels that are essential for normal cell function.[68] In the sarcoplasmic reticulum of muscle cells, the Ca pump promotes reuptake of Ca^{2+} that has been released to the cytoplasm in the course of the stimulation of muscular contraction.

Whereas ATPases carrying out active transport of H^+, Na^+, K^+, and Ca^{2+} are well characterized, the existence of a Cl^- transport ATPase is less certain.[69]

B. BASIC FEATURES OF ION PUMPS WHICH MAY PLAY A ROLE IN THEIR SPECIFICITY

Ion pumps are unique among proteins in their ability to couple a scalar reaction, such as ATP hydrolysis, to transmembrane ion translocation. Regarding this capacity some questions arise:

1. What conditions must a protein meet in order to perform this task?
2. To what extent is free energy transiently stored in the form of conformational energy of the pump protein?
3. Is the free energy used mainly for switching the ion binding sites from an inward-facing to an outward-facing configuration or for creating a difference in the affinity of ion binding?

As channels, ion pumps also have some common properties with the enzymes. They operate in a cyclic fashion, binding and releasing substrates in a functionally ordered sequence.

The Na,K-ATPase found in the plasma membrane of virtually all animal cells is responsible for the low sodium concentration and high potassium concentration in the cytoplasm.[70] The reaction mechanism of the Na,K pump (the Post-Albers scheme)[70a,b] is based on the notion that the pump protein can assume two principal conformations, E_1 and E_2, with inward-facing (E_1) and outward-facing (E_2) binding sites for Na^+ and K^+. E_1 prefers to bind Na^+

and/or ATP and is stabilized by these ligands; E_2 binds K^+ and/or P_i and is stabilized by these ligands. In the presence of Na^+ at the cytoplasmic side, the protein is phosphorylated by ATP, whereby the bound Na^+ ions are *occluded,* i.e., trapped inside the protein ($Na_3 \cdot E_1 \cdot ATP \rightarrow (Na)_3\ E_1$-P). State $(Na_3)E_1$-P undergoes a conformational transition to state E_2 in which the ion-binding sites assume an outward-facing configuration. Release of Na^+ to the extracellular side and binding of K^+ is followed by occlusion of K^+ and dephosphorylation (P-$E_2 \cdot Na_3 \rightarrow$ P-$E_2 \rightarrow$ P-$E_2 \cdot K_2 \rightarrow E_2[K_2]$). Binding of ATP then shifts the conformation back to state E_1 from which K^+ is released to the cytoplasm ($E_2[K_2] \rightarrow [K_2]E_2 \cdot ATP \rightarrow K_2 \cdot E_1 \cdot ATP \rightarrow E_1 \cdot ATP$). After binding of Na^+ at the cytoplasmic side, the pump is ready for a new cycle (Figure 9).

Recent experiments indicate that the Post-Albers scheme has to be supplemented by at least one further state E^x (or E $\sim$ P) of elevated free energy.[71,72] This state of the phosphoenzyme is thought to contain two occluded Na^+ ions. It is formed by release of a Na^+, from state $(Na_3)E_1$-P to the extracellular side:

$$(Na_3)E_1 - P \rightleftharpoons (Na_2)E \sim P + Na^+_{ext}$$

$$(Na_2)E \sim P \rightleftharpoons P - E_2Na_2 \rightleftharpoons P - E_2 + 2Na^+_{ext}$$

The step in which $(Na_2)E \sim P$ interconverts into P-$E_2 \cdot Na_2$ means transformation of a high energy acylphosphate (capable of donating its phosphate to ADP) into a low energy form (capable only of being hydrolyzed by water to yield inorganic phosphate). This step is apparently associated with the occlusion of the sodium ions. In some way that we still have to understand, occlusion of sodium ions leads to destabilization of the acylphosphate bond, rendering it available for the phosphorylation of ADP. The movement of the sodium charges is presumably associated with this destabilization.[73]

On the other hand, ion pumps are clearly different from enzymes in that one of the substrates, the transported ion, remains chemically unchanged. In enzymatic reactions, close contact between reactants is usually an advantage, but it is not obvious that this should also be true in active transport, in which an otherwise unaltered solute is driven from low to high electrochemical potential.[74] For this reason, mechanistic models for ion pumps based on direct interaction between transported ion and chemical reactants such as ATP or phosphate have not found wide acceptance.[75-77] Instead, it is commonly thought that coupling between chemical reaction and ion transport is indirect.[58] Applied to an ATP-driven pump, this conjecture means that ATP hydrolysis and ion translocation may take place in spatially separate parts of the protein. It is obvious that indirect coupling requires conformational changes as a link between the chemical reaction and translocation event.

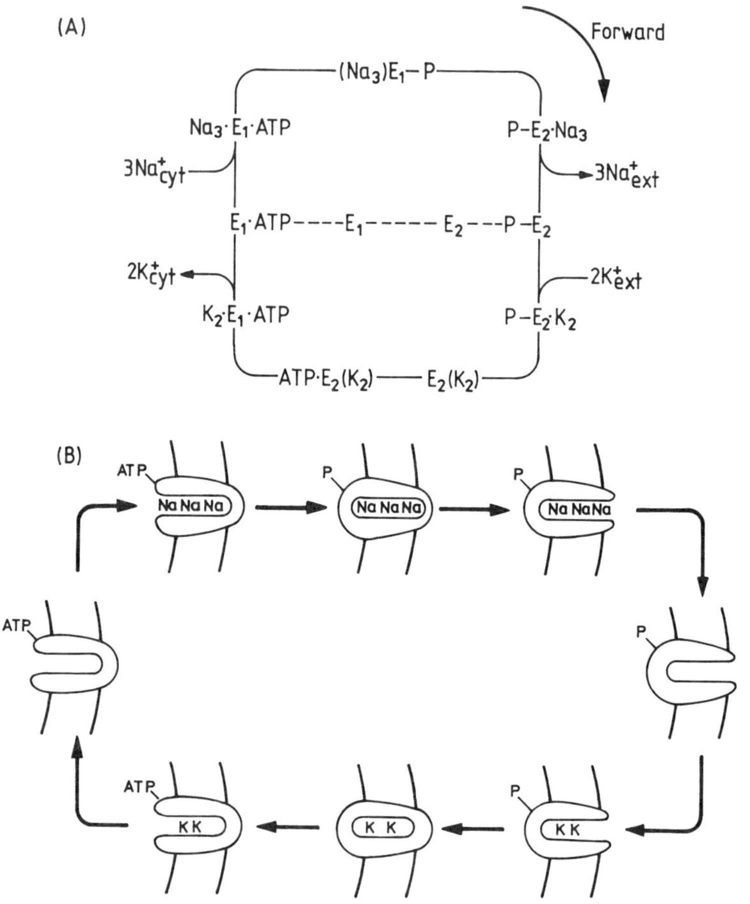

FIGURE 9. (A): Post-Albers scheme for the pumping cycle of Na, K-ATPase. E_1 and E_2 are conformations of the enzyme with ion binding sites exposed to the cytoplasm and the extracellular medium, respectively. In the "occluded" states (Na_3)-E_1-P and $E_2(K_2)$, the bound ions are unable to exchange with the aqueous phase. (B): Pictorial representation of the reaction scheme in part A, showing the sidedness of the ion binding sites. (From Läuger, P., *Electrogenic Ion Pumps,* Sinauer Associates, Inc., Sunderland, MA, 1991. With permission.)

Current models of energy-driven ion translocation are almost exclusively based on the alternating-access mechanism (Figure 10),[74,78-80] which accounts for the fact that ion pumps are medium- to large-sized membrane-spanning proteins that are unlikely to function by a shuttle-type mechanism. Rather, the pump protein is considered as a gated channel in which the height and position of energy barriers restricting the movement of ions are transiently modified by the energy-supplying reaction. The actual motion of the protein matrix associated with the gating event may be minor, involving reorientation of a small number of amino acid residues.

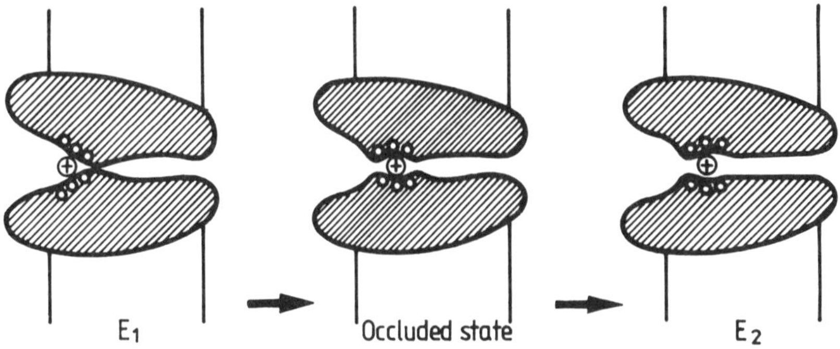

FIGURE 10. Alternating-access mechanism of ion pumps. E_1 and E_2 are states with inward- and outward-facing ion binding sites. In the occluded state, the ion is unable to exchange with either aqueous medium. Open circles represent ligand groups of the protein. (From Läuger, P., *Electrogenic Ion Pumps,* Sinauer Associates, Inc., Sunderland, MA, 1991. With permission.)

The distance over which bound ions are translocated in the E_1E_2 conformational transition is not known for any ion pump. It is unlikely, however, that the amplitude of movement is as large as the entire membrane thickness. It is therefore thought that pump proteins have access channels on one or both faces, connecting the binding sites with the adjacent aqueous medium (Figure 10).[58] Within the access channel, ions may migrate by a diffusion-type mechanism. Accordingly, an access channel can be defined as a pathway through which ions can diffuse without the aid of a transport-related conformational change in the protein.[81] For the structure of access channels, two limiting cases may be distinguished:

1. The channel may consist of a wide opening (or vestibule) into which water and all kinds of ions are allowed to enter. Under this condition the electrical conductance of the access channel is large, meaning that the field strength resulting from a transmembrane voltage is low within the channel.
2. In the other extreme case, the channel is narrow and specific for the transported ions, so that part of the transmembrane voltage decreases across the length of the channel.[58]

These two limiting cases may be referred to as low-field and high-field access channels. In a high-field access channel, or ion well,[82] the effective affinity of ion-binding sites becomes a function of voltage.

Studies of the kinetic behavior of Na,K-ATPase and Ca-ATPase indicate that transitions between conformations E_1 and E_2 with inward- and outward-facing binding sites proceeds through an intermediate state in which ions are trapped inside the protein.[83] Such *occluded states,* in which the bound ions

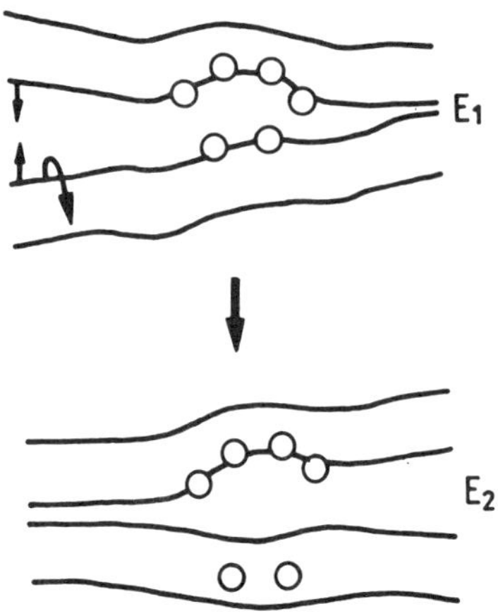

FIGURE 11. Possible mechanism leading to conformation-dependent affinity changes of bound ions. Polar groups (circles) are located on adjacent alpha-helixes, forming an ion-binding site. A conformational transition ($E_1 \rightarrow E_2$) leads to rotation of one of the helixes, so that part of the liganding groups moves to a greater distance. (From Läuger, P., *Electrogenic Ion Pumps,* Sinauer Associates, Inc., Sunderland, MA, 1991. With permission.)

are unable to exchange with the aqueous medium, can occur when gates on either side of the binding site are closed (Figure 10). Thus, occlusion of ions does not necessarily mean that the binding site has an unusually high affinity, but rather that release of the bound ions is inhibited by high activation barriers.

C. CHANGES OF BINDING AFFINITY

As discussed, the transition from state E_1 to E_2 is usually associated with a large change of ion affinity. A simple mechanism by which this affinity change can be brought about has been proposed by Tanford.[84] His model (Figure 11) is based on the notion that the transmembrane portion of an ion pump consists of a bundle of several more or less parallel alpha helixes as has been found for bacteriorhodopsin.[85] A row of polar groups may be assumed to run down the center of the bundle, providing a pathway for ion translocation. Polar groups from one helix that would be in close proximity to a site on the second helix in conformation E_1 can be imagined as moving away to a greater distance in conformation E_2 by rotation of the helix. This movement would lead to a less favorable complexation of the ion and to a decrease of binding affinity.

All known pumping mechanisms have in common that the free energy supplied by the external source is transferred to the pump protein in a single, discrete step.[73] In a light-driven pump, this energy-donating step is the absorption of a photon,[86] and in a P-ATPase it is the transfer of a phosphate residue from ATP to an aspartyl group of the protein.[73] At the end of the pumping cycle, part of the supplied energy appears in the form of an electrochemical gradient. This implies that free energy is transiently stored in the pump. At least two different possibilities exist namely that free energy can be stored in an energy-rich ligand bond or in a particular macromolecular conformation. In P-ATPases, E $\sim$ P is a state of elevated free energy in which the bound phosphate has an energetic status comparable to that of the terminal phosphate residue in ATP.[73] A strained conformation of the pump protein, on the other hand, may occur as a metastable state in the cycle, which has a tendency to relax to a more stable conformation.

Strained macromolecular states may be involved in conformational coupling between an energy-supplying reaction and ion translocation, as depicted in Figure 12. The mechanism of conformational coupling shown in Figure 12 represents a two-step process in a P-type protein ATPase. State $HE_1 \sim P$, in which the phosphate bond is of the high-energy type, first undergoes transformation to state HE^x-P, in which a strained conformation has been induced by formation of additional noncovalent bonds between the phosphate residue and the protein. In a second step, the strained state relaxes to the more stable conformation HE_2-P; this relaxation is associated with reorientation of the ion-binding site and a decrease of binding strength. Mechanistic models involving strained protein conformations have been proposed, for instance, for the F_0F_1-ATPase[87] and for bacteriorhodopsin.[88] An alternative possibility for conformational coupling[74] consists of a single-step mechanism in which state $E_1 \sim P$ undergoes direct transformation to state E_2-P (dashed arrow in Figure 12). In both mechanisms, an internal conversion takes place in which the transition from low-affinity to high-affinity phosphate binding energetically compensates for the transition from high-affinity to low-affinity ion binding.

D. CATION SPECIFICITY AND CATION BINDING SITES IN P-ATPASES

The Na,K pump exhibits a high ion specificity in the sodium limb of the transport cycle, but a remarkably low specificity in the potassium limb of the cycle. As far as is known, only H^+ and Li^+ can substitute for Na^+ under steady state turnover conditions, and only to a limited extent. On the other hand, K^+ can be replaced at the extracellular side by other monovalent cations such as Rb^+, Cs^+, Tl^+, NH^+_4, and even by Na^+, Li^+, and H^+.[89–91] Rb^+, Cs^+, and Tl^+ are almost equally well transported as K^+, whereas Na^+, Li^+, and H^+ are poor substitutes for K^+.

A characteristic feature of the Post-Albers model is the change of binding affinities in the course of the reaction cycle: in conformation E_1 the transport site prefer Na^+ over K^+, in conformation E_2 they prefer K^+ over Na^+. Numerous attempts have been made to determine affinities from binding studies with radioactive ions.[92] These attempts have met with limited success, however, because of the difficulty of discriminating between specific and

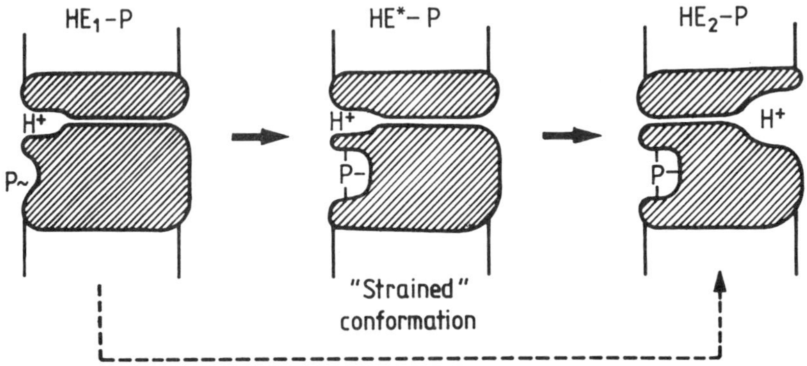

FIGURE 12. Two-step mechanism for conformational coupling in a P-type proton ATPase. State $HE_1 \sim P$, in which the phosphate bond is of the high-energy type, first undergoes transformation to a strained conformational state H^x-P, in which the phosphate group forms additional noncovalent bonds with the protein. In a second step, the strained state relaxes to a more stable conformation HE_2-P; this relaxation is associated with reorientation of the ion binding site and a decrease of binding strength. (From Läuger, P., *Electrogenic Ion Pumps,* Sinauer Associates, Inc., Sunderland, MA, 1991. With permission.)

nonspecific binding. For this reason, binding sites are usually characterized by apparent affinities which are obtained from the ion-concentration dependence of transport rates or enzymatic reaction rates. (The apparent affinity is defined as the reciprocal of the concentration of half-maximal activation of a given process.) The apparent affinities for Na^+ and K^+ change 1000-fold and 50-fold, respectively, in the E_1/E_2 conformational transition, corresponding to a high affinity at the uptake side and a low affinity at the release side.[93,94]

The existence of multiple binding sites for Na^+ and K^+ has been taken into account for the interpretation of apparent affinities.[95] If, for instance, two Na^+ ions bind with high affinity and the third Na^+ ion binds with low affinity, the two high-affinity sites will have already been occupied at moderate Na^+ concentration, so that the apparent affinity essentially reflects binding to the last, low-affinity, site.

Several lines of evidence indicate that, in addition to the low-affinity release sites for Na^+, a high-affinity site for Na^+ exists at the extracellular face of the pump.[96-98] Binding of Na^+ to this site with an apparent affinity of 0.5 mM^{-1} leads to inhibition of the slow spontaneous dephosphorylation that is observed in the absence of K^+ ($P\text{-}E_2 \rightarrow E_2 + P_i$) and, accordingly, leads to inhibition of uncoupled ATP-driven Na extrusion.[96] This high-affinity site may be identical with one of the three transport sites for sodium, or may represent an additional regulatory site.

Furthermore, extracellular Na^+ seems to have allosteric transmembrane effects on cation-binding affinities at the cytoplasmic side.[99] The relation

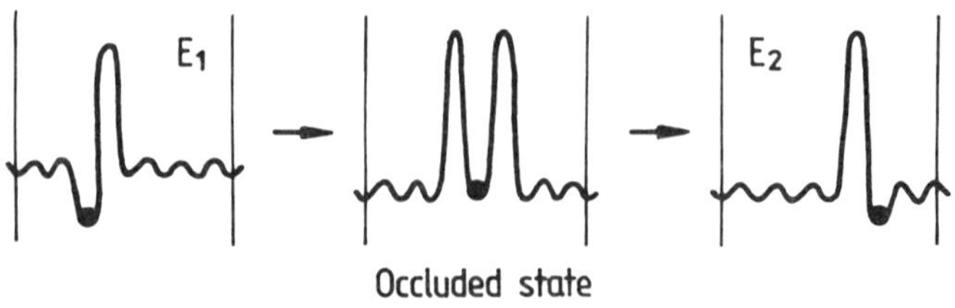

FIGURE 13. Ion occlusion as an intermediate step in the $E_1 \rightarrow E_2$ conformational transition. In the occluded state, equilibration between binding site and medium is blocked by high-energy barriers. (From Läuger, P., *Electrogenic Ion Pumps*, Sinauer Associates, Inc., Sunderland, MA, 1991. With permission.)

between this trans effect and the inhibitory effect of extracellular Na^+ is not clear so far.

Little is known about the nature of the binding sites for Na^+ and K^+. It has been proposed that the protein contains a single binding pocket that can accommodate either three Na^+ ions or two K^+ ions.[100]

Ion occlusion does not necessarily mean that the binding sites in the occluded state have an unusually high affinity for the bound ions. In fact, the apparent affinities of the ion occlusion sites on the Na,K-ATPase are not remarkably high.[101]

By definition, the occluded state is a state in which high activation barriers prevent rapid equilibration between binding sites and the medium. The transport pathway of the ion through the protein is represented by a potential profile consisting of a series of barriers and wells. Potential wells are located at places where the ion interacts in an energetically favorable way with the liganding groups of the protein; binding sites correspond to deep potential wells (Figure 13). In conformation E_1 the ion in the binding site has easy access to the cytoplasmic side, but is prevented from escaping to the extracellular side by a high energy barrier. In conformation E_2, the position of the high barrier is inverted. In the occluded state, equilibration between the binding site and the medium is blocked by two energy barriers. In either conformational state, barriers must then exist preventing escape of the bound ions to the "wrong" side. Experimentally determined rates of release of ions from the occluded state lie in the range of 0.001 to 100 s^{-1};[100] these rates are much smaller than the diffusion-limited rate of ion dissociation from a binding site exposed to water (about 10^5 s^{-1} for K_d = 0.1 m*M*). The access channels connecting the binding sites with the adjacent aqueous phase are represented by a number of smaller barriers.

The hypothesis of a single binding pocket is supported by experiments. Shani-Sekler et al. have studied the effects of chemical modification of carboxyl residues in the Na,K-ATPase by dicyclohexylcarbodiimide.[102] From the results of these experiments, the authors proposed that occlusion of Na^+

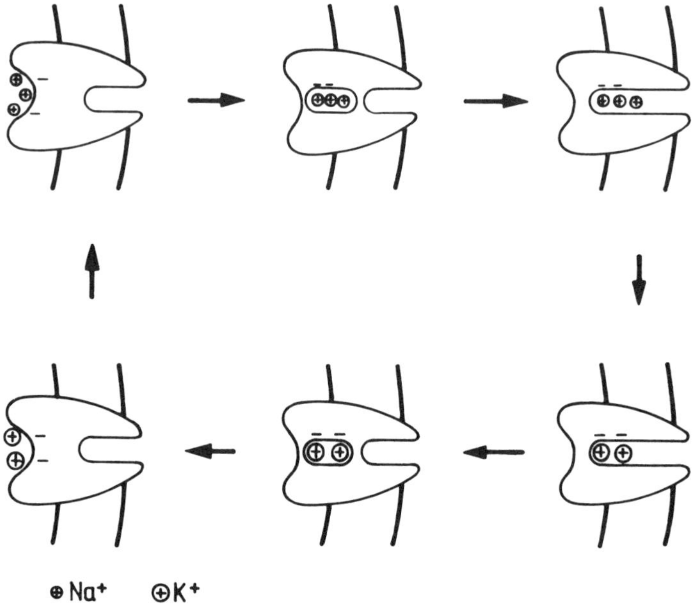

FIGURE 14. Electrostatic model of the Na, K pump. The ligand system of the cation binding pocket is assumed to bear two fixed negative charges and to accommodate either three Na^+ ions or two K^+ ions. A wide-access channel is assumed to be present at the cytoplasmic side and a narrow-access channel (ion well) at the extracellular side (From Läuger, P., *Electrogenic Ion Pumps,* Sinauer Associates, Inc., Sunderland, MA, 1991. With permission.)

and K^+ involves the same carboxyl groups in the transmembrane portion of the protein (Figure 14).

Karlish et al. have recently shown that tryptic cleavage in the presence of Rb^+ removes a large part of the cytoplasmic portion of the alpha chain of the enzyme, containing the sites for ATP binding and phosphorylation.[103] The remaining membrane-embedded part of the protein retains the capacity for occlusion of Na^+ and Rb^+. These interesting findings indicate that the cation binding sites are physically separate from the ATP binding sites and lie within transmembrane segments of the protein.

In the absence of Na^+, the Na,K-ATPase can sustain H^+/K^+ exchange coupled to ATP hydrolysis, as was shown in experiments with reconstituted vesicles and inside-out membrane vesicles derived from erythrocytes.[91,104] The proton affinity of the pump is remarkably high, since H^+/K^+ exchange can be detected when the proton concentration is as low as 10^{-6} *M*. It is feasible that the transported protons bind to ionized ligand groups that normally accept Na^+ ions. Another possibility is that the Na^+ sites accept protons in the form of hydronium ions (H_3O^+).[105] In analogous experiments, it was demonstrated that in the absence of extracellular K^+, the Na,K pump can perform ATP-driven N^+/H^+ exchange.[91]

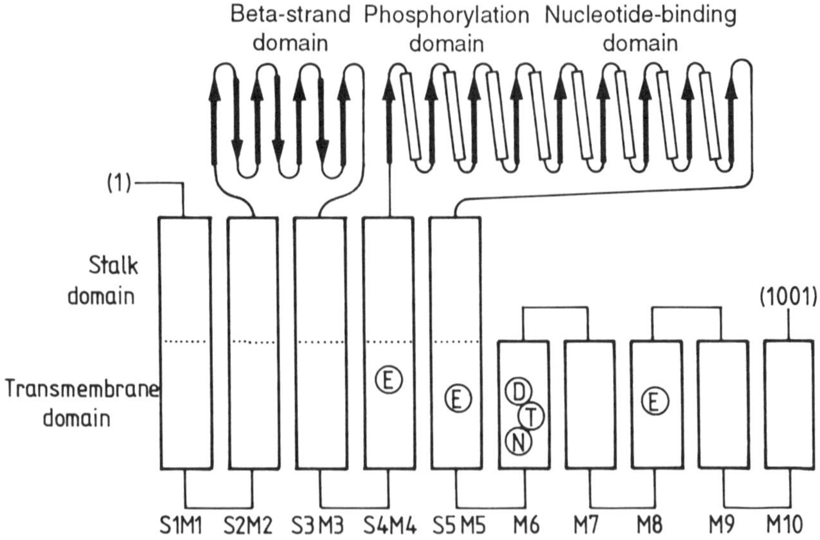

FIGURE 15. Structural model of the Ca-ATPase of sarcoplasmic reticulum. The stalk and transmembrane regions are shown as boxes to indicate alpha helixes. In the globular domains alpha helixes are shown as boxes, and antiparallel or parallel beta-sheet domains are indicated by thick arrows. Six amino acids, which are thought to be involved in high-affinity calcium binding, are indicated in helixes M_4, M_5, M_6, and M_8 (M4: Glu-309; M5: Glu-771; M6: Asn-796, Thr-799, and Asp-800; and M8: Glu-908). (From Läuger, P., *Electrogenic Ion Pumps*, Sinauer Associates, Inc., Sunderland, MA, 1991. With permission.)

Calcium ATPases are found also in the plasma membrane of virtually all animal cells, as well as in the membranes of sarcoplasmic reticulum. A structural model of the Ca-ATPase is given in Figure 15: it contains ten alpha-helical transmembrane segments and two large cytoplasmic domains. The cytoplasmic domains are connected with the transmembrane segments by a "stalk" consisting of amphipathic alpha helixes. The presence of many negatively charged amino acids in the stalk segments has led to the speculation that the high-affinity transport sites for Ca^{2+} may be located in the stalk domain.[65,106] One could identify six amino acid residues that seem to be involved in high-affinity calcium binding.[107] Alteration of Glu-309, Glu-771, Asn-796, Thr-799, Asp-800, or Glu-908, each of which is predicted to lie near the center of the transmembrane domain in the putative transmembrane sequences M_4, M_5, M_6, and M_{10} resulted in complete loss of Ca^{2+} transport and of Ca^{2+}-dependent phosphorylation of ATP. These results suggest that at least six polar, oxygen-containing residues lying near the center of the transmembrane domain provide ligands for one or both of the two high-affinity Ca^{2+} binding sites. The six amino acid residues are involved in the activation of the catalytic site by Ca^{2+}, permitting utilization of ATP.[68]

In turn, the bound calcium is displaced into the lumen of the channel-like structure formed by clustering of some transmembrane alpha helixes of the Ca pump upon phosphorylation of the catalytic site by ATP. The distance between the catalytic and calcium binding domain is quite large, and the allosteric mechanism of transduction most likely involves a segment of the ATPase sequence that includes the phosphorylation site and the transmembrane helix M_4 (Figure 15).

E. BINDING SITES FOR ATP AND PHOSPHATE

The location of the ATP binding site on the alpha subunit of Na,K-ATPase (Figure 16) has been probed by covalent attachment of reactive ATP analogues.[108] An indication of the functional groups required to form a nucleotide binding site may be obtained from examining sites in other ATP binding proteins of known structure, such as adenylate cyclase[72] or sites in different transport ATPases.[109]

An important feature is a hydrophobic pocket for accommodation of the adenine and ribose moieties, which is formed by Ile, Val, His, and Leu residues. The triphosphate moiety is flanked by a hydrophobic strand of a parallel beta-pleated sheet terminated by aspartate. Both sequence homologies and covalent labeling indicate that at least three different segments of the alpha chain participate in ATP binding, namely, residues 543–561, 655–664, and 704–722.[72]

As in the other P-type ATPases, the acceptor of the phosphate group donated by ATP is an aspartate residue (Figure 16). In the Na,K-ATPase, the phosphorylation site is Asp-369. The segment containing the phosphorylation site is highly conserved in all P-type ATPases.[66,67]

IV. TIME SCALES OF ION TRANSPORT

Compared to ion channels, ion pumps are extremely slow transport systems. The rate of ion translocation through a channel may be as high as 10^8 s^{-1}.[11] On the other hand, maximal turnover rates of known ion pumps are in the range of 10 to 1000 s^{-1}, corresponding to cycle times of between 1 and 100 ms.[58] In ion pumps, the turnover rate is limited, as far as we know, by the rate of conformational changes associated with transitions between the inward- and outward facing orientation of the ion binding sites. In ionic channels, transport rates are thought to be limited by the frequency of thermally activated jumps of the ions over energy barriers, which may not require changes of protein conformation.

Apart from these differences in the overall transport rate, interesting similarities exist between the kinetic properties of ion channels and ion pumps. Ion migration in the access channel of a pump is likely to be as fast as in an ordinary transmembrane ion channel.

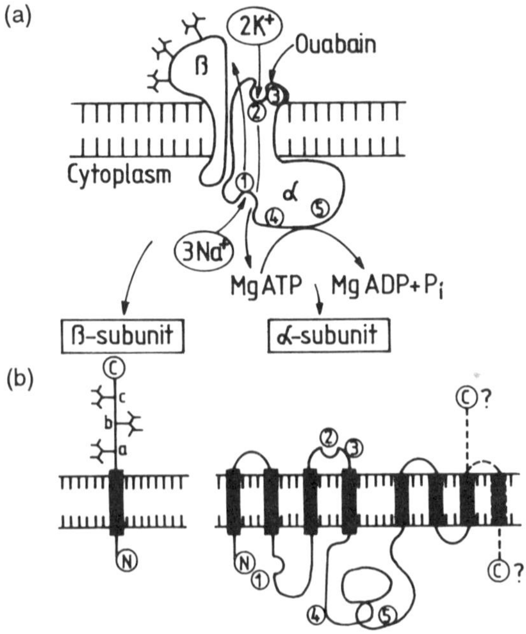

FIGURE 16. (a): General model for Na,K-ATPase: an alpha-beta heterodimer is schematically represented with five binding sites: the Na binding site (1), the K binding site (2), the ouabain binding site (3), the phosphorylation site (4), and the ATP binding site (5). (b): Model for the membrane insertion of the beta subunit (left) and the alpha subunit (right). (From Rossier, B. C., Geering, K., and Kraehenbuhl, J. P., *Trends Biochem. Sci.*, 12, 483, 1987. With permission.)

The slow conformational transitions of pump proteins have their counterpart in the gating processes of ionic channels. Many ion channels exhibit open-closed transitions in the time range of milliseconds. It is possible that the processes responsible for occlusion and deocclusion of ions in a pump are mechanistically similar to the closing and opening of a gate in a flickering ion channel.[58]

The large activation energy barriers, the erection and dismantling of which is controlled by the binding and debinding of specific ions, control the paths along which the cycle of pump activities occur. The movement of the ions opens and closes the barriers that control the chemical reactions of the phosphate ligands, while the progress of these reactions opens and closes the barriers to ion movements.[73]

V. CONCLUDING REMARKS

One can conclude that the specificity of channels and pumps is provided, beyond the steric reasons, by the ability of ions to open or close the energetic barriers which are present on their pathway through the membrane. A specific interaction between an ion and a transporter means that an energetically favorable process occurs.

In channels, the energetic barriers are lowered by ions which can interact with the selectivity filter, leading to a rearrangement of the protein, which allows diffusion to proceed. As a final result the free energy of the system: channel-ion diminishes.

In pumps which work against the transmembrane electrochemical gradients, two sorts of events make this antientropic transport possible. Energy (chemical, light, or redox) is used to change the affinity of the ion binding sites by modifying the various energetic barriers which control these sites. These barriers prevent ion escape from the wrong face. Reciprocally, the chemical reactions of the phosphate ligands (for P-ATPases) of the photocycle (bacteriorhodopsin) or of the electron transfer (cytochrome oxidase) are controlled by the presence of occluded ions in the pump.

REFERENCES

1. **Eisenberg, R. S.,** Channels as enzymes, *J. Membr. Biol.,* 115, 1, 1990.
2. **Flonta, M. L., Nistor, M., and Frangopol, P. T.,** Evaluation of the stimulatory capacity of procaine on Na transport through frog skin, *Arch. Int. Physiol. Biochem. Biophys.,* 99, 335, 1991.
3. **Li, J. H. Y., Palmer, G., Edelman, I. S., and Lindemann, B.,** The role of sodium-channel density in the natriferic response of the toad urinary bladder to an antidiuretic hormone, *J. Membr. Biol.,* 64, 77, 1982.
4. **Dixon, M. and Webb, E. C.,** *Enzymes,* Academic Press, New York, 1979, chap. 6.
5. **Koshland, D. E.,** Mechanism of transfer enzymes, in *The Enzymes,* Vol. 1, Boyer, P. D., Lardy, H., and Myrback, K., Eds., Academic Press, New York, 1959.
6. **Bone, R., Silen, J. L., and Agard, D. A.,** Structural plasticity broadens the specificity of an engineered protease, *Nature (London),* 339, 191, 1989.
7. **Mc-Cleskey, E. W. and Almers, W.,** The calcium channel in skeletal muscle is a large pore, *Proc. Natl. Acad. Sci. U.S.A.,* 82, 7149, 1985.
8. **Eisenman, G. and Alvarez, O.,** Structure and function of channels and channelogs as studied by computational chemistry, *J. Membr. Biol.,* 119, 109, 1991.
9. **Schoffeniels, E. and Mărgineanu, D. G.,** *Molecular Basis and Thermodynamics of Bioelectrogenesis,* Kluwer Academic Publishers, Dordrecht, 1990, chap. 2.
10. **Hodgkin, A. L.,** The ionic basis of electrical activity in nerve and muscle, *Biol. Rev.,* 26, 339, 1951.
11. **Läuger, P.,** Ion transport through pores: a rate-theory analysis *Biochem. Biophys. Acta,* 311, 423, 1973.
12. **Hille, B.,** *Ionic Channels of Excitable Membranes,* Sinauer Associates Inc., Sunderland, MA, 1984, chap. 10.
13. **French, R. J. and Adelman, W. Y.,** Competition, saturation and inhibition — ionic interactions shown by membrane ionic currents in nerve, muscle and bilayer systems, *Curr. Top. Membr. Transp.,* 8, 161, 1976.
14. **Eisenman, G. and Horn, R.,** Ionic selectivity revisited: the role of kinetic and equilibrium processes in ion permeation through channels, *J. Membr. Biol.,* 76, 197, 1983.

15. **Hille, B.,** Ionic selectivity, saturation and block in sodium channels. A four-barrier model, *J. Gen. Physiol.*, 66, 535, 1975.
16. **Heckmann, K.,** Single-file diffusion, in *Biomembranes,* Vol. 3, Passive Permeability of Cell Membranes, Kreuzer, F. and Slegers, J. F. C., Eds., Plenum Press, New York, 1972, 127.
17. **Läuger, P.,** Ion transport through pores. A rate-theory analysis, *Biochim. Biophys. Acta,* 311, 423, 1973.
18. **Hille, B., Courtney, K., and Dum, R.,** Rate and site of action of local anesthetics in myelinated nerve fibers, in *Molecular Mechanisms of Anesthesia, Progr. in Anesthesiology,* Vol. 1, Raven Press, New York, 1975.
19. **Eisenman, G.,** Cation selective glass electrodes and their mode of operation, *Biophys. J.*, 2(2), 259, 1962.
20. **Hille, B.,** Ionic selectivity of Na and K channels of nerve membranes, in *Membranes — A Series of Advances,* Vol. 3, Lipid Bilayer and Biological Membranes: Dynamic Properties, Eisenman, G., Ed., Marcel Dekker, New York, 1975, 255.
21. **Armstrong, C. M. and Benzanilla, F.,** Currents related to movement of the gating particles of the sodium channels, *Nature (London),* 242, 459, 1973.
22. **Läuger, P., Benz, R., Stark, G., Hamberg, E., Jordan, P. C., Fahr, A., and Brock, W.,** Relaxation studies of ion transport systems in lipid bilayer membranes, *Q. Rev. Biophys.*, 14, 513, 1981.
23. **Läuger, P.,** Microscopic calculation of ion-transport rates in membrane channels, *Biophys. Chem.*, 15, 89, 1982.
24. **Karplus, M. and Petsko, G. A.,** Molecular dynamics simulations in biology, *Nature,* 347, 631, 1990.
25. **Jones, T. A. and Liljas, L.,** Structure of satellite tobacco necrosis virus after crystallographic refinement at 2.5 Å resolution, *J. Mol. Biol.*, 177, 735, 1984.
26. **Stauffacher, C. V., Usha, R., Harrington, M., Schmidt, T., Hosur, M. V., and Johnson, J. E.,** The structure of cowpea mosaic virus at 3.5 Å resolution, in *Crystallography in Molecular Biology,* Moras, D., Drenth, J., Strandberg, B., Suck, D., and Wilson, K., Eds., Plenum Press, New York, 1987, 293.
27. **Yellen, G.,** Permeation in potassium channels: implications for channel structure, *Annu. Rev. Biophys. Chem.*, 16, 227, 1987.
28. **Montelius, I., Liljas, L., and Unge, T.,** Structure of EDTA-treated satellite tobacco necrosis virus at pH 6.5, *J. Mol. Biol.*, 201, 353, 1988.
29. **Lear, J. D., Wasserman, Z. R., and De Grado, W. F.,** Synthetic amphiphilic peptide models for protein ion channels, *Science,* 240, 1177, 1988.
30. **Oiki, S., Danho, W., Madison, V., and Montal, M.,** M2S, a candidate for the structure lining the ionic channel of the nicotinic cholinergic receptor, *Proc. Natl. Acad. Sci. U.S.A.*, 85, 8703, 1988.
31. **Ermishkin, L. N., Kasumov, K. M., and Potseluyev, V. M.,** Properties of amphotericin B channels in a lipid bilayer, *Biochim. Biophys. Acta,* 470, 357, 1977.
32. **Perutz, M. F.,** Mechanisms of cooperativity and allosteric regulation in proteins, *Q. Rev. Biophys.*, 22, 139, 1989.
33. **Barnard, E. A., Darlison, M. G., and Seeburg, P.,** Molecular biology of the $GABA_A$ receptor: the receptor/channel superfamily, *Trends Neurosci.*, 10, 502, 1987.
34. **Bormann, J., Hamil, O. O., and Sakmann, B.,** Mechanism of anion permeation through channels gated by glycine and γ-aminobutyric acid in mouse cultured spinal neurons, *J. Physiol. (London),* 385, 243, 1987.
35. **Dani, J. A. and Eisenman, G.,** Monovalent and divalent cation permeation in acetylcholine receptor channels. Ion transport related to structure, *J. Gen. Physiol.*, 89, 959, 1987.
36. **Aqvist, J. and Warshel, A.,** Energetics of ion permeation through membrane channels — solvation of Na^+ by Gramicidin-A, *Biophys. J.*, 56, 171, 1989.

37. **Furois-Corbain, S. and Pullman, A.,** A possible model for the inner wall of the acetylcholine receptor channel, *Biochim. Biophys. Acta,* 984, 339, 1989.
38. **Eisenman, G.,** Cation selective glass electrodes and their mode of operation, *Biophys. J.,* 2, 259, 1962.
39. **Quiocho, F. A., Sack, J. S., and Vyas, N. K.,** Stabilization of charges on isolated ionic groups sequestered in proteins by polarized peptide units, *Nature (London),* 329, 561, 1987.
40. **Montal, M.,** Channel protein engineering, in *Ion Channels, Vol. 2,* Narahashi, T., Ed., Plenum Press, New York, 1990, 251.
41. **Palmer, L. G.,** Ion selectivity of the apical membrane Na^+ channel in the toad urinary bladder, *J. Membr. Biol.,* 67, 91, 1987.
42. **Smith, P. R. and Benos, D. J.,** Epithelial Na^+ channels, *Annu. Rev. Physiol.,* 53, 509, 1991.
43. **Palmer, L. G.,** Epithelial Na^+ channels: the nature of the conducting pore, *Renal Physiol. Biochem.,* 13, 51, 1990.
44. **Garty, H. and Benos, D. J.,** Characteristics and regulatory mechanisms of the amiloride-blockage Na^+ channel, *Physiol. Rev.,* 68, 309, 1988.
45. **Hamilton, K. L. and Eaton, D. C.,** Single-channel recordings from amiloride-sensitive epithelial sodium channel, *Am. J. Physiol.,* 249(C), 200, 1985.
46. **Eaton, D. and Hamilton, K. L.,** The amiloride-blockable sodium channel of epithelial tissue, in *Ion Channels,* Vol. 1, Narahashi, T., Ed., Plenum Press, New York, 1988.
47. **Wills, N. K. and Zweifach, A.,** Recent advances in the characterization of epithelial ionic channels, *Biochim. Biophys. Acta,* 906, 1, 1987.
48. **Turnheim, K.,** Intrinsic regulation of apical sodium entry in epithelia, *Physiol. Rev.,* 71, 429, 1991.
49. **Flonta, M. L., Galter, D., Frangopol, P. T., and Mărgineanu, D. G.,** Procaine has opposite effects on passive Na and K permeabilities in frog skin, *Eur. J. Physiol.,* 408, 215, 1987.
50. **Flonta, M. L., Endstrasser, W., Kirmeyer, I., and Nagel, W.,** Procaine effects on sodium and chloride transport in frog skin, *Eur. J. Physiol.,* 411, 386, 1988.
51. **Lewis, S. A. and Donaldson, P. J.,** Sodium dependence of cation permeabilities and transport, *Curr. Top. Membr. Transp.,* 34, 83, 1989.
52. **Van Driesshe, W. and Zeiske, W.,** Ionic channels in epithelial cell membranes, *Physiol. Rev.,* 65, 833, 1985.
53. **Hess, P. and Tsien, R. W.,** Mechanism of ion permeation through calcium channels, *Nature (London),* 309, 453, 1984.
54. **Hosey, M. M. and Lazdunski, M.,** Calcium channels: molecular pharmacology, structure and regulation, *J. Membr. Biol.,* 104, 81, 1988.
55. **Stoeckenius, W. and Bogomolni, R. A.,** Bacteriorhodopsin and related pigments of halobacteria, *Annu. Rev. Biochem.,* 52, 587, 1982.
56. **Oesterhelt, D. and Tittor, J.,** Two pumps, one principle: light-driven ion transport in halobacteria, *Trends Biochem. Sci.,* 14, 57, 1989.
57. **Krab, K. and Wikstrom, M.,** Principles of coupling between electron-transfer and proton translocation with special reference to proton-translocation mechanisms in cytochrome oxidase, *Biochim. Biophys. Acta,* 895, 25, 1987.
58. **Läuger, P.,** *Electrogenic Ion Pumps,* Sinauer Associates, Sunderland, MA, 1991, chap. 2.
59. **Senior, A. E.,** The proton-translocating ATPase of *Escherichia coli, Annu. Rev. Biophys. Chem.,* 19, 7, 1990.
60. **Penefsky, H. S.,** Mechanism of action of the mitochondrial proton pumping ATPase in ATP synthesis and hydrolysis, in *The Ion Pumps, Structure, Function and Regulation,* Stein, W. D., Ed., Alan R. Liss, New York, 1988, 261.

61. **Stein, W. D. and Läuger, P.,** Kinetic properties of F_0F_1-ATPases: theoretical predictions from alternating-site models, *Biophys. J.*, 57, 255, 1990.
62. **Forgac, M.,** Structure and function of vacuolar class of ATP-driven proton pumps, *Physiol. Rev.*, 69, 765, 1989.
63. **Pedersen, P. L. and Carafoli, E.,** Ion motive ATPases. I. Ubiquity, properties and significance to cell function, *Trends Biochem. Sci.*, 12, 146, 1987.
64. **Pedersen, P. L. and Carafoli, E.,** Ion motive ATPases. II. Energy coupling and work output, *Trends Biochem. Sci.*, 12, 186, 1987.
65. **Inesi, G. and Kirtley, M. E.,** Coupling of catalytic and channel function in the Ca^{2+} transport ATPase, *J. Membr. Biol.*, 116, 1, 1990.
66. **Horisberger, J. D., Lemas, V., Kraehenbuhl, J. P., and Rossier, B. C.,** Structure-function relationship of Na,KATPase, *Annu. Rev. Physiol.*, 53, 565, 1991.
67. **Geering, K.,** Subunit assembly and functional maturation of Na,-K-ATPase, *J. Membr. Biol.*, 115, 109, 1990.
68. **Inesi, G., Sumbilla, C., and Kirtley, M. E.,** Relationship of molecular structure and function in Ca^{2+}transport ATPase, *Physiol. Rev.*, 70, 749, 1990.
69. **Gerencser, G. A., White, J. F., Gradmann, D., and Bonting, S. L.,** Is there a Cl^- pump?, *Am. J. Physiol.*, 255, R677, 1988.
70. **Rossier, B. C., Geering, K., and Kraehenbuhl, J. P.,** Regulation of the sodium pump: how and why?, *Trends Biochem. Sci.*, 12, 483, 1987.
70a. **Post, R. L., Hegyvary, C., and Kume, S.,** Activation by adenosine triphosphate in the phosphorylation kinetics of sodium and potassium ion transport adenosine triphosphatase, *J. Biol. Chem.*, 247, 6530, 1972.
70b. **Albers, R. W.,** Biochemical aspects of active transport, *Annu. Rev. Biochem.*, 36, 727, 1967.
71. **Yoda, A. and Yoda, S.,** Two different phosphorylation-dephosphorylation cycles of Na,K-ATPase proteoliposomes accompanying Na^+ transport in the absence of K^+, *J. Biol. Chem.*, 262, 110, 1987.
72. **Jørgensen, P. L. and Andersen, J. P.,** Structural basis for E_1-E_2 conformational transitions in Na,K-pump and Ca-pump proteins, *J. Membr. Biol.*, 103, 95, 1988.
73. **Stein, W. D.,** Energetics and the design principles of the Na/K-ATPase, *J. Theor. Biol.*, 147, 145, 1990.
74. **Tanford, C.,** Mechanism of free energy coupling in active transport, *Annu. Rev. Biochem.*, 52, 379, 1983.
75. **Mitchell, P.,** Molecule, group and electron translocation through natural membranes, *Biochem. Soc. Symp.*, 22, 142, 1963.
76. **Mitchell, P.,** Compartimentation and communication in living systems. Ligand conduction: a general catalytic principle in chemical, osmotic and chemiosmotic reaction systems, *Eur. J. Biochem.*, 95, 1, 1979.
77. **Scarborough, G. A.,** A chemically explicit model for the molecular mechanism of the $F_0F_1H^+$-ATPase/ATP synthases, *Proc. Natl. Acad. Sci. U.S.A.*, 83, 3688, 1986.
78. **Läuger, P.,** A channel mechanism for electrogenic ion pumps, *Biochim. Biophys. Acta*, 552, 143, 1979.
79. **Klingenberg, M.,** Membrane protein oligomeric structure and transport function, *Nature*, 290, 449, 1981.
80. **Tanford, C.,** Translocation pathway in the catalysis of active transport, *Proc. Natl. Acad. Sci. U.S.A.*, 80, 3701, 1983.
81. **Jennings, M. L.,** Evidence for an access channel leading to the outward-facing substrate site in human red blood cell band 3, in, *Anion Transport Protein of the Red Blood Cell Membrane,* Hamasaki, N. and Jennings, M. L., Eds., Elsevier, Amsterdam, 1989, 59.
82. **Mitchell, P. and Moyle, J.,** The mechanism of proton translocation in reversible proton-translocating adenosine triphosphatases, *Biochem. Soc. Spec. Publ.*, 4, 91, 1974.

83. **Glynn, I. M. and Karlish, S. J. D.,** Occluded cations in active transport, *Annu. Rev. Biochem.,* 59, 171, 1990.
84. **Tanford, C.,** Simple model for the chemical potential change of a transported ion in active transport, *Proc. Natl. Acad. Sci. U.S.A.,* 79, 2882, 1982.
85. **Henderson, R.,** The purple membrane from *Halobacterium halobium, Annu. Rev. Biophys. Bioeng.,* 6, 87, 1977.
86. **Ottolenghi, M. and Sheves, M.,** Synthetic retinals as probes for the binding site and photoreaction in rhodopsin, *J. Membr. Biol.,* 112, 193, 1989.
87. **Boyer, P. D., Chance, B., Ernster, L., Mitchell, P., Racker, E., and Slater, E. C.,** Oxidative phosphorylation and photophosphorylation, *Annu. Rev. Biochem.,* 46, 955, 1977.
88. **Fodor, S. P. A., Ames, J. B., Gebhard, R., van der Berg, E. M. M., Stoeckenius, W., Lugtenburg, J., and Mathies, R. A.,** Chromophore structure of bacteriorhodopsin's N intermediate: implications for the proton-pumping mechanism, *Biochemistry,* 27, 7097, 1988.
89. **Dunham, P. B. and Hoffman, J. F.,** Na and K transport in red blood cells in *Physiology of Membrane Disorders,* Andreoli, T. E., Hoffman, J., F., and Fanestil, D. D., Eds., Plenum Press, New York, 1978, 255.
90. **Gache, C., Rossi, B., Leone, F. A., and Lazdunski, M.,** Pseudo-substrates to analyze the reaction mechanism of the Na,KATPase, in *Na, K-ATPase. Structure and Kinetics,* Skou, J. C. and Nørby, J.-G., Eds., Academic Press, London, 1979, 301.
91. **Polvani, C. and Blostein, R.,** Protons as substitutes for sodium and potassium in the sodium pump reaction, *J. Biol. Chem.,* 263, 16757, 1988.
92. **Homareda, H., Nozaki, T., and Matsui, H.,** Interaction of sodium and potassium ions with Na,K-ATPase. III. Cooperative effects of ATP and Na^+ on complete release of K^+ from E_2K, *J. Biochem.,* 101, 789, 1987.
93. **Stein, W. D.,** *Transport and Diffusion Across Cell Membranes,* Academic Press, Orlando, FL, 1986.
94. **Simons, T. J. B.,** Potassium:potassium exchange catalyzed by the sodium pump in human red cells, *J. Physiol. (London),* 237, 123, 1974.
95. **Mezele, M., Lewitzki, E., Ruf, H., and Grell, E.,** Cation selectivity of membrane proteins, *Ber. Bunsenges. Phys. Chem.,* 92, 988, 1988.
96. **Glynn, I. M.,** The Na^+,K^+-transporting adenosine triphosphatase, in *The Enzymes of Biological Membranes,* Vol. 3, 2nd ed., Martonosi, A. N., Ed. Plenum Press, New York, 1985, 35.
97. **Glynn, I. M.,** The coupling of enzymatic steps to the translocation of sodium and potassium, in *The Na^+,K^+-Pump, Part A: Molecular Aspects,* Skou, J. C., Nørby, J. G., Maunsbach, A. B., and Esmann, M., Eds., Alan R. Liss, New York, 1988, 435.
98. **Pedemonte, C. H.,** Kinetic mechanism of inhibition of the Na^+-pump and some of its partial reactions by external Na^+ (Na^+_0), *J. Theor. Biol.,* 134, 165, 1988.
99. **Karlish, S. J. D. and Stein, W. D.,** Cation activation of the pig kidney sodium pump: transmembrane allosteric effects of sodium, *J. Physiol. (London),* 359, 119, 1985.
100. **Forbush, B., III,** Occluded ions and Na,K-ATPase, in *The Na^+,K^+-Pump, Part A: Molecular Aspects,* Skou, J. C., Nørby, J. G., Maunsbach, A. B., and Esmann, M., Eds., Alan R. Liss, New York, 1988, 229.
101. **Forbush, B., III,** Na^+,K^+ and Rb^+ movements in a single turnover of the Na/K pump, *Curr. Top. Membr. Transp.,* 28, 19, 1987.
102. **Shani-Sekler, M., Goldshleger, R., Tal, D. M., and Karlish, S. J. D.,** Inactivation of Rb^+ and Na^+ occlusion on (Na^+,K^+)-ATPase by modification of carboxyl groups, *J. Biol. Chem.,* 263, 19331, 1988.
103. **Karlish, S. J. D., Goldshleger, R., and Stein, W. D.,** A 19-kDa C-terminal tryptic fragment of the alpha chain of Na/K-ATPase essential for occlusion and transport of cations, *Proc. Natl. Acad. Sci. U.S.A.,* 87, 4566, 1990.

104. **Hara, Y. and Nakao, M.,** ATP-dependent proton uptake by proteoliposomes reconstituted with purified Na^+,K^+-ATPase, *J. Biol. Chem.*, 261, 12655, 1986.
105. **Boyer, P. D.,** Bioenergetic coupling to proton motive force: should we be considering hydronium ion coordination and not group protonation?, *Trends Biochem. Sci.*, 13, 5, 1988.
106. **Green, N. M.,** Ions, gates, and channels, *Nature,* 339, 424, 1989.
107. **Clarke, D. M., Loo, T. W., Inesi, G., and MacLennan, D. H.,** Location of high affinity Ca^{2+}-binding sites within the predicted transmembrane domain of the sarcoplasmic reticulum Ca^{2+}ATPase, *Nature,* 339, 476, 1989.
108. **Pedemonte, C. H. and Kaplan, J. H.,** Chemical modification as an approach to elucidation of sodium pump structurefunction relations, *Annu. J. Physiol.*, 258, Cl, 1990.
109. **Taylor, W. R. and Green, N. M.,** The predicted secondary structures of the nucleotide-binding sites of six cation-transporting ATPases lead to a probable tertiary fold, *Eur. J. Biochem.*, 179, 241, 1989.

Chapter 8

MOLECULAR RECOGNITION DURING LIGAND-RECEPTOR INTERACTIONS. APPLICATIONS IN RECEPTOR MAPPING

Tudor Ionel Oprea and Ludovic Kurunczi

TABLE OF CONTENTS

0-8493-5398-X/93/$0.00 + $.50

I. INTRODUCTION

Ligand-receptor interaction can be studied using thermodynamics and pharmacology (macroscopic methods), but such methods do not give information about individual molecules. There are no experimental data about the intimate behavior of ligands and receptors during molecular recognition processes and docking at the receptor binding site. It is difficult to study experimentally the manner by which the ligand approaches the receptor, is specifically recognized, and eventually anchored at the binding site. Technical limits do not allow observation of processes such as:

1. The precise effect of the ligand upon the receptor as the sole investigated phenomenon, excluding connected processes such as *in situ* catabolization of the ligand, nonspecific binding or passage of the ligand through the membrane (if it is lipid soluble), and even the interaction of the ligand with other receptor (sub)types (e.g., cyproheptadine[1] is an equipotent blocker on three different membrane receptors: histaminic H_1, serotoninic 5-HT_2, and muscarinic M_1, respectively).
2. *In vitro* pharmacological techniques that partly overcome such problems measure some, but not all biological effects triggered by ligand binding. It is technically difficult to monitor all intracellular changes, but it is known that agonists do trigger more than one effector pathway. Stimulation of beta adrenergic receptors is followed by intracellular cAMP (cyclic adenosine monophosphate) accumulation, yet chlorine secretion is a non-cAMP dependent beta adrenergic mediated process.[2]
3. Structural data accumulated via X-ray crystallography[3] and NMR (with different techniques,[4] such as 2D-NMR, nOe-NMR, *in vivo* NMR) for both ligands and receptors are not yet able to investigate ligand-receptor interactions at the atomic level. Data available on cocrystallized ligands and enzymes offer some static information, but dynamic processes are not yet detectable.

Therefore, the study of ligand receptor interactions at the molecular level is possible only with theoretical methods, using available experimental data to hypothesize and speculate upon such processes. This chapter aims to investigate molecular recognition processes during ligand receptor interaction. Later, we briefly discuss the approach of molecular modeling techniques in an attempt to define the receptor binding site (receptor mapping). A model for the dipole-dipole interaction is proposed, and its use in receptor mapping is shown.

For pharmacological aspects, the reader is referred to basic pharmacology textbooks (e.g., Taylor and Insel[5]). Thermodynamic (theoretical and practical) aspects of the ligand-receptor-transducer complex are investigated elsewhere (Leff and Harper[6]). Electronic interactions between the ligand and the receptor at the binding site were thoroughly examined (Menekyan and Bonchev[7]).

II. DEFINITIONS

The receptor is a macromolecule specialized for the recognition of one and only one molecular pattern[8] from the large number of molecular species with which it could interact. The term receptor[3] in biology also comprises in its definition receptor stimulation by its specific signal molecule. This is followed by a specific biological response of the cell (organ) containing the receptor. The mass of the receptor may reach 10^5 Da (or more) with a diameter of about 20 to 50 Å.

The signal molecule is generically called a ligand. Ligands are usually micromolecules[9] with masses up to 10^3 Da and diameters of 10 Å, excepting some polypeptidic hormones that are dimensionally comparable with the receptor. Ligands can be of endogenous origin (physiological mediators produced by specialized structures of the organism, e.g., hormones, prostaglandins, etc.), or exogenous (e.g., pharmacological agents).

Molecular recognition[10] is the process that occurs when any two molecules approach each other, before any chemical reaction takes place. This process involves alterations in the movement of both electrons and nuclei from any given molecule, enforced by the external perturbative field of the other molecule. Molecular recognition becomes very important during stereoselective processes.

It should be emphasized that receptors are stereoselective with respect to the approaching ligand,[5,8] although the subject is not addressed in this chapter. For example, optical isomers of known mediators like adrenaline have different pharmacological effects, R(−) adrenaline being 100 times more potent than S(+) adrenaline. The explanation of a stereoisomeric binding site was initially suggested by Easson and Steadman in 1933, (cited in Reference 5), and even receptors for natural ligands with no optical activity (e.g., dopamine and acetylcholine) are stereoselective for configuration isomers of synthetic compounds.

III. STAGES OF THE MOLECULAR RECOGNITION PROCESS

From the molecular point of view, the interaction of the ligand with the receptor could be described in several steps, which occur in a relatively well-delimited succession.

A. THE ELECTROSTATIC INTERACTION

The ligand approaches the receptor — the first step in molecular recognition — when the electrostatic attractive/repulsive Coulomb forces dominate and the ligand is reorientated about the receptor. For distances from 200 to 10 Å these long-range forces are decisive, being the first type of force chronologically involved in molecular recognition. Two examples that suggest the importance of coulomb forces are shown below.

A molecular modeling analysis of the electronic volume of the phosphodiesterase inhibitor SKF 95305 and of the natural substrate cAMP (structurally dissimilar compounds) shows that the inhibitor displays similar electrostatic minima, concerning intensity and position, with one of the most stable conformers of cAMP. This can explain the inhibitory activity of the compound SKF 95305 (Davis et al.[11]).

A molecular modeling study of the natural substrate 7,8-dihydrofolate and the synthetic inhibitor methotrexate, both active ligands on DHFR (dihydrofolate reductase), revealed that in spite of the good structural fit obtained by superposition of the nitrogen atoms of the two compounds, a detailed analysis of the distribution of the electronic volume of the molecules suggests the opposite structural fit in the binding site. By matching two-dimensional electrostatic isopotential maps across the aromatic planes, the position of methotrexate in the enzymatic binding site is upside down compared to the natural substrate (overlay with minima upon minima positive and negative regions). This was confirmed later by X-ray crystallographic studies (Vinter[12]).

B. THE LOSS OF HYDRATION SPHERES

The progressive loss of hydration spheres (of the more or less ordered water molecules surrounding the ligand) as the ligand approaches closer to the receptor (from 100 to 10 Å) is the second step in molecular recognition, when probably a conformational filtering process takes place under the influence of the receptor electrostatic potential (magnitude[13] of about 10^6 V/m) and of the permanent dipole moment of the receptor (about 10^3 D). The filtration step is described in more detail in Section VI.

The loss of hydration speres is considered a hydrophobic interaction[14] by which the exhibited contact surface between water and the nonpolar regions of the ligand and the receptor molecules is minimized. For the interaction equilibrium, this phenomenon presents a decisive entropic component that can be semiquantitatively estimated using the approach proposed by Williams et al.[15]

C. THE STERIC SELECTION

In a simultaneous phase the steric selection of the ligands (which may or may not fit the receptor cavity) is accomplished — this being the third step in molecular recognition. The steric interaction could actually occur at any moment during the approach of the ligand to the receptor (probably, more than one steric barrier with filtration function exists).

In ligand-receptor interaction the decisive[16] role of steric effects have been recognized, and in the QSAR and drug design techniques they are used as steric parameters. The MTD — Minimal Steric (Topologic) Difference[17] — method is an example of such a steric parameter, which can be applied to quantify steric effects.

In the docking process of the ligand on the specific receptor site there are "forbidden zones"[18] imposed by the nature of the receptor itself. The ligand molecule is (partly) included in the receptor cavity, being forced to match the receptor using only certain regions of its configurational space to contact the receptor atoms. The MTD parameter quantifies only steric interaction for series of ligands. More complex is the comparative molecular field analysis (CoMFA) method, which analyzes both steric and van der Waals interactions of the ligand with the receptor.[19] CoMFA is also used for series of ligands.

D. ELECTRONIC AND CONFORMATIONAL REARRANGEMENTS

The electronic redistribution at the level of the ligand which adapts itself to the receptor microenvironment is the final step in molecular recognition during which the ligand gradually binds to the receptor. The atoms of the ligand lose degrees of freedom, and these modifications have an impact upon the molecular orbitals. The polarization[20] phenomenon occurs when valence electrons redistribute due to perturbations induced by the atoms of the receptor and by the lack of contact with the water (resulting in alterations of electronic densities in the contact areas and modifications in the zero and superior electric moments). Conformational changes in the ligand occur under the influence of electrostatic and steric effects exerted by the receptor itself, and these are variable according to the nature of the ligand.

Theoretical chemistry studies concerning ligand-receptor electronic interaction[7] correlated the reaction centers (monocentric . . . multicentric) with biological effects, showing that multicentric interactions are more frequent and that they proceed in more steps.

Bacteriorhodopsin[21] (bR) — a photosensitive membrane protein with proton pump function, is used as a model structure[22,23] for a transmembrane G-protein coupled receptor (GPCR), although light is the natural agonist of bR as opposed to other GPCRs, whereas chemical compounds are agonists to other GPCRs. Two-photon double resonance spectroscopy studies[24] had revealed that due to the transition of bR from the inactive to the activated state a dipole moment alteration (between 9.1 and 13.5 D) is produced — namely a redistribution of the electrons from the high-energy orbitals. A strong electrostatic (likely ionic) interaction was detected between the chromophore binding site of bR and the β-ionylidine moiety (part of the retinal implicated in the transduction mechanism). The intervention of hydrogen bonds was excluded in this region as probably not implicated in the proton pumping activity.

The proton transfer mechanism along the hydrogen bonds of bR has been hypothetically analyzed[25] with *ab initio* methods, by calculating the proton migration in supermolecular systems (e.g., the tetramer cation $[H_2OH_2O\text{-}H\text{-}OH_2OH_2]^+$) with a rigorous quantum-chemical formalism. The results have revealed the sensitivity of the energy barrier for proton transfer to the precise geometry of the hydrogen bond. This might be used to modulate the rate of proton transfer by conformational changes of bR. The proton pump function might even have directionality due to the differences in basicity between the hydroxyl and amino groups (also preventing proton "leaks").

The nature of the hydrogen bonds has been understood, but their effects upon the structured water (a feature of other macroscopic systems as well) had not been completely elucidated. Thus, among other effects[26] of hydrogen bonding, the stretching of the O-H bond (by approximately 2.3% of the bond length found in gaseous water) and alteration of the bond dipole moment orientation (dipole moment vector deviated by 25.2° from the H-O axis) can be observed.

At the stage of ligand docking in the binding site, "partial" transition states of the ligand may occur.[27] An alteration of the frontier orbital electronic density of the ligand could be obtained without reaching a *true* transition state (i.e., no chemical reaction with the receptor occurs). Thus a high electronic density localized in a small volume could produce, via electrostatic interactions, conformational alterations of the receptor — triggering, for instance, the achievement of the active receptor conformation by setting up the activated ligand-receptor complex (agonism).

IV. INTERACTIONS BETWEEN THE LIGAND AND THE RECEPTOR BINDING SITE

The following section presents a strictly qualitative discussion of the forces involved in the binding of the ligand to the receptor. We make a distinction[28] between long-range and short-range forces. The distance dependence defines the radius of action: the energies which are functions of r^{-p} (r being the distance) correspond to long-range forces if $p \leq 3$ (e.g., ion-ion or ion-dipole interactions) and to short-range forces if $p > 3$ (e.g., Lennard-Jones attractions and repulsions), respectively. This "natural" division is obtained by integrating the energy (for simple pure media), which diverges for $p \leq 3$, and converges for $p > 3$, respectively (Hill, 1960, cited by Dill[28]). Besides this distinction, the differentiation between local and nonlocal forces is also useful. For macromolecular systems local forces are interactions which occur between chemically bonded atoms (1–2 vicinity) and between immediate neighbors (1–3, 1–4, and 1–5 interactions). The nonlocal forces are interactions between significantly remote but topologically close groups in the three-dimensional macromolecular structure (e.g., in a globular protein, atoms from different amino acid residues which are hundreds of residue-units apart in the primary sequence, could, in the tertiary structure, be within the hydrogen bond range, less than 3 Å in space). Local and nonlocal forces can be short-range or long-range as well. During the ligand-receptor interactions nonlocal forces could play an important role in delimitation of the binding site (from the pharmacophoric point of view).

The van der Waals forces are not discussed here because of their ubiquitous and transient character. They appear between all the atoms of the molecular systems involved in the discussed process: between ligand and solvent, receptor and solvent, and ligand and receptor. The van der Waals forces (permanent dipole-induced dipole and induced dipole-induced dipole interactions) are dependent on r^{-6} (short-range forces) and thus they might be important only in the close contact phase.

A. THE IONIC FORCES

The interaction between the ligand net charge and preexisting counterions in the receptor site are ionic in nature. These forces are present in the case of biogenic amines due to the amino groups, protonated at the physiologic pH of 7.4. The positive charges probably interact with the aspartate[29] or glutamate carboxyl moieties that carry an overall unit negative charge, although changes in the pK of the intrareceptor carboxyl moieties have been reported (Hiley, 1974, cited by Burgen[29]), since they are exposed to a hydrophobic environment unlike those at the interface with water molecules. Structured water (with a substantially reduced dielectric constant compared to disorganized water[30]) is involved in the weaker ionization of the aspartate.

Most likely, these ionic interactions do not have the strength of a crystal Na^+Cl^- interaction. In the discussed case electric charges are lower (they have subunitary values) because the electrons are redistributed among the surrounding molecular orbitals of the vicinal atoms (both in the ligand and in the receptor). Thus, charges are more diffuse. At the same time, they are also "shielded" through nonlocal interactions (with the topologic neighbors), whose orbitals are close by and effectively contribute to the minimization of the perturbation produced by the charges.

Molecular biology studies have directly confirmed the ionic interaction (previously noted also for bR[24]), showing by site-directed mutagenesis studies the requirement for the aspartate residue in this kind of interaction[22] (Asp^{113} on the α-helix III of the human β_2 adrenergic receptor). As this amino acid residue is conserved in the GPCR, it was selected for mutagenesis and replaced by asparagine. The agonist efficacy of the mutant receptors dropped 10^5 times, being partially restored after the reappearance of the negative charge[31] by replacing asparagine with glutamate (yet being 10^3 times lower).

Molecular modeling studies performed after 1988 used this result in the simulation of the ligand (protonated amine) receptor (represented by Asp^{113} or by another potentially ionic amino acid residue) interaction (e.g., Linschoten et al.[32]). Asp^{113} was modeled as a key residue both in the agonist[33] and antagonist[34] β-adrenergic receptor binding site — which suggests a common binding region for both categories of ligands.

B. THE COVALENT BONDS

These are short-range forces ($r \leq 2$ Å in the discussed type of compounds) which make the entire process *reversible* (if they are missing) or *irreversible* (if they are formed). The use of irreversible ligands as a technique for controlled sequential inactivation has become a classical pharmacological method to measure intrinsic efficacy.[8] By using such ligands (usually blockers), "permanent" chemical bonds are established with the receptor — this method is being used in the exploration of the receptor binding site.

For example, the irreversible β_2 adrenergic blocker ^{125}I-*para*(bromoacetamido)-benzyl-1-iodocarazolol, is covalently bound to an α helix II site, Asp^{79} being probably implicated in agonist binding (Dohlman et al., 1987, cited in Reference 22).

C. THE HYDROGEN BONDS

Hydrogen bonds are short-range forces which require several conditions:[35] (1) *electronic* — ground state and closed shell systems for both molecules; (2) *chemical* — the donor atom must be quite electronegative and of low volume, e.g., O or F, while the acceptor must be electronegative as well, with an available nonbonding pair of electrons; and (3) *geometric* — acceptor-hydrogen-donor angles of about 120° and hydrogen-acceptor distances between 1.8 and 3 Å. The electronic perturbations in both molecules are much lower compared to the covalent bond.

Stouten and Kroon[36] statistically investigated 45 different compounds containing alcoholic O-H moieties (196 analyzed hydrogen bonds, in which each O-H moiety is only once a donor and only once an acceptor). The resultant oxygen-oxygen distance histogram was well interpreted by using the same molecular pairs considered as dimers in the gaseous state, with a Boltzmann distribution at $T = 298$ K. The results were successfully simulated by molecular dynamics using Lennard-Jones type potentials. The results (in concordance with experimental data) show that during the simulation (closer to the dynamic reality), hydrogen bonds have no fixed character (the H-acceptor distance varied between broader limits than in the crystals — between 2.6 and 3.2 Å — but the mean value is within that observed in the crystal, i.e., 2.8 Å), they are not permanent (the hydrogen bonds are formed and broken with a temperature-dependent mean lifetime), and that single bonds are preferred for the given simulation (90% of the potential acceptors are implicated in only one hydrogen bond, 1% in two simultaneous bonds, and 9% are not involved — as observed in ice).

Hydrogen bonds between the ligand and receptor have a reciprocal character, being most frequently invoked in anchoring the ligand at the receptor binding site. They appear between the hydroxyl, fluoro-, etc., moieties of the ligand, and similar moieties (hydroxyl and amino) on the receptor side (e.g., serine residues). The energetically optimum hydrogen acceptor distance is between 1.8 and 2 Å. For the computer-assisted modeling of these interactions much broader limits are used:[37] hydrogen-acceptor distances $\leq$3.25 Å and donor-H-acceptor angles above 90°, directed towards the nonbonding electron pair.

D. DIPOLE-DIPOLE INTERACTIONS

Such forces[3] appear between the ligand permanent dipole moment and the total dipole moment of the receptor (this interaction diminishes with the inverse third power of the distance,[20] being effective only at very short distances). Dipole-dipole interactions are likely to occur between the ligand and the "local" dipole of some clearly delimited groups from the receptor site. Because of their short-range action, these forces might influence the orientation of the ligand in the receptor site and the mobility of the amino acid residues present at the contact zone. A hypothetic dipole-dipole interaction model is proposed in Section VI.

E. THE HYDROPHOBIC FORCES

These are interactions which protect the nonpolar groups of the both ligand and receptor against the influence of the surrounding water (outside the receptor). The hydrophobic effect means[38] transfer of nonpolar solutes into an aqueous solution when characteristic temperature dependence is observed, or in a larger sense, any transfer of a nonpolar solute to any aqueous solution. The nonpolar compound solvation is opposed by entropy (unlike the solvation of polar products, which is dominated by enthalpy). The most distinctive characteristic[39] of hydrophobicity is the unusually large heat exchange capacity. This feature probably offers the energy required for protein folding,[28] by burying nonpolar fragments in nonpolar cores (of hydrophobic nature).

In a theoretical study of applied thermodynamics[40] it is stated that the nature of hydrophobic forces is not well established, and that the belief that hydrophobic interactions are dominant in biology (understanding by this the "tendency of nonpolar groups to adhere to each other in polar solvents") is exaggerated. The author examined the hydrophobic and hydrophilic interactions and argues that the hydrophobic one — as a solvent-induced interaction — is practically negligible, especially during intermolecular processes. On the other hand, the hydrophilic interaction (which includes hydrogen bonds and solvent-induced interactions between polar groups in polar solvents) seems to be an important contributor to the total energy of the system, particularly in the case of intramolecular forces. These later forces might be more important in biologic systems than the hydrophobic interactions. The "aromatic cage" interaction (*vide infra*) seems to have a stronger influence in anchoring the ligand (polar) in the nonpolar surroundings.

The reason for overestimating hydrophobic forces probably has emerged from the studies initiated by Overton and Meyer (cited in Reference 5), who demonstrated in 1901 that the anesthetic power can be directly correlated with the water/oil partition coefficient of the anesthetic compound. This discovery revealed the importance of hydrophobic forces at the interface between the polar solvent (water) and the nonpolar one (lipid membrane), but has undoubtedly influenced the approach to the nature and contribution of the forces involved during ligand-receptor interaction.

The correlational model has been extended by Hansch and Leo,[41] who introduced the hydrophobicity constant π (the octanol/water partition coefficient) in the QSAR approach. The major physical factors contributing to the π constant[42] are (1) a bulk factor (the molecular weight or the van der Waals volume of the ligand); (2) an electronic factor (the fragmental dipole moment, introduced by Lien et al.[43]); and (3) a factor that estimates the hydrogen bonds (through correlation with the mean strength of the hydrogen bonds between the groups and the number of these bonds offered/accepted by a ligand). Since the first two factors can be measured exactly, but the third one can only be estimated, the authors consider that the results of the correlational equations ($r^2 \doteq 0.8$) suffered. Nevertheless, the negative coefficient of the dipole moment and of the hydrogen bond term show the detrimental contribution of these two types of interactions to the hydrophobic forces.

F. THE AROMATIC CAGE

These are cation–pi-electron interactions between positively ionized groups of the ligand and groups with an aromatic character from the receptor site. NMR studies[44] have revealed that compounds of the $R\text{-}N^+Me_3$ type are preferentially included in the cavity of a synthetic receptor (in which six aromatic groups are disposed in a cylinder-like fashion), unlike the isosteric $R\text{-}CMe_3$ compounds, in spite of better solvation energy for quaternary ammonium ions than for neutral tert-butyl compounds in the water phase, and in spite of considerable desolvation involved by this interaction. A Monte Carlo-type calculation based on the perturbation theory estimated the magnitude of this effect to be $\Delta\Delta G^0_{sol} = -65$ kcal/mol (with R as phenyl or methyl). The aromatic cage forces seem powerful enough to compete with the ionic forces, and in the above-mentioned study[44] they are seen as the dominant interaction in the bonding of acetylcholine to the receptor site. (The aromatic cage is constituted by aromatic moieties rich in electrons — like tryptophan and tyrosine residues — which completely surround the quaternary ammonium ion.)

This arrangement has been experimentally confirmed by Sussman et al.,[45] who determined the atomic structure of the acetylcholine esterase cocrystallized with acetylcholine. In this crystal structure the ammonium ion is surrounded by nine aromatic rings (and not by negatively charged groups) which "protect" the cation from other interactions.

G. DIRECT MEASUREMENTS

Assessments of the forces between the ligand and the receptor on the streptavidine (receptor model)/biotin (ligand model) system, using a surface force transducer, have recently been summarized.[46] The transducer has been constructed with reference to the possibility of measuring the surface tension between two macroscopic curved and smooth surfaces in a liquid (with a resolution of 10 n*N*, then normalized to the radius of the curved surface) as a function of the distance r measured through optical interference techniques (with a resolution of 1 Å).

Streptavidine is a homotetrameric nonglycosylated protein with four binding sites for biotin, disposed in pairs on the two faces of the protein (allowing formation of monolayers). The two examined molecules have been trapped on lipid bilayers and the attractive forces measured at pH 7.4 and 33°C in a 0.2 m*M* solution of sodium phosphate and 0.1 m*M* sodium chloride solution. The $r = 0$ position has been considered when the two lipid bilayers were in contact.

The results revealed a gradual increase of the electrostatic repulsive forces between the two electrically charged bilayer surfaces at large distances (between $r < 85$ Å and $r = 4$ Å), reaching approximately 1.5 mN/m at $r = 4$ Å. For $r < 4$ Å a short-range attractive force of -36 mN/m appeared (this value actually being the force necessary to separate the two surfaces, after they adhered) that was ten times greater than the typical van der Waals attraction forces measured between lipid bilayers. This adhesive force also persisted after a pH increase, thus being of nonelectrostatic origin (hydrophobic, according to the authors).

The resumption of the experiments using inactivated biotin and streptavidine (thus measuring only the repulsion between the receptor and the ligand which appeared during the process of scanning of biotin for the binding site) demonstrated that both repulsive (long-range) and attractive (shortrange) effects were present, the latter increasing exponentially with the decrease of the distance. The effect of temperature upon the experiment has revealed that below 30°C there is not enough lateral mobility (the lipid bilayer solidified) and fewer bonds occur (the adhesive force diminished).

By a rapid approach of biotin to streptavidine, a steric repulsion was detected at $r < 20$ Å — which corresponds to the length of the biotin, 16 Å — the steric force being associated with the reorientation of biotin in the configurational space. Approaching at slow rate (below 1 Å s^{-1}) this repulsion disappeared, because biotin had time enough to reorientate and detect the binding site.

The contact forces from the binding site were detected by crystallization of the complex followed by molecular modeling analyses: there are numerous hydrogen bonds, van der Waals, and hydrophobic interactions.

The interactions measured in these studies have practical value, yet despite their elegance some comments are required. In the interior of the binding site they refer to the interaction among nonpolar groups, which could be strictly of van der Waals nature, and to the interaction between aromatic rings and polar groups (neither being hydrophobic in nature, nor possessing proper aromatic cage forces). In the exterior of the binding site, repulsive forces are probably Coulombian repulsions between the two molecular lipid bilayers (negatively charged at the surface). However the extrapolated biotin-streptavidine interaction curves gives useful information that can be considered in future studies.

The fact that the pH-independent adhesive forces at $r < 4$ Å are considered hydrophobic in nature is not well established because at a pH of 7.4 both molecules have a certain "physiological" configurational space, altered at a basic pH. Alterations in the status of the protonated moieties are likely to happen — by losing their protons, these become nonpolar, thus promoting hydrophobic interactions in regions where other types of interactions normally occur.

It is interesting to note the experimental proof for the steric interaction, which has not been discussed above because of its circumstantial character. The steric repulsion is of electrostatic and van der Waals nature (see Section III.C).

H. ADDITIONAL COMMENTS

The nature of the discussed forces, and their role in the approach of the ligand and its anchoring at the binding site, allow us to state that at long distances the electrostatic forces are dominant (for the orientation of the ligand in the receptor micromedium). At the receptor site the critical interactions are the hydrogen bonds and the aromatic cage (and possibly ionic) forces in the case of ionized ligands, and the hydrophobic and van der Waals forces when uncharged ligands are involved.

The dipole-dipole forces (especially those of higher-order dipole-induced dipole and induced dipole-induced dipole) are of importance only in the van der Waals contact zone, most probably being negligible at distances greater than 3.5 Å. Covalent forces are not involved in physiological binding, this being a reversible process. Hydrophobic interactions seem to play a minor role in the binding of ionized ligands to the receptor, being more important in the pre-receptor stages (the crossing of the intestinal mucous membrane barriers and of the hematoencephalic barriers) and not specifically in the arrangement of the ligand in the binding site. Steric interactions are undoubtedly involved in the pharmacological selectivity of the receptors, but as a result of the reciprocal conformational rearrangements, their importance is probably diminished after the passage of the "steric barrier" (they are not involved when a ligand is already fitted in the binding pocket).

The agonist[5] type of ligand-receptor interaction is followed at a molecular scale by electronic rearrangements and conformational transitions both at the ligand and, especially, at the receptor level. Through this process the activated ligand-receptor complex is formed, which initiates the second stage of the signal transduction mechanism by stimulation of the transducer molecule, when the intracellular launching of the second messenger takes place (for membrane receptors), or other specific processes occur which end in intranuclear penetration of the ligand-receptor complex (e.g., for steroid hormones) followed by the activation of protein synthesis.

The antagonist[5] (or blocker) type of ligand-receptor interaction, pharmacologically speaking, is the failure of an agonist to activate the receptor when another ligand (the blocker) is already present. Antagonists occupy the receptor agonist-binding site, blocking the interaction of (endogenous) agonists with their receptors and thus the physio-pharmacological effects fail to occur. Another blocking mechanism is to perturbate the molecular events which occur during the ligand-receptor activated complex formation (a necessary step in agonism), or to perturbate the activated ligand-receptor complex/transducer molecule coupling. Other blockers interact irreversibly with the receptors; thus, agonists have no place to bind.

The most controverted type of interaction is partial agonism.[5] The pharmacological definition is that these ligands produce a certain degree of agonism, but no matter how high the dose, the effect is below 30% of the effect of a full agonist. In strictly molecular terms, to activate a receptor a ligand must be present in the binding site for a certain lapse of time, Δt. The reduced effect of the partial agonist might be explained through: (1) abandonment of the receptor sooner than Δt (agonism would appear only by successive entrance of two partial-agonist ligands — a process with more reduced probability), or (2) the persistence of the ligand in the receptor binding site longer than Δt — thus the partial-agonist ligand exerting blocker properties as well.

For the "$<\Delta t$" hypothesis, we take as example the first discovered β blocker, R-dichloroisoprenaline (DCI). In this compound the two hydroxyl moieties of the catechol group are replaced by two chlorine atoms. This modification allows hydrogen bond formation (similar to R-isoprenaline) — therefore, the molecule would behave like an agonist, but these bonds are probably weaker (they do not offer reciprocity to the receptor like the H-O moieties in isoprenaline), and thus the ligand abandons the receptor earlier than isoprenaline. It would probably take more than one ligand in the binding site to trigger agonism. Yet, DCI competes with a full agonist to occupy the receptor (thus, the occupancy time is probably not too small). DCI was not clinically tested as a blocker due to its partial agonist character.

For the ''>**Δ**t'' hypothesis, we take the example of pindolol, a β blocker with partial agonism (intrinsic) activity. Structurally, *S*-pindolol resembles *S*-propranolol (antagonist), and only the indole moiety could offer some ''agonist'' activity (closer to the catechol moiety of R-isoprenaline, as compared to the α-naphthyl moiety of propranolol, since it offers the possibility of a hydrogen bond). It is possible that besides triggering agonism in a weaker fashion, pindolol fails to leave the receptor in due time, thus the onset of antagonism occurs. Pindolol is used clinically partly due to its intrinsic activity property.

If a true agonist activates spare receptors as well (thus enhancing its effect), then a partial agonist might simply fail to activate such receptors and the above hypothesis may be no longer valid.

V. RECEPTOR MAPPING IN DRUG DESIGN

Direct[9] molecular modeling techniques start from the known structure of both receptor and ligand, and search for the binding site and for the improvement of the ligand structure. Indirect[9] molecular modeling techniques start mainly from biological data on series of ligands, QSAR analyses, and a search for common patterns in the structural and electrostatic features of the ligands, in order to define a ''receptor binding site'', or a pharmacophore.[47]

Common knowledge about receptor structure at the atomic level is still scant,[48] the description of receptors being stated mainly in terms of their function, not structure. Crystallographic data are difficult to obtain — although the receptors were purified; the (mono)crystallization stability of the membrane arrangement remains very difficult. This explains the attempts to theoretically characterize at least the receptor binding site by using indirect molecular modeling techniques.

The belief that agonist-induced conformational modifications at the receptor level trigger the intracellular signal is still widespread, but there arc reports[49] that a receptor contains all the necessary key factors to release the intracellular signal. Moreover, there are receptors that achieve active conformation in the absence of any ligand. Site-directed mutagenesis studies identify key amino acid residues[49] for the receptor function, allowing molecular modeling studies to focus on small receptor fragments (see Section VII).

The nature of ligand-receptor binding outruns the simple steric fit. Chemical alterations within the ligand may be involved, e.g., histamine[50] tautomerism, transprotonation (the exchange of a proton between the ligand ammonium moiety and a carboxyl moiety of the β adrenoreceptor[32]), electrostatic field-induced conformational changes within the ligand, etc.

The processing[48] of crystallographic data on proteins and enzymes and on their contact with the ligands, respectively, furnished by the Brookhaven Protein Data Bank, made possible a general study concerning the geometry of some functional groups from the enzyme-ligand complexes. The following conclusions[48] about the interaction of functional groups with the surrounding atomic microenvironment are useful to any receptor mapping process. It should be outlined that this study considered neighbor atoms located at distances between 3.2 and 3.8 Å — lesser distances implying interatomic repulsion forces, while at greater distances the interactions were neglected.

Phenyl groups (-C_6H_5) — These are not predominantly surrounded by nonpolar residues; the paucity of the surrounding atoms shows weak interactions between phenyls and these atoms; there is a preference for the interaction of the oxygen atoms with the edge of the aromatic nucleus. The authors did not look for the "aromatic cage". **Carboxyl groups (-COO^-)** are mostly surrounded by polar atoms; there are two types of interactions with the environment — either with one oxygen alone, or with both of them (chelate type, in which the arginine guanidino group is invariably the only partner) — via hydrogen bonds, occasionally intermediated by water molecules; the carboxyl-arginine interaction seems to be generalized, this result being confirmed by various studies concerning the hydrogen bonds in proteins, and the intramolecular geometry of the proteins, respectively.

Carbonyl groups (−C=O) — The investigation of the area at distances of 3.8 to 3.2 Å from the group shows a hemispheric cloud around the oxygen, without the planar directionality expected for hydrogen bonds; previous studies of almost 1000 different molecules, based on X-ray crystal data, showed the planarity of the R-C=O · · · H-N-R hydrogen bond,[51] but these studies were performed on small molecules; the absence of concordance of these data most likely reflects the difference between chemical functions (amide, aldehyde, ketone).

Hydroxyl groups (-OH) — Directionality is even less obvious than in the case of the carbonyl environment; this result confirms the theory, since the freedom of rotation is large, both for the acceptor and for the donor; however, the problem remains open because the hydrogen atom cannot be observed through usual crystallographic methods. **The amino group (-C-NR_2)** shows greater preference for directionality as compared to the carbonyl group; the neighboring groups always contain oxygen, either glutamate or aspartate residues, or even fragments from the peptide skeleton. A second, less marked oxygen atom cloud is observed in the same plane, but shifted 90° from to the first.

Although the starting assumptions of QSAR studies for receptor map constructions are subject to debate,[48,52] correct usage of the available data could lead to useful receptor models which would provide the possibility for predicting the behavior of chemical compounds. In the case of β adrenergic receptors, for instance, through the analysis of the pharmacophoric groups many receptor models have been postulated (some are elegantly reviewed by Linschoten[53]). These models do not superimpose, first because both agonists and antagonists were used in modeling (presupposing different pharmacologic processes), and second, because of inherent experimental errors, especially in the case of antagonists (antagonists do not differentiate the active/inactive/spare states of the receptors, while agonists initially bind on the active, or high-affinity receptors).

Indirect molecular modeling methods in drug design are based on a comparison of the similarities and differences between various compounds that operate at the same receptor — a comparison grounded on structural and physico-chemical properties. Study of the molecular vicinity around the receptor site allows probing for auxiliary sites, thus obtaining entirely new ligands with fundamentally different pharmacologic properties.[54] Receptor mapping studies are based on the hypothesis that the structure and properties of the ligand are complementary to those of the receptor binding site. An often used assumption is to consider that both agonists and antagonists have the same binding site at the receptor; this assumption can be false for some categories of antagonists that may have even two binding sites, one partially joint with the agonists, the other entirely different (e.g., the case of β blockers tertatolol and bopindolol[55]).

The principles applied in receptor mapping have been defined by Marshall.[47,56] Some general comments are presented below:

1. Each considered compound must be critically examined by viewing its activity and/or inactivity regarding each target-receptor subtype, where the existence of the receptor was pharmacologically (e.g., radioligand binding studies) and physiologically (functionally active receptors) certified (several tissues, if data are available).
2. Pharmacophoric groups should be selected to concomitantly allow ionic, hydrogen bond type, aromatic cage, and/or hydrophobic interactions; these groups should be precisely defined on the computer, using molecular graphics; the choice of those groups which allow the achievement of aromatic cage(s) should have priority when ionized ligands are studied.
3. On the basis of docking studies and through parameterization of the physico-chemical properties of the ligands from the test series (part of the data should be retained as control series), the hypothetic receptor site (pharmacophore) is defined, specifying the (steric) forbidden regions, the possible hydrogen bonds, and the electrostatic and hydrophobic type interactions.
4. An alternate approach is to investigate "binding pockets" and use such information for the receptor chart (e.g., for the angiotensin-converting

enzyme using a computer-assisted cavity search[57] algorithm); steric information would be given by any QSAR technique using steric techniques (e.g., both MTD and CoMFA offer such details when a good correlation is found).

5. For the obtained receptor map the ligand-receptor interaction should be calculated using the control series — the exactitude of the previously obtained model(s) being tested; discovered errors are to be corrected on a new hypothetic pharmacophore; and preferably, more than one hypothetic pharmacophore should be tested.[47]

A comparative study[58] on natural and synthetic steroids using the COSMIC force field and the MTD method as compared with the CoMFA method — based on previously published data[19] — show that similar results can be obtained with the MTD method, a less CPU-intensive technique, yet the obtained receptor maps are not entirely superimposable.

Although limitary (such studies must take into account the more active structures, thus they require conformational analysis for each compound), the above-stated principles are useful in the attempt to design new compounds with higher selectivity and activity for the examined receptor.

VI. ROLE OF THE DIPOLE MOMENT DURING LIGAND-RECEPTOR INTERACTION. A HYPOTHETIC STATIC MODEL

A. THEORETICAL CONSIDERATIONS

1. Physical Chemistry Data

Intermolecular interactions can be studied using either the static or the dynamic approach. The static approach[20] deals with pure electrostatic interactions between molecules, considering only the charge distribution for each isolated molecule in the absence of any external perturbation. These methods consider the interactions through multipole expansion: zero order (monopole) moment (the proper net electric charge), first order (dipole) moment, second order (quadrupole) moment, third order (octupole) moment, etc. Because the distribution of the electrons in molecules is not uniform, these moments are helpful to explain the behavior of molecules in the electrical and/or magnetic field.

Simple examples from chemistry: $(NH_4)^+$ has a zero order moment (unitary positive charge), but the dipole and quadrupole moments cancel because of the internal symmetry; H_2O has a null zero order moment (being an uncharged molecule), but it has a dipole moment (through vectorial addition the H-O bond dipole moments do not cancel) — this dipole moment being extremely important in the explanation of the physical and chemical properties of water; CO_2 has a null zero order moment (the molecule has no electrical charge), a null first order moment (through vectorial addition the C=O bond dipoles cancel), but a nonzero second order moment — the quadrupole moment which explains the properties of CO_2 in a magnetic field.

A general treatment of this theory for the multipole interactions has been proposed by Applequist,[59] who applied the method of Cartesian polytensors and treated the permanent and induced moment fields including multipole moments of all orders and polarizabilities.

An electric dipole is defined as an ensemble of two equal charges (**q** and −**q**) separated by a distance **l**, and is represented by a vector $\boldsymbol{\mu}$ that points from the negative charge to the positive one (Atkins[20]). By convention, organic chemists define its direction from the positive toward the negative end (Exner[60]). The magnitude of the vector $\boldsymbol{\mu}$ is **ql** and is called the electric dipole moment. Dipole moments are reported in units called Debyes (D):

$$1\ \mathrm{D} = 3.336 \cdot 10^{-30}\mathrm{C} \cdot \mathrm{m} \tag{1}$$

This unit is named after P. Debye, a pioneer in dipole moment studies. For instance a molecule with one positive (−e) and one negative (e) electron charge ($e = -1.6 \cdot 10^{-19}$ C) separated by 1 Å would have a dipole moment of $1.6 \cdot {}^{-29}$ C · m, or 4.8 D. Usually, small molecules have dipole moments about 1 D. For more details about polarizability and hyperpolarizability, permanent and induced dipole moments, and the equations (Clausius-Mossotti and Debye) linking these notions, the reader is referred to physical chemistry textbooks.[20]

The simple physical meaning of the molecular dipole moment is in contrast with the striking difficulties encountered when this moment is colligated with the charge distribution in organic molecules — the problem of dipole moment modeling, especially with quantum chemical methods, being a subject of debate.

In the calculation[61] of molecular dipole moments through quantum mechanics this quantity is expressed as the sum over the net charges and first order moment of every atom in the molecule. Two contributions, $\boldsymbol{\mu}_s$ and $\boldsymbol{\mu}_p$, result. The term determined by atomic charge transfer is $\boldsymbol{\mu}_s = \Sigma_A\ \mathbf{q}(A)X_A$, where **q**(A) represents the atomic charge (sum of the nuclear and average electronic charges) and X_A the position vector of the nucleus of atom A (measured from some arbitrary origin). The first order moment contribution arising from the polarization of the atomic charge densities is expressed as $\boldsymbol{\mu}_p = \Sigma_A \boldsymbol{\mu}_A$, where $\boldsymbol{\mu}_A$ is the first order moment of atom A. Although both terms are of equal importance, some of the currently used molecular modeling methods neglect the second one, thus leading to erroneous results.

Another problem in discussion is the moment of the C–H bond.[62] Apparently, the use of the partial atomic charges calculated by quantum-chemical methods leads to the assignment C^--H^+, which contradicts vibrational spectroscopy measurements. Taking into account the explanation proposed by Coulson, a bond $c_Ah_A + c_Bh_B$ will have a "homopolar" contribution

$<h_A|er|h_B>$ to the dipole toward the more contracted atomic hybrid orbital, even if h_A and h_B have no net moments (i.e., spherically symmetric orbitals) and the polarization coefficients c_A and c_B are equal (i.e., in the simplest picture zero net charges on each atom). In the C-H bond, the "atomic dipole" $<c_C h_C|er|c_H h_H>$ is of C^+-H^- polarity.

The dynamic approach to intermolecular interactions is based on quantum chemistry (the Hartree-Fock-Roothan[63] equation), using the perturbation theory (Schrödinger). Such methods add to the Hamiltonian a time-dependent term (useful to explain spectroscopic properties), or a time-independent term (the Schrödinger equation being time independent). Time-independent methods deal with external perturbations (same degrees of freedom) or internal perturbations (one extra degree of freedom) — with applications in the assessment of short-range intermolecular forces. Recent approaches to the calculation of partial charges and molecular dipole moments via *ab initio* methods are reviewed elsewhere.[64]

There are some difficulties encountered: (1) molecules move (thermal motion influences the electrostatic interactions); (2) the static approximation is valid only for large intermolecular distances compared to the diameter of the molecules involved; (3) the dynamic approach becomes too complicated as the number of basis sets (implicitly the number of atoms involved) increases — a general problem in quantum chemistry. There is no accurate physical model to approximate the change in dielectric constant for distances below 50 Å (when molecules are rather organized, and such properties are different from those measurable at macroscopic scale) when the microenvironment shifts from hydrophilic (water phase) to hydrophobic (lipid membrane phase) and vice versa.

2. Biological Data

Direct measurements (by optical and electrode techniques) have shown without doubt that unstimulated cells have a net negative charge, caused by the accumulation of trapped organic anions inside the cell. A typical transmembrane potential is -60 mV, which causes a local electric field of 10^6 V/m within the lipid bilayer.[13] Such an electric field has nonlinear oscillations that directly influence the conformation and dielectric properties of transmembrane proteins. However, for the GPCR the permanent dipole moment rotation rate of a given conformation can be approximated to zero — in other words, for an unstimulated cell with unoccupied receptors in the equilibrium state, the permanent dipole of the receptor can be considered as having a fixed character.

The mean density of the net negative charge on the membranal surface is 0.008 to 0.035 C/m^2, which means one electron charge per 2000 to 450 Å^2 (Barber, 1980, cited in Reference 65). GPCRs are glycosylated transmembranar proteins[66] and the sialic acid residues are also negatively charged, being disposed on the N-terminal chain outside the cell.

3. Bacteriorhodopsin as a Model for the Dipole Moment

bR has been used as a model for the three-dimensional structure of GPCR, as it belongs to this protein gene family, and is the only one whose structure has been established through X-ray crystallography by Henderson et al.[67] (with a resolution of 3.5 Å in plane and 10 Å in depth, but refined[68] by heavy atom labeling and computerized analysis of the high resolution electron diffraction patterns 3.5 Å in depth). The putative structure[22] for GPCR consists of seven transmembrane α helixes organized inside the lipid bilayer (with a 'hydrophobic type' inner arrangement[69] — i.e., with the polar groups captured in hydrogen bonds and masked towards the interior of the helix), with the N-terminal segment outside the cell, with three extracellular and three intracellular loops, and with the C-terminal segment in the cytoplasm. The GPCR structure is entirely based on structural homology with bR, although this has 248 amino acid residues, while the majority of GPCR have over 350 residues.

The dielectric properties of bR have been studied *in extenso,* and some of these properties[70] are a net negative charge (−3 . . . −4) located inside the cell, according to the model proposed by Henderson et al.[67] and a ground state permanent dipole moment of 720 D (the measurement results being between 100 and 1000 D). Even in excited (transition) states the dipole moment variation of light-adapted bR does not exceed 15 D (as measured by two-photon double resonance spectroscopy[24]).

We have used[71] the permanent dipole moment of bR to model the GPCR dipole moment, the above-mentioned results being assumed as such. From the electric dipole definition

$$\mathbf{q} = \boldsymbol{\mu}/\mathbf{l} \qquad (2)$$

Assuming a dipole length of 40 Å (which is the mean length of one α helix from the bR crystal), it follows that the point charge is $\mathbf{q} = 3.75$ electrons. Since the net negative charge (zeroth moment charge) is predominantly inside the cell for bR, we assumed that the first order moments charges are similarly disposed and the dipole moment vector (placed in the center of the receptor) is pointing towards the extracellular space. So, in the linear dipole approximation, the dipole moment of both bR and GPCR can be seen as:

(extracellular) ⊕————⊖ (intracellular)
(−3.75e)—— 40 Å ——(+3.75e)

Using the approach described by Hol et al.[72] for α helices and subsequently developed for soluble proteins,[73] we assumed that this approximation describes the dipole moment of the GPCR, and that the negative end of this dipole is shielded by the lipid bilayer. Therefore, the field created by such a line dipole is equal to the field of a positive charge of 3.75 e at the extracellular surface of the GPCR.

The influence of such a point charge upon the approaching ligand can be approximated by a Coulomb (charge-charge) interaction and a charge-dipole interaction, using the following equations:

$$DV = \mathbf{q}_{GPCR} \cdot \mathbf{q}_L / 4\pi\epsilon_0 \cdot r \tag{3}$$

$$DV = -\mathbf{q}_{GPCR} \cdot \boldsymbol{\mu}_L \cdot (4\pi\epsilon_0 \cdot \mathbf{l} \cdot r)^{-1}\{[1 - \mathbf{l} \cdot (r \cdot \cos\phi - \mathbf{l}/4)/r^2]^{-1/2} - [1 + \mathbf{l} \cdot (r \cdot \cos\phi + \mathbf{l}/4)/r^2]^{-1/2}\} \tag{4}$$

where D is the dielectric constant, V is the potential energy, ϵ_0 is vacuum permittivity [$\epsilon_0 = (1/36\pi) \cdot 10^{-9}\ J^{-1}C^2m^{-1}$], $\mathbf{q}$ are charges, $\boldsymbol{\mu}_L$ is the ligand dipole moment, $\mathbf{r}$ is the distance between the point charge $\mathbf{q}_{GPCR}$ and the center of the dipole $\boldsymbol{\mu}_L$, $\mathbf{l}$ is the length of $\boldsymbol{\mu}_L$ and ϕ the angle between the dipole axis $\boldsymbol{\mu}_L$ and the line which unites the center of $\boldsymbol{\mu}_L$ and the receptor point charge $\mathbf{q}_{GPCR}$.

B. CALCULATIONS FOR THE DIPOLE-DIPOLE INTERACTION

The following simplifications were made:

- The ligand is seen as a protonated flexible molecule with a hydrophobic side (such as biogenic amines at physiological pH).
- Hydration shells of both ligand and receptor are ignored, as well as the glycosylated residues outside the receptor; the role of the glycosylated residues, because of their numerous contained O-H groups, could be to take over the hydration water around the ligand and via some ''molecular peristalsis'' mechanism (term proposed by Vinter[74]) to take the ligand progressively into the receptor site (by successively forming and breaking hydrogen bonds).
- The dipole moment of GPCR is considered to be linear, fixed, and centered on the receptor for the lapse of time in which the first ligand approaches and binds to the receptor (the receptor dipole is probably further modified, in the case of agonists, when the activated ligand-receptor complex is formed via conformational changes in the receptor).
- The higher order interactions (e.g., dipole/induced-dipole and induced-dipole/induced-dipole) are not calculated.
- The conformation of the approaching ligand is considered to be stable, and its permanent dipole moment fixed during the interaction, for the sake of simplicity.
- The variation of the dielectric constant is calculated using the approach of Kollman et al.[75] in which D is numerically equal with r (the distance

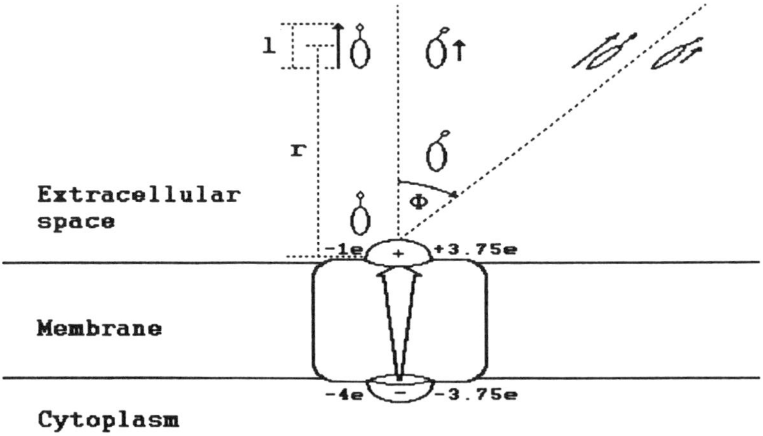

FIGURE 1. Interaction between the two hypothetic conformers and the receptor dipole. The receptor is depicted with the zero moment charges (left) and first moment charges (right).

between the two electrically charged particles) in angstroms, for $r <$ 80 Å (for $r \geq 80$ Å, $D = 80$, the value for the water dielectric constant).

- The influence of the thermal motion is not calculated.

The calculations were performed for the approach of two hypothetical conformers of the same ligand, with equal energy (thus, equal Boltzmann distribution in the solution). The first one with $\boldsymbol{\mu}_L = 17$ D and a more lengthened conformation of $\mathbf{l} = 7$ Å length, and the other with $\boldsymbol{\mu}_L = 9$ D and a more folded conformation of $\mathbf{l} = 4$ Å length.

Equations 3 and 4 were used, considering $\cos\phi = 1$ (i.e., the ligand approach having the most favorable orientation, perpendicular to the receptor). The results are summarized in Table 1. The computations were carried out by considering the receptor negatively charged at the external side ($\mathbf{q}_{GPCR} = -1e$), i.e., taking into account the bR model of Henderson et al.[67] in which at the contact zone between the membrane and the extracellular space there are three electropositive and four electronegative residues.

The potential energy (depending on the inverse second power of the distance) becomes more and more important, according to our model, as the ligand approaches the receptor. Analyzing the results, one can conclude that the interaction of the ligand with the higher dipole moment is energetically favored. An artistic impression is proposed in Figure 1, and a graphical representation of the results from Table 1 is outlined in Figure 2.

One must note that the above-mentioned formulas are not trustworthy for smaller distances (and in fact, even at 40 Å the results represent approximate values) because the equation does not include the polarization and hence the contribution of induced dipole moment. These results are used merely to

TABLE 1
Potential Energy Values (in kcal/mol) for Charge (Ligand)-Charge (Receptor) [q_L-q_{GPCR}], Charge (Ligand)-Dipole (Receptor) [$q_L\mu_{GPCR}$], and Dipole (Ligand)-Dipole (Receptor) [$\mu_L\mu_{GPCR}$] Interactions

Distance r (Å)	q_Lq_{GPCR} (+1/−1e)	$q_L\mu_{GPCR}$ (+1/+3.75e)	$\mu_L\mu_{GPCR}$ (Dipole 9/4)	Total potential for 9/4	$\mu_L\mu_{GPCR}$ (Dipole 17/7)	Total potential for 17/7
100.0	−0.041	0.154	−0.051	0.062	−0.097	0.016
80.0	−0.052	0.195	−0.080	0.063	−0.151	−0.008
60.0	−0.091	0.341	−0.190	0.060	−0.359	−0.109
40.0	−0.210	0.780	−0.641	−0.071	−1.216	−0.646
30.0	−0.370	1.390	−1.522	−0.502	−2.901	−1.881
20.0	−0.830	3.110	−5.164	−2.884	−9.963	−7.683

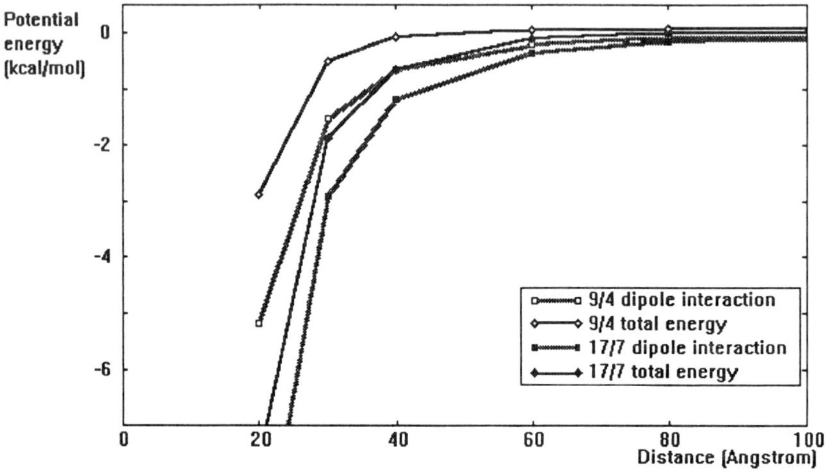

FIGURE 2. Dipole-dipole [μ_L-q_{GPCR}] and total electrostatic interaction energies.

illustrate a qualitative trend, not a real approach. These figures show that the ''receptor dipole-ligand dipole'' interaction are within the same order of magnitude as the charge-charge interaction, thus not at all negligible (it may influence the conformer's approach toward the binding site) — although in some molecular modeling programs it is neglected, considering that the strictly mathematic r^{-3} dependence (r = distance) justifies this.

C. SOME SPECULATIVE REMARKS

- The dipole moment of the receptor acts as a conformational filter during the docking process, and from more conformations with the same Boltzmann probability (equienergetical), the higher dipole conformer is preferred.
- In the process of ligand ''selection'', even at Boltzmann probabilities lower by 40%, a conformer with double dipole moment will be preferred when compared to the lowest energy conformer (with half the value of the dipole).
- During agonist-induced conformational changes, through reorientation of its permanent dipole, the receptor may ''expel'' the ligand (providing, for instance, a virtual negative first order moment charge at the extracellular margin). The mechanism by which a ligand leaves the receptor is not yet known.
- Assuming different orientations and values for the dipole moment according to the receptor subtypes, a particular conformation of the same ligand may occupy a certain receptor subtype. There are three major receptor types (α_1, α_2, and β) and ten minor subtypes only in the case

of adrenergic receptors, and all these respond to noradrenaline stimulation. The ligand being the same, there must be a mechanism by which receptors are ''selectively'' activated (not in pharmacological, but in informational sense). This process could involve the receptor dipole: a certain receptor dipole (negative first moment charge on the extracellular surface) could easily dock the low-energy conformers (consequently, an activation through low ligand concentrations). A sign and/or direction change in the dipole could permit only the access of some higher-energy conformers — thus, the ligand could activate the receptor only at higher concentrations for a particular subtype of receptor, when more receptor subtypes are present on the same tissue/cell.

D. PRACTICAL CONSIDERATIONS AND CONCLUSIONS

In the conformational equilibrium, the following thermodynamic process is involved:

$L_i \rightarrow L_0$ with the Gibbs equation:

$$\Delta G = -RT \ln([L_1]/[L_0]) \qquad (5)$$

where [] stands for concentrations, L_i is a given conformation of the ligand, L_0 is the lowest-energy conformer (each conformer having its own dipole moment) available in the solution at the temperature T (in Kelvin), R is the gas constant (R = 1.987 cal/mol) and ΔG is the Gibbs energy (free enthalpy).

Without knowing the precise amount of L_0, one can consider $[L_0] = 1$. Thus, the relative amount ($[L_i]/[L_0]$) of all other considered conformers can be determined using the Gibbs equation. As ΔG one can utilize ΔU values (the difference between potential energies calculated in the molecular mechanics approximation, e.g., the COSMIC[76] force field) or ΔH_f (the difference between the enthalpies of formation calculated by semiempirical quantum-chemical methods, e.g., MOPAC[77]). In both cases the entropic variation is approximated to zero (the variation of the degree of disorder following the modification of some torsion angles is insignificant, and conformational modifications in micromolecules are unlikely to induce important changes in the hydration shell's order). Therefore the ΔU or ΔH_f values can be approximated, within certain limit, as free energy ΔG.

Based on the above-described simplifications, the probability of a given conformer of the ligand to interact with the receptor would be directly proportional with its dipole moment and its Boltzmann distribution probability. Supposing that there are N conformations in study with the corresponding calculated dipole moment values μ_i, respectively, and that the energy differences are compared to the conformation with the lowest energy, $\Delta G_i = G_i - G_0$, then one can define:

$$\Delta p_i = \left| \pm \sqrt{\{\mu_i^2 \cdot [\exp(-\Delta G/RT]\}} \right| \quad (6)$$

$$P_T = \Sigma_N(\Delta p_i) \quad (7)$$

$$P_i = (\Delta p_i/P_T) \cdot 100 \quad (8)$$

Equation 6 represents the contribution of the conformer L_i to the effective dipole moment of the considered ligand, and at same time Δp_i is a relative measure of the contribution of the conformer L_i to the total probability P_T defined by Equation 7 (Δp_i being proportional to the probability of conformer "i" to occupy the receptor binding site). Thus, one may calculate the absolute probability P_i of a given conformer "i" to occupy the receptor via Equation 8.

The above-proposed method attempts to also introduce, besides energy — so far the only parameter used in the conformational analysis for receptor mapping (besides geometric data) — a modulating parameter, which connects the Boltzmann distribution and the dipole moment of the conformer, assessing its contribution to the pharmacological activity of the given ligand.

We briefly comment on the method:

1. When more metastable and geometrically different conformers are analyzed (i.e., meaningful conformers in respect to their lifetime and concentration), better results are expressed in P_i values;
2. The more extensive the conformational analysis, the better the chance to find the global minimum, therefore the computation will be closer to the correct answer — yet, geometry optimizations are usually CPU time-consuming;
3. The method does not consider ways of transition from one conformer into another (i.e., it is not dependent on the activation energy $\Delta G^{\#}$ corresponding to the conformational transition) and thus one may include conformers which do not really exist at the room temperature because of high transition barriers;
4. The method is tributary to the correctness by which one calculates the dipole moment μ_i of the studied conformers; it has already been stated that quantum chemical methods hardly cope with the calculation of dipole moments, especially in the case of ionized states — we recommend those empirical methods which account not only for electronegativities, but also for polarizabilities (e.g., CHARGE3[78]);
5. Using the effective dipole moment, our method does not consider the orientation of the dipole with respect to the receptor (angle ϕ from Equation 4), and thus we strictly refer to the conformational analysis.

The described method is based on the bR model. If future experimental results will give a GPCR dipole moment of the same order of magnitude as for bR, then the above-proposed method can be introduced as a "filter" parameter in receptor mapping by virtue of the speculative calculations presented in Section VI.B and D.

The advantage of the method consists in the fact that we propose a fast semiquantitative assessment for conformational analysis, and thus offer the opportunity to automatically test all available conformers at room temperature. A computer program could calculate the corresponding P_i values for all the conformers produced by a conformational analysis program, and then separately store those with significant P_i. Thus, from millions of possible conformers, the modeler can select only a few tens for a thorough individual analysis using classical methods.

VII. PERSPECTIVES

To avoid human (subjective) factors in the hunt for the best pharmacophore, Dean and co-workers developed an automated site-directed drug design technique,[79-82] which uses given protein structures searching for all possible binding sites. The procedure generates molecular templates which are eventually transformed into chemical structures — candidates for synthesis and further biological testing. This method is a direct technique, yet it uses receptor mapping principles. The computer algorithm is not fully operational yet, but with the emerging technical improvements it is likely that it will become an important method.

Progress is made in the field of computer-assisted receptor modeling, based on bR as a starting structure for GPCR. These studies involve structural homology building,[22] primary structure alignment using the SPECTRAMAP (Lewi[83]) technique with cluster analysis using the variable gap penalty,[84] or three dimension receptor building assisted by molecular graphics[23] using results from indirect molecular modeling studies. In conformational analysis, geometry optimization techniques are currently developed: in the *ab initio* formalism,[85] new parameterized Hamiltonians for semiempirical calculations,[86] or new parameterized force fields for molecular mechanics calculations.[87]

By combining such tools, one is likely to identify the *real* binding site in future. Cooperation with molecular biology laboratories allows the targeting of site-directed mutagenesis studies followed by direct evidence of the involvement of single amino acid residues. Thus, one can build a receptor map and experimentally check the results, refining the model until the correct answer is found.

ACKNOWLEDGMENTS

The authors wish to express their gratitude to the following persons for stimulating discussions and advice: Dr. R. R. Birge (Syracuse, NY), Dr.

P. M. Dean (Cambridge, U.K.), Prof. Dr. L. H. M. Janssen (Utrecht, the Netherlands), Dr. H. Moerels (Beerse, Belgium), Dr. E. E. Moret (Utrecht, the Netherlands), Dr. M. Mracec (Timişoara, Romania), Dr. Z. Szabadai (Timişoara, Romania), Prof. Dr. J. P. Tollenaere (Utrecht, the Netherlands), Dr. J. G. Vinter (Cambridge, U.K.), Dr. J. Wilting (Utrecht, the Netherlands). Prof. Dr. Z. Simon and Prof. Dr. Fr. Schneider (both from Timişoara) are acknowledged for revising the manuscript. This work was in part supported by a grant offered by the Faculty of Pharmacy, State University of Utrecht (T.I.O). The Romanian version of this work is part of the Ph.D. thesis of Dr. T. I. Oprea.

REFERENCES

1. **Sadek, M., Craik, D. J., Hall, J. G., and Andrews, P. R.,** Conformational analysis of cyproheptadine hydrochloride, *J. Med. Chem.,* 33, 1098, 1990.
2. **Feldman, R. D., Brotherton, A., and Welsh, M. J.,** β-Adrenergic-mediated Cl secretion: evidence for additional non-cAMP-dependent pathway of effect, *Am. J. Physiol.,* 259, L426, 1990.
3. **Dean, P. M.,** *Molecular Foundations of DrugReceptor Interaction,* Cambridge University Press, Cambridge, U.K., 1987.
4. **Bax, A.,** Two-dimensional NMR and protein structure, *Annu. Rev. Biochem.,* 58, 223, 1989.
5. **Taylor, P. and Insel, P. A.,** Molecular basis of pharmacologic selectivity, in *Principles of Drug Action,* 3rd ed., Pratt, W. B. and Taylor, P., Eds., Churchill Livingstone, London, 1990.
6. **Leff, P. and Harper, D.,** Do pharmacological methods for the quantification of agonists work when the ternary complex mechanism operates?, *J. Theor. Biol.,* 140, 381, 1989.
7. **Menekyan, O. and Bonchev, D.,** On the possible types of electronic drug-receptor interactions, *Eur. J. Med. Chem.,* 24, 397, 1989.
8. **Simon, Z.,** *Quantum Biochemistry and Specific Interactions,* Abacus Press, Tunbridge Wells, Kent, U.K., 1976.
9. **Cohen, N. C., Blaney, J. M., Humblet, C., Gund, P., and Barry, D. C.,** Molecular modelling software and methods for medicinal chemistry, *J. Med. Chem.,* 33, 883, 1990.
10. **Rebeck, J.,** Molecular recognition with model systems, *Angew. Chem. Int. Ed. Engl.,* 29, 245, 1990.
11. **Davis, A., Warrington, B. H., and Vinter, J. G.,** Strategic approaches to drug design. II. Modelling studies on phosphodiesterase substrates and inhibitors, *J. Comput.-Aided Mol. Design,* 1, 97, 1987.
12. **Vinter, J. G.,** Understanding Molecular Modelling, course held at the State University of Utrecht, 1991.
13. **Astumian, R. D. and Robertson, B.,** Nonlinear effect of an oscillating electric field on membrane proteins, *J. Chem. Phys.,* 91, 4891, 1989.
14. **Moţoc, I.,** *Molecular Structure and Biologic Activity,* 'Facla' Publishing House, Timişoara, 1980.
15. **Williams, D. H., Cox, J. P. L., Doig, A. J., Gardner, M., Gerhard, U., Kaye, P. T., Lal, A. R., Nicholls, I. A., Salter, C. J., and Mitchell, R. C.,** Toward the semiquantitative estimation of binding constants. Guides for peptidepeptide binding in aqueous solution, *J. Am. Chem. Soc.,* 113, 7020, 1991.

16. **Balaban, A. T., Chiriac, A., Moţoc, I., and Simon, Z.,** *Steric Fit in Quantitative Structure-Activity Relations,* SpringerVerlag, Heidelberg, 1980.
17. **Simon, Z., Chiriac, A., Holban, S., Ciubotaru, D., and Mihalaş, G. I.,** *Minimum Steric Difference. The MTD Method for QSAR Studies,* John Wiley & Sons, Chichester, U.K., 1984.
18. **Mayer, D., Naylor, C. B., Moţoc, I., and Marshall, G. R.,** A unique geometry of the active site of angiotensin-converting enzyme consistent with structure-activity studies, *J. Comput.-Aided Mol. Design,* 1, 3, 1987.
19. **Cramer, R. D., III, Patterson, D. E., and Bunce, J. D.,** Comparative molecular field analysis (CoMFA). 1. Effect of shape on binding of steroids to carrier proteins, *J. Am. Chem. Soc.,* 110, 5959, 1988.
20. **Atkins, P. W.,** *Physical Chemistry,* 4th ed., Oxford University Press, Oxford, 1990, 645.
21. **Rees, D. C., Komyia, H., Yeates, T. O., Allen, J. P., and Feher, G.,** The bacterial photosynthetic reaction center as a model for membrane proteins, *Annu. Rev. Biochem.,* 58, 607, 1989.
22. **Dohlman, H. G., Thorner, J., Caron, M. G., and Lefkowitz, R. J.,** Model systems for the study of seven transmembrane segment receptors, *Annu. Rev. Biochem.,* 60, 653, 1991.
23. **Hibert, M. F., Trumpp-Kallmeyer, S., Bruinvels, A., and Hoflack, J.,** Three-dimensional models of neurotransmitter G-binding protein-coupled receptors, *Mol. Pharmacol.,* 40, 8, 1991.
24. **Birge, R. R. and Zhang, C.-F.,** Two-photon double resonance spectroscopy of bacteriorhodopsin. Assignment of the electronic and dipolar properties of the low-lying $^1A'^-{}_g$-like and $^1B'^+\pi$-like π,π' states, *J. Chem. Phys.,* 92, 7178, 1990.
25. **Scheiner, S.,** Quantum chemical approach to study the mechanisms of proton translocation across membranes through protein molecules, in *Structure and Properties of Cell Membranes,* Vol. 3, Benga, G., Ed., CRC Press, Boca Raton, FL, 1985.
26. **Whalley, E. and Klug, D. D.,** Effect of hydrogen-bonding on the direction of the dipole moment derivative of the O-H bond in the water molecule, *J. Chem. Phys.,* 84, 78, 1986.
27. **Vinter, J. G.,** personal communication, 1991.
28. **Dill, K. A.,** Dominant forces in protein folding, *Biochemistry,* 29, 7133, 1990.
29. **Burgen, A. S. V.,** Some considerations of receptor specificity, *Trends Pharmacol. Sci.,* Suppl. 10, 1, 1989.
30. **Cole, R. H.,** Dielectrics in physical chemistry, *Annu. Rev. Phys. Chem.,* 40, 1, 1989.
31. **Strader, C. D., Sigal, I. S., Candelore, M. R., Rands, E., Hill, W. S., and Dixon, R. A. F.,** Conserved aspartic acid residues 79 and 113 of the β adrenergic receptor have different roles in receptor function, *J. Biol. Chem.,* 263, 10267, 1988.
32. **Linschoten, M. R., Kranenbarg, G. W. K., de Kimpe, S. J., Wilting, J., Janssen, L. H. M., and van Lenthe, J. H.,** A computer modelling study of hydrogen bonds in ligandβ-adrenoceptor complexes: its implications in the deduction of a receptor map, *J. Mol. Struct.,* 237, 339, 1990.
33. **Mitchell, T. J., Tute, M. S., and Webb, G. A.,** A molecular modelling study of the interaction of noradrenaline with the β_2-adrenergic receptor, *J. Comput.-Aided Mol. Design,* 3, 211, 1989.
34. **van Vlijmen, H. W. T. and IJzerman, A. P.,** Molecular modelling of a putative antagonist binding site on helix III of the β-adrenoceptor, *J. Comput.-Aided Mol. Design,* 3, 165, 1989.
35. **Scheiner, S.,** Calculating the properties of hydrogen bonds by *ab initio* methods, in *Reviews In Computational Chemistry II,* Lipkowitz, K. B. and Boyd, D. B., Eds., VCH Publishers, New York, 1991, 165.
36. **Stouten, P. F. W. and Kroon, J.,** Hydrogen bonds *in concreto* and *in computro, J. Mol. Struct.,* 177, 467, 1988.

37. **Vedani, A. and Dunitz, J. D.,** Lone-pair directionally in hydrogen bond potential functions for molecular mechanics calculations: the inhibition of human carbonic anhydrase II by sulfonamides, *J. Am. Chem. Soc.*, 107, 7653, 1985.
38. **Dill, K. A.,** The meaning of hydrophobicity, *Science,* 250, 297, 1990.
39. **Privalov, P. L., Gill, S. J., and Murphy, K. P.,** Answer to ''The meaning of hydrophobicity'', *Science,* 250, 297, 1990.
40. **Ben-Naim, A.,** Solvent-induced interactions: hydrophobic and hydrophilic phenomena, *J. Chem. Phys.*, 90, 7412, 1989.
41. **Hansch, C. and Leo, A.,** *Substituent Constants for Correlation Analysis in Chemistry and Biology,* Wiley Interscience, New York, 1979.
42. **Yang, G.-Z., Lien, E. J., and Guo, Z.-R.,** Physical factors contributing to hydrophobic constant π, *Quant. Struct.-Act. Relat.*, 5, 12, 1986.
43. **Lien, E. J., Guo, Z. R., Li, R.-L., and Su, C.-T.,** Use of the dipole moment as a parameter in drug receptor interaction and QSAR studies, *J. Pharm. Sci.*, 71, 641, 1982.
44. **Dougherty, D. A. and Stauffer, D. A.,** Acetylcholine binding by a synthetic receptor: implications for biological recognition, *Science,* 250, 1558, 1990.
45. **Sussman, J. L., Harel, M., Frolow, F., Oefner, C., Goldman, A., Toker, L., and Silman, I.,** Atomic structure of acetylcholinesterase from *Torpedo californica:* A prototypic acetylcholine-binding protein, *Science,* 253, 872, 1991.
46. **Leckband, D. E., Israelachvili, J. N., Schmitt, F.-J., and Knoll, W.,** Long range attraction and molecular rearrangements in receptor-ligand interaction, *Science,* 255, 1419, 1992.
47. **Marshall, G. R.,** Computer-aided drug design, in *Computer-Aided Molecular Design,* Richards, W. G., Ed., IBC Technical Services, London, 1989, 91.
48. **Tintelnot, M. and Andrews, P.,** Geometries of functional group interactions in enzyme-ligand complexes: guides for receptor modelling, *J. Comput.-Aided Mol. Design,* 3, 67, 1989.
49. **Hollenberg, M. D.,** Receptor triggering and receptor regulation: structure-activity relationships from the receptor's point of view, *J. Med. Chem.*, 33, 1275, 1990.
50. **Worth, G. A., King, P. M., and Richards, W. G.,** Histamine tautomerism and its mode of action, *Biochim. Biophys. Acta,* 1036, 158, 1990.
51. **Taylor, R., Kennard, O., and Versichel, W.,** Geometry of the NH-OC hydrogen bond. 1. Lone pair directionality, *J. Am. Chem. Soc.*, 105, 5761, 1983.
52. **Mandema, J. W.,** Conformational Requirements for Adrenoceptor Activity. Master thesis, State University of Utrecht, Utrecht, 1987.
53. **Linschoten, M. R.,** A Ligand's View of the β-Adrenoceptor, PhD thesis, State University of Utrecht, Utrecht, 1991.
54. **Tallarida, R. J. and Freeman, K. A.,** Control and stability of ligand-receptor interaction in the presence of a competitive compound, *Life Sci.*, 48, PL-19, 1991.
55. **De Blasi, A., Fratelli, M., and Marasco, O.,** Certain β-blockers can decrease β-adrenergic receptor number: I. Acute reduction in receptor number by tertatolol and bopindolol, *Circ. Res.*, 63, 273, 1988.
56. **Marshall, G. R.,** Computer-aided drug design, *Annu. Rev. Pharmacol. Toxicol.*, 27, 193, 1987.
57. **Ho, C. W. J. and Marshall, G. R.,** Cavity search: an algorithm for the isolation and display of cavity-like binding regions, *J. Comput.-Aided Mol. Design,* 4, 337, 1990.
58. **Oprea, T. I., Ciubotariu, D., Sulea, T. I., and Simon, Z.,** Comparative analysis by Minimal Steric Difference (MTD) and Comparative Molecular Field Analysis (CoMFA) methods for binding of steroids to carrier proteins, *Quant. Struct.-Act. Relat.*, 1993 (submitted).
59. **Applequist, J.,** A multipole interaction theory of electric polarization of atomic and molecular assemblies, *J. Chem. Phys.*, 83, 809, 1985.
60. **Exner, O.,** *Dipole Moments in Organic Chemistry,* Georg Thieme Verlag, Stuttgart, 1975, 2.

61. **Bader, R. F. W., Larouche, A., Gatti, C., Carroll, M. T., MacDougall, P. J., and Wiberg, K. B.,** Properties of atoms in molecules: dipole moments and transferability of properties, *J. Chem. Phys.,* 87, 1142, 1987.
62. **Reed, A. E. and Weinhold, F.,** Some remarks on the C-H bond dipole moment, *J. Chem. Phys.,* 84, 2428, 1986.
63. **Davidson, E. R.,** Perspectives on *ab initio* calculations, in *Reviews in Computational Chemistry,* Lipkowitz, K. B. and Boyd, D. B., Eds., VCH Publishers, New York, 1990, 373.
64. **Williams, D. E.,** Net atomic charge and multipole models for the *ab initio* molecular electric potential, in *Reviews In Computational Chemistry II,* Lipkowitz, K. B. and Boyd, D. B., Eds., VCH Publishers, New York, 1991, 219.
65. **Wojtczak, L. and Nalecz, M. J.,** The surface potential of membranes: its effect on membrane-bound enzymes and transport processes, in *Structure and Properties of Cell Membranes,* Vol. 2, Benga, G., Ed., CRC Press, Boca Raton, FL, 1985, 215.
66. **Vassilev, P. M. and Tien, H. T.,** Planar lipid bilayers in relation to biomembranes, in *Structure and Properties of Cell Membranes,* Vol. 3, Benga, G., Ed., CRC Press, Boca Raton, FL, 1985, 63.
67. **Henderson, R., Baldwin, J. M., Ceska, T. A., Zemlin, F., Beckmann, E., and Downing, K. H.,** Model for the structure of Bacteriorhodopsin based on high-resolution electron cryo-microscopy, *J. Mol. Biol.,* 213, 899, 1990.
68. **Ceska, T. A. and Henderson, R.,** Analysis of high-resolution electron diffraction patterns for purple membrane labelled with heavy atoms, *J. Mol. Biol.,* 213, 539, 1990.
69. **Jennings, M. L.,** Topography of membrane proteins, *Annu. Rev. Biochem.,* 58, 999, 1989.
70. **Birge, R. R.,** Photophysics and molecular electronic applications of the rhodopsins, *Annu. Rev. Phys. Chem.,* 41, 683, 1990.
71. **Oprea, T. I., Kurunczi, L., and Moret, E. E.,** Role of the dipole moment during ligand receptor interaction — a hypothetic static model, IX Eur. Symp. QSAR Molec. Model., Strasbourg, 1992.
72. **Hol, W. G. J., van Duijnen, P. T., and Berendsen, H. J. C.,** The α-helix dipole and the properties of proteins, *Nature,* 273, 443, 1978.
73. **Hol, W. G. J., Halie, L. M., and Sander, C.,** Dipoles of the α-helix and β-sheet: their role in protein folding, *Nature,* 294, 532, 1981.
74. **Vinter, J. G.,** personal communication, 1991.
75. **Daggett, V., Kollmann, P. A., and Kuntz, I. D.,** Molecular dynamics simulation of small peptides: dependence on dielectric model and pH, *Biopolymers,* 31, 285, 1991.
76. **Morley, S. D., Abraham, R. J., Haworth, I. S., Jackson, D. E., Saunders, M. R., and Vinter, J. G.,** COSMIC(90): An improved molecular mechanics treatment of hydrocarbons and conjugated systems, *J. Comput.-Aided Mol. Design,* 5, 475, 1991.
77. **Stewart, J. P.,** MOPAC: a semiempirical molecular orbital program, *J. Comput.-Aided Mol. Design,* 4, 1, 1990.
78. **Abraham, R. J. and Grant, G. H.,** Charge calculations in molecular mechanics. IX. A general parameterisation of the scheme for saturated halogen, oxygen and nitrogen compounds, *J. Comput.-Aided Mol. Design,* 6, 273, 1992.
79. **Danziger, D. J. and Dean, P. M.,** Automated site-directed drug design: a general algorithm for knowledge acquisition about hydrogen-bonding regions at protein surfaces, *Proc. R. Soc. Lond.,* B236, 101, 1989.
80. **Danziger, D. J., and Dean, P. M.,** Automated site directed drug design: the prediction and observation of ligand point positions at hydrogen-bonding regions on protein surfaces, *Proc. R. Soc. London,* B236, 115, 1989.
81. **Lewis, R. A. and Dean, P. M.,** Automated site-directed drug design: the concept of spacer skeletons for primary structure generation, *Proc. R. Soc. London,* B236, 125, 1989.

82. **Lewis, R. A. and Dean, P. M.,** Automated site-directed drug design: the formation of molecular templates in primary structure generation, *Proc. R. Soc. London,* B236, 141, 1989.
83. **Lewi, P. J.,** Spectral map analysis: factorial analysis of contrasts, especially from log ratios, *Chemometrics Intelligent Lab. Sys.,* 5, 105, 1989.
84. **Lewi, P. J. and Moereels, H.,** Latent variables and cluster analysis of receptor protein sequences, *Trends Anal. Chem.,* 10, 283, 1991.
85. **Potter, T. and Klessinger, M.,** Geometry optimization in semiempirical SCF-MO-CI calculations, *J. Comp. Chem.,* 12, 167, 1991.
86. **Zerner, M. C.,** Semiempirical molecular orbital methods, in *Reviews In Computational Chemistry II,* Lipkowitz, K. B. and Boyd, D. B., Eds., VCH Publishers, New York, 1991, 313.
87. **Dinur, U. and Hagler, A. T.,** New approaches to empirical force fields, in *Reviews In Computational Chemistry II,* Lipkowitz, K. B. and Boyd, D. B., Eds., VCH Publishers, New York, 1991, 99.

INDEX

A

B

C

D

F

G

I

N

Q

R

S